MORE ARDUINO
for HAM RADIO

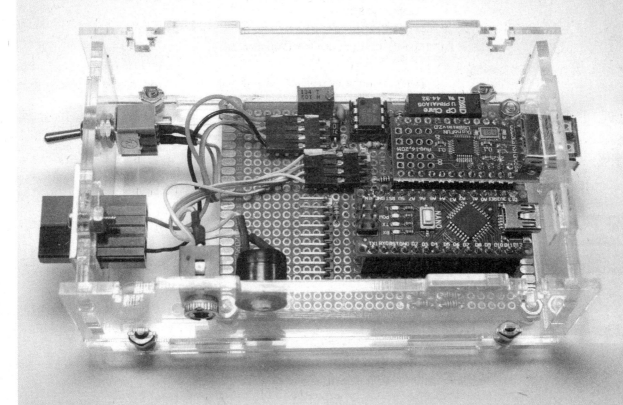

 ARRL
The National Association for
Amateur Radio®

Glen Popiel, KW5GP

Production

Jodi Morin, KA1JPA
Michelle Bloom, WB1ENT
David F. Pingree, N1NAS
Maty Weinberg, KB1EIB

Copyright © 2021 by
The American Radio Relay League, Inc.

Copyright secured under the Pan-American Convention

All rights reserved. No part of this work may be reproduced in any form except by written permission of the publisher. All rights of translation are reserved.

Printed in the USA

Quedan reservados todos los derechos

ISBN: 978-1-62595-147-2

First Edition

We strive to produce books without errors. Sometimes mistakes do occur, however. When we become aware of problems in our books (other than obvious typographical errors), we post corrections on the ARRL website. If you think you have found an error, please check **www.arrl.org/product-notes** for corrections. If you don't find a correction there, please let us know by sending email to **pubsfdbk@arrl.org**.

Sketches, libraries, Fritzing (project layout) diagrams, and other files for the projects in this book are available online from **www.arrl.org/arduino3**, **www.sunriseinnovators.com/Arduino3** and **www.kw5gp.com/Arduino3**.

Contents

Preface
Acknowledgements
About the Author
About This Book
How This Book Is Organized
About ARRL

1. The Arduino — An Update
2. Arduino Processor Boards — An Update
3. Arduino Shields, Modules, and Devices
4. Arduino I/O Methods
5. Creating Arduino Sketches
6. Tools, Construction Techniques, and Troubleshooting
7. Yaesu FH-2 Keypad
8. Peltier Cooler Controller
9. Rotator Turn Indicator
10. Rotator Position Indicator
11. Build Your Own AR-40 Rotator Controller
12. Modified AR-40 Rotator Controller
13. USB CW Keyboard
14. Yaesu CAT Display
15. CDE/HyGain Keypad Entry Rotator Controller
16. RTTY Reader
17. In Conclusion

Index

Preface

I've often said that working with the Arduino is more of an adventure than anything else. Based on my own experiences with the Arduino, nothing could be more true. I started out on my Arduino adventure more than nine years ago, and here we are, three books later, and I'm still having fun. What has been exciting and amazing to me has been the sheer number of projects planned for this book that had to be left on the drawing board. They weren't left there because they couldn't be made to work — actually it's quite the opposite. There simply wasn't enough room to fit them all in this book. Just because this book is finished doesn't mean that my own personal Arduino adventure has come to an end. More than likely, as soon as this book heads off to print, I'll start playing with a half-dozen new ideas and who knows what can happen from there.

Having stretched the limits of what could be done with the simple, inexpensive Arduino boards such as the Uno and Nano in my previous books, this time around I chose to focus on inexpensive and more versatile projects that lend themselves to more enhancement and customization on your part. Several of the projects in this book could easily be combined and rolled into one all-inclusive project you could be proud of. As part of this, I have also chosen some areas of the Arduino that are not well documented and remain relatively unexplored territory, such as USB communication with the Arduino.

Since my last Arduino book, there have been a number of new Arduino-compatible CPU boards, modules, and components. Part of my own personal Arduino adventure has been to take some of these newer parts and develop a project to demonstrate their usefulness in ham radio-related projects. Because a primary goal has been to make each book a standalone Arduino project guide, some material from the first two books has been carried over into this book, such as the Open Source licensing models and the older, basic Arduino components such as the Uno and Nano. From there, we'll quickly move into some of the new modules and components and put them to use in some fun ham-related projects.

As with my previous books, every project in this book is complete and functional, but along with each project I have included some ideas and suggestions on how you can enhance and customize your finished project. There have been a number of times where it has been very difficult for me to choose a stopping point in order to leave something for you to expand upon.

Quite often, I'll go back and put my own finishing touches on these projects and put them to use in my own station. But what has also been part of the fun of this whole process for me has been to see others pick up where I leave off and create some truly wonderful projects of their own. To me, this is what Open Source and the Arduino are all about, sharing ideas and projects for others to expand upon. And there's absolutely nothing preventing me from taking the ideas of others and expanding on them myself, and then sharing them back out there for others, in a never-ending cycle. To me, that's what Open Source is really all about.

In keeping with the Open Source philosophy of sharing knowledge, in this book, I have focused on how to interface to real-world devices such as antenna rotators and transceivers. We'll add Arduino power to your rotator controller, or use an Arduino to talk to your rig's CAT (Computer Assisted Transceiver) interface. In many cases, you'd be surprised to find out just how easy it is to inexpensively add some very powerful and fun features to your existing equipment.

It has been a wonderful experience to see the art of homebrewing making a comeback in the ham radio world, with the Arduino being a large part of that resurgence. When I first started out homebrewing way back in the Dark Ages before cell phones and the internet, we didn't have devices like the Arduino. Instead, we had to bury ourselves in the manufacturer's component databooks, then design, solder (or wirewrap) everything from scratch, with no guarantee of success. Now, with the Arduino, projects that used to take weeks, or even months, can be built in a matter of days for a fraction of the cost.

So who knows where your next Arduino project can take you? Let your imagination be your guide. My favorite words throughout this adventure have been "Wouldn't it be cool if…?" That has led to the creation of a lot of fun projects for me. Hopefully, it will be the same way for you.

And whatever you do, please, share your ideas, projects, and knowledge with the rest of the Open Source community. Without the philosophy of freely sharing knowledge and ideas, none of the things in the Arduino universe would have been possible.

73,
Glen Popiel, KW5GP
kw5gp@arrl.net

Acknowledgements

To my Dad, thanks for all the encouragement and teaching me how rewarding it can be to learn how to do things yourself.

I would also like to thank you, my fellow Makers and hams, for making my personal Arduino adventure so much fun. Without your support, this third ARRL Arduino book would never have happened. I would also like to thank the late Tim Billingsley, KD5CKP, for introducing me to the Arduino, providing ideas and key insights during my Arduino learning process, and providing invaluable help, advice, and trying to keep me on track and focused on what I was actually trying to accomplish. The late Craig Behrens, NM4T, also helped make this whole series of books a reality.

I would also like to thank Martin Jue and the folks at MFJ Enterprises for loaning me an AR-40 rotator for the AR-40 projects. Thanks also go out to Bill Brown, WB8ELK, and Tom Medlin, W5KUB, for providing photos and Picoballooning information. There are many others who have helped along the way, and I apologize in advance to anyone I may have omitted.

Thanks also to the Olive Branch (Mississippi) and Helena (Arkansas) Amateur Radio Clubs, the Chickasaw Amateur Radio Association, along with the Orlando Hamcation and the Huntsville Hamfest for allowing me to present Arduino forums to help spread the word on the Arduino and how it can be applied to amateur radio.

I also need to thank former ARRL Publications Manager, Steve Ford, WB8IMY; my editor, Mark Wilson, K1RO; ARRL Publications Manager Becky Schoenfeld, W1BXY, and the staff at ARRL for the opportunity to work with them. Don't let anyone tell you otherwise — putting together a book like this is an arduous task, and they've helped to make the whole process enjoyable.

And a special thanks to the Open Source community. Their sharing of knowledge and ideas helped make this book become a reality. It is through their efforts that the Arduino has become the popular microcontroller platform that it is today and we all owe them a huge debt of gratitude. Thank you.

About the Author

Glen Popiel, KW5GP, is the author of ARRL's *Arduino for Ham Radio*, *More Arduino Projects for Ham Radio*, and *High Speed Multimedia for Amateur Radio*. By day, he is a Network Engineer and Technology Consultant, specializing in Open Source technology solutions. First published in *Kilobaud Microcomputing* in 1979 for circuits he designed for the RCA 1802 microprocessor, he continues to work with microcontrollers and their uses in amateur radio, and has written numerous articles on computers and amateur radio.

Always taking things apart (and sometimes even getting them to work again afterwards), he discovered electronics in high school and has never looked back. As a teenager, he had one of the first true "home computers," a Digital Equipment (DEC) PDP-8 minicomputer (complete with state-of-the-art Model 35 Teletype) in his bedroom that he and his friends salvaged from the scrap heap. Over his 40+ year career, he has worked for various aerospace and computer manufacturers on radio and military turbojet research data acquisition and control systems.

Since discovering the Arduino more than a decade ago, he has developed a passion for this powerful, inexpensive microcontroller and has given a number of seminars and hamfest forums on the subject of the Arduino and Open Source. He is a member of the Olive Branch Amateur Radio Club (OBARC), the Chickasaw Amateur Radio Association, QRP Amateur Radio Club International (QRP-ARCI), and the QRP Skunkwerks, a design team of fellow hams and Arduino enthusiasts.

Glen is also a former cat show judge and has exhibited Maine Coon cats all over the country, with the highlight being a Best in Show at Madison Square Garden in 1989. He now lives in Southaven, Mississippi, where he continues to create fun and exciting new Arduino projects for amateur radio with his two champion Maine Coon show cats and lab companions, Angel and Shadow.

About This Book

When I started out on my adventures with the Arduino, I never thought in my wildest dreams that I'd end up here. Welcome to the third of my Arduino project series of books. I hope that you find the projects in this book as much fun for you as it was for me to create them.

When I finished my second Arduino book, *More Arduino Projects for Ham Radio* (ISBN 978-1-6295-070-3, out of print), I thought that would be the end of it. I was all but out of ham radio Arduino project ideas and never thought I'd have enough to even think about writing a third book. But, new components and modules became available, and with them, the ideas began to come. By time I got to the point of actually writing this book, I had more project ideas than I could fit into these pages. Based on feedback I received from my fellow hams, Makers, and Arduino enthusiasts, this time around I decided to focus on easier, inexpensive, fun, and highly expandable projects.

As with my other Arduino books, a primary goal is not to provide a collection of fully finished, gold-plated Arduino projects for you to copy and build, but instead, provide functional and usable projects with room for you to add your own personal touches and expand on each of these projects. To help you understand the projects and the concepts behind them, we'll start with a review of the Arduino and cover some of the newer Arduino boards, shields, and modules that can be used for ham radio projects. Then we'll move into practical projects, gradually increasing in complexity as we go. The projects in this book were chosen to use as wide of an array of different modules and components as possible to help you learn how to apply all these various components in your own ham radio projects. On a personal note, part of the project selection process was based on uniqueness, simplicity, innovative approach, and an all-around "cool" factor. Many of my Arduino projects start out from the simple thought of "Wouldn't it be cool if…?"

A major focus of this book was also to peel away the layers of complexity on how to interface with the various rotator controllers, transceivers, communication protocols, and specialized Arduino components and functions, while at the same time trying to keep each project interesting, fun, easy to build, and as inexpensive as possible. Just to let you in on a little secret, I create and build these projects as much for myself and my ham station as I do for you and the book. What I end up with is more of a chronicling of my own Arduino adventures in a format that allows others to join in on the fun as well as a bunch of fun projects to play with. And, for those of you that have been here before with my other books, isn't it fun to look back and see how much we've all painlessly learned in this process?

As with my other Arduino books, this book assumes that you have a basic working knowledge of electronics, components, and construction techniques. You don't have to be an expert, as most of these projects require only a handful of wires and components and are easy to build. You will, however, need to know how to solder and feel comfortable building electronic projects.

For those of you new to electronics, there are a number of good books and online tutorials to help you along the way. Among the books that I recommend are *Understanding Basic Electronics* by Walter Banzhaf, WB1ANE (ISBN 978-087259-082-3, ARRL order no. 0823) and *The ARRL Handbook*, which ARRL publishes annually. Every ham should have a copy of *The ARRL Handbook* in their library. I find myself constantly referring to mine for research or when I want to learn some new aspect of electronics or ham radio. Many of these books and other useful resources are available from amateur radio dealers, **www.arrl.org/shop**, or online booksellers.

You will also need a working knowledge of the Arduino and the Arduino Integrated Development Environment (IDE). Entire books have been written about how to learn and use the Arduino, and there is simply not enough space to cover those basics and all of the projects in one book. This book does include discussions on some specific areas such as installing and using libraries, and using the Arduino IDE in general, but it's best that you take the time to become familiar with the Arduino. In particular, concentrate on the various methods of installing libraries. Library installation issues account for nearly 90% of all of the Arduino questions I get. Installing libraries is really not that difficult, but is an often misunderstood and overthought process. The more you can work to understand Arduino libraries, the more it will help you further on down the road.

There are several excellent introductory books and online tutorials to help you along the way. The ones I have found most useful include *Beginning Arduino* by Michael McRoberts (ISBN 978-1430232407) and *Arduino Cookbook* by Michael Margolis (ISBN 978-1449313876). Another excellent book for learning the Arduino, especially the hardware interfacing aspect, is *Exploring Arduino: Tools and Techniques for Engineering Wizardry* by Jeremy Blum (ISBN 978-1-118-54936-0). There are also some outstanding Arduino tutorials available online at **www.arduino.cc**, **www.learn.adafruit.com**, and **learn.sparkfun.com**. Recently, I have discovered several other very good online course and tutorials available for the Arduino. The Programming Learning Academy (**www.programmingelectronics.com**) offers a free 12-lesson Arduino Video Course and an inexpensive e-book PDF download of the *Arduino Course for Absolute Beginners* by Michael James. Both are excellent Arduino learning tools.

Because one or two of the projects in this book may require that you have a ham radio operator's license, now would be a good time get or upgrade your license, if you haven't already. There is something for everyone in the ham radio community, and there is absolutely no reason for you to go it alone as you start out on your Arduino adventures. For more information on how to become a radio amateur, check out the *ARRL Ham Radio License Manual* (ISBN 978-1-62595-087-1, ARRL order no. 00871). I recommend that you find a local ham radio club at **www.arrl.org/find-a-club** and attend a few meetings. That's what got me started out on the Arduino, and you will not meet a friendlier, more helpful group of people anywhere. The odds are that, like me, you will meet other Arduino enthusiasts with whom you can collaborate on your own Arduino projects.

I also urge you to attend some of the forums presented at many hamfests. Several of the best ideas for the projects in this book have come from my Arduino forums and club presentations. Forums are also a great place to meet others who share similar interests and to share knowledge and ideas.

All of the Arduino project sketches in this book were written and compiled on Windows 7, 8, and 10, using version 1.8.5 of the Arduino IDE. If you have any problems compiling the sketches, please try compiling using the same version of the IDE that I used, and verify that your libraries are properly installed. While sometimes a bit on the vague side, the error messages you may receive when experiencing problems with your sketches can point you in the right direct as to where the problem is. When in doubt, do an internet search for the error message itself. Quite often, someone else has experienced the same issue and has posted an explanation or possible solution.

All of the sketch files, libraries, and any errata information for the book projects are available online at **www.sunriseinnovators.com/Arduino3** or at **www.kw5gp.com/Arduino3**. While every effort has been made to eliminate any errors in this book, they always seem to have a way to creep in and cause all kinds of havoc, even in the most carefully prepared book. So please, check the Errata first if you have any problems. And by all means, feel free to contact me if you need help, I'll be glad to help as much as I can.

How This Book is Organized

This book builds on the ideas and concepts presented in my previous books, *Arduino for Ham Radio* and *More Arduino Projects for Ham Radio*. As with the earlier books, the projects here are arranged to begin with simpler designs and concepts, gradually increasing in complexity and functionality. Each new component, concept, and programming technique is described in detail as it appears in a project. As presented here, each project is fully functional and usable as is, but each was designed and constructed to allow you to add your own personal touches for expansion and enhancement. Each chapter includes some enhancement ideas to help you take your project and Arduino prowess to the next level.

While I understand your desire to skip ahead and dive right into the projects, I urge you to take the time to read the chapters leading up to the projects. These preliminary chapters will provide you with vital information and insights into the various modules, components, and technologies used in the projects that follow. They will also provide you with information and guidance on creating a good, functional work area to build and troubleshoot your Arduino projects. The more prepared and organized you are in these areas, the better things will often turn out. This doesn't necessarily mean your work area will stay clean and uncluttered for very long, but it helps to at least start out that way. After 20 minutes on a project, my work area tends to look like a complete and total mess, but at least I had a good start.

Chapter 1, *The Arduino — An Update*, provides an introduction and update to the world of the Arduino, its history, and provides a basic understanding of the concepts of Open Source and the various Open Source licenses.

Chapter 2, *Arduino Processor Boards — An Update*, discusses the most common Arduino boards with an emphasis on the new Arduino and Arduino compatible boards that have been introduced since the last book.

Chapter 3, *Arduino Shields, Modules, and Devices*, covers the various boards and components that can be used to interface with the Arduino, allowing the Arduino to sense and communicate with the outside world, again, with an emphasis placed on the newer devices and those that lend themselves for use in ham radio-related projects.

Chapter 4, *Arduino I/O Methods*, discusses in detail the I/O capabilities of the Arduino, including some of the newer methods now available, which method is best for communicating with the various shields and components, and how to best implement each I/O method. This chapter also discusses the Bluetooth, USB, and serial RS-232 communication methods with the Arduino, including the use Human Interface Devices (HID) such as keyboards and mice with the Arduino.

Chapter 5, *Creating Arduino Sketches*, introduces the Arduino Integrated Development Environment (IDE), creating sketches (programs), outlining and flowcharting sketches, and organizing your projects using sketchbooks. It also discusses Arduino simulators and alternatives to the Arduino IDE, as well as how to install and manage your Arduino libraries.

Chapter 6, *Tools, Construction Techniques, and Troubleshooting*, covers tools, test equipment and methods on how to troubleshoot your sketches, library problems, hardware, and other issues you may encounter while building Arduino projects.

Chapter 7, *Yaesu FH-2 Keypad*, demonstrates how to use programmable digital potentiometer chips to build your own version of the Yaesu FH-2 keypad used with many recent Yaesu transceivers, including the FT-950, FTDX1200, FT-2000, and others.

Chapter 8, *Peltier Cooler Controller*, shows you how you can construct your own Peltier cooler module controller using a touchless infrared temperature sensor.

Chapter 9, *Antenna Rotation Indicator*, introduces the individually addressable NeoPixel RGB LED ring and how to interface it to your CDE/HyGain or Yaesu rotator controller to provide an animated LED display while your antenna rotates.

Chapter 10, *Antenna Direction Indicator*, demonstrates how to interface to your HyGain or Yaesu antenna rotator controller to use the individually addressable NeoPixel RGB LED ring to display your current antenna bearing.

Chapter 11, *Build Your Own AR-40 Rotator Controller*, shows you how to build your own HyGain AR-40 rotator controller and interface it your PC for control by *Ham Radio Deluxe* and other rotator control software.

Chapter 12, *Modified AR-40 Rotator Controller*, allows you to modify your existing AR-40 antenna rotator control unit, add a color TFT display module, and interface it to your PC for control by *Ham Radio Deluxe* and other rotator control software.

Chapter 13, *USB CW Keyboard*, revisits the popular CW keyboard from my first book, *Arduino Projects for Ham Radio* and updates it for use with a USB keyboard and color TFT display. This project also demonstrates how to interface USB Human Interface Devices (HID) to the Arduino using a small, inexpensive USB host module.

Chapter 14, *Yaesu CAT Display*, demonstrates how to interface an Arduino and a touchscreen color graphic TFT display to the CAT (Computer Assisted Transceiver) interface for many Yaesu transceivers such as the FT-450D, FT-950, FTDX1200, and others. This project could easily be adapted to other transceiver brands and models.

Chapter 15, *CDE/HyGain Keypad Entry Rotator Controller*, interfaces an inexpensive analog 4×4 keypad to your CDE/HyGain rotator controller for direct bearing entry and rotation control without the need for a PC or holding down the control unit's front panel switches.

Chapter 16, *RTTY Reader*, demonstrates how you can use an Arduino to decode and display a standard 45-baud RTTY signal on a color graphic TFT display.

Chapter 17, *In Conclusion*, discusses projects, ideas, and concepts not included in this book, in order to provide you with concepts and ideas for other projects to encourage going beyond the scope of this book.

WHAT DO YOU WANT TO DO WITH AMATEUR RADIO?

ARRL is the national association for amateur radio in the US. We provide opportunities to discover radio, develop skills, and serve your local community.

Membership in ARRL can help you:

Discover New Interests

Whether you're interested in radiosport, new technologies, project building, emergency preparedness, or public service, ARRL has resources to help you learn, get active, and get on the air.

Your membership provides digital access to all four ARRL publications, with offerings for beginners as well as advanced hams. They include *QST*, the membership journal of ARRL; *On the Air*, an introduction to the world of amateur radio; *QEX*, covering topics related to radio communications experimentation; and *National Contest Journal (NCJ)*, covering radio contesting.

Build & Share Your Knowledge

With online learning courses, members-only web content, and leadership opportunities, you can grow your skills and interest in amateur radio through the many ARRL programs available to members.

Shape the Future

Your membership dollars help to preserve and protect access to frequencies allocated to the Amateur Radio Service.

Anyone who is active in amateur radio or who wishes to get more involved to pursue technological interests, public service, or personal enjoyment will benefit from ARRL Membership.

Benefits
To get you involved and keep you up to date with all that amateur radio has to offer!

INFORMATION

As a member, you will gain access to all four digital magazines, several special interest e-newsletters, & personalized answers to your technical and operating questions.

LEARNING

From licensing exam prep, to live training forums; to online training courses for new hams, emergency communicators, and more.

PROGRAMS & SERVICES

License renewal, member recognition programs, contesting opportunities, advocacy efforts, and an active local club system.

Two Easy Ways to Join or Renew

ONLINE at arrl.org/join

CALL toll free at **1-888-277-5289**

About ARRL

We're the American Radio Relay League, Inc. — better known as ARRL. We're the largest membership association for the amateur radio hobby and service in the US. For over 100 years, we have been the primary source of information about amateur radio, offering a variety of benefits and services to our members, as well as the larger amateur radio community. We publish books on amateur radio, as well as four magazines covering a variety of radio communication interests. In addition, we provide technical advice and assistance to amateur radio enthusiasts, support several education programs, and sponsor a variety of operating events.

One of the primary benefits we offer to the ham radio community is in representing the interests of amateur radio operators before federal regulatory bodies advocating for meaningful access to the radio spectrum. ARRL also serves as the international secretariat of the International Amateur Radio Union, which performs a similar role internationally, advocating for amateur radio interests before the International Telecommunication Union and the World Radiocommunication Conference.

Today, we proudly serve nearly 160,000 members, both in the US and internationally, through our national headquarters and flagship amateur radio station, W1AW, in Newington, Connecticut. Every year we welcome thousands of new licensees to our membership, and we hope you will join us. Let us be a part of your amateur radio journey. Visit www.arrl.org/join for more information.

225 Main Street
Newington, CT 06111-1400 USA
Tel: 860-594-0200
FAX: 860-594-0259
Email: membership@arrl.org
www.arrl.org

1 The Arduino — An Update

The Arduino has come a long way since its humble beginning in Ivrea, Italy, back in 2005. The Arduino-compatible family has grown tremendously since then and continues to expand today, with new, faster 600-MHz processors eclipsing the original 16-MHz Arduino Uno. And yet, the Arduino Uno and its smaller twin, the Nano, are still the primary boards used in the typical Arduino project. The Uno and Nano have always been my first choices when designing a project, simply because they're inexpensive and easy to work with. Quite often, they have all the I/O (input/output connections) and horsepower that I'll need for the project that I have in mind. But it's always nice to know that there are other members in the Arduino family to choose from that have the horsepower, memory, and I/O for just about any project I can think up.

Differences Between an Arduino and Raspberry Pi

We'll take a look at some of the newer Arduino and Arduino-compatible boards in a bit, but first, let's fall back and cover some Arduino basics. One of the most common questions I am asked about the Arduino is "What is the difference between the Arduino and a Raspberry Pi?" This is often followed up with "How do I know which one of the two to use for a project?" And of course, the follow-up question to that is usually "How do I know which Arduino to use in my projects?"

An easy way to visualize the differences between the Arduino (**Figure 1.1**) and the Raspberry Pi (**Figure 1.2**) is to think of the Arduino as the systems management computer in your car. While your engine is running, your car's embedded computer is constantly sampling things such as accelerator pedal position, engine temperature, air intake volume and temperature, and a dozen other sensors in order to keep your car's engine running properly and efficiently. The computer then can then use all of the information provided by the sensors to control things such as the fuel-to-air mixture and ignition timing, and to ensure that the exhaust emissions are within allowable limits. If the computer senses anything out of whack, you get to see the ominous "Check Engine" light.

Figure 1.1 — An Arduino Uno and Nano.

Figure 1.2 — Raspberry Pi 3.

In the case of a sensor failure or other major issue, your car's computer can switch into a secondary operation configuration, often called "Limp mode," which will alter the engine control settings to allow the engine to continue functioning in a degraded manner until you can safely stop and get it fixed. Now, add in those dreaded wireless tire pressure sensors, and you see that your car's computer is capable of sensing and controlling a lot of things based on the sensor inputs. The Arduino is pretty much the same way, only you get to choose what is sensed and controlled, and what operations occur based on those inputs.

The Raspberry Pi is actually a small single-board, general-purpose computer, running a customized version of the Linux operating system (OS). Its function and operation are nearly identical to a full-size PC, except instead of a hard drive, the Pi uses an SD (Secure Digital) flash memory card. The Raspberry Pi also typically has an Ethernet and/or Wi-Fi port, Bluetooth, USB, and HDMI video interface onboard. The Raspberry Pi also has a significantly faster CPU and more RAM memory than the Arduino.

With the advancement in the two technologies, there is some crossover between the Arduino and the Pi. With the newer, faster Arduino boards having more memory, and the Raspberry Pi's general-purpose I/O pins, choosing which board to use can be confusing. A simple way to differentiate between the two is that the Arduino is best used when you want to sense and control things, while the Raspberry Pi is a mini-PC that runs Linux programs. So for applications such as an antenna rotator controller, you'd most likely want to go with an Arduino, but if you want to build an SDR (software defined radio), you'd probably want to use the Raspberry Pi.

Why Choose the Arduino?

Now that you've decided your project needs a microcontroller, why should you choose the Arduino? For starters, the Arduino is inexpensive and easy to work with. Arduino programs, or *sketches*, as they are called in the Arduino world, are written using a variant of the C++ programming language that is easy to understand and program with.

Released under the Open Source Creative Commons Attribution Share-Alike license, the Arduino is totally Open Source. While Open Source is often associated with being free, the primary meaning to Open Source is that the design, concepts, and ideas are free of patent and copyright restrictions, as long you follow the philosophy of Open Source, share your Arduino work with the community, and properly attribute the people whose work you used to create yours.

For example, while this book itself is protected by copyright law, the projects, designs, and sketches are all Open Source. This means that under copyright law, you can't just copy this book and sell it, but you are permitted to use and distribute the concepts, projects, designs, ideas, and the programs (sketches), even if you intend to sell your project commercially, as long as you follow the rules of Open Source licensing. We'll talk more on the specifics of the types of Open Source licenses later in this chapter.

In a nutshell, you have a worldwide network of people developing both Arduino hardware and software, and sharing their ideas and designs with others in the Arduino community. The components generally used in the vast majority of Arduino projects are, for the most part, inexpensive, and easy to interface. The Arduino Integrated Development Environment (IDE) software that you use to program and communicate with the Arduino and the Arduino-compatible processors is free. And, as you will see, there are literally hundreds of Arduino processors, shields, modules, and components that you can use in your projects. We'll discuss some of this hardware in later chapters.

Everything about the Arduino itself is Open Source. The board designs and schematic files are Open Source, meaning that anyone can create their own version of the Arduino free of charge. The Creative Commons licensing agreement allows for unrestricted personal and commercial derivatives as long as the developer gives credit to Arduino, and releases the work under the same license.

You may see the Arduino also called a "Genuino." This is the result of a trademark dispute among the Arduino developers that has since been resolved and both names are now under just the one Arduino umbrella. The names Arduino and Genuino are trademarked, which is why the various Arduino-compatible boards have names like Iduino, Ardweeny, Boarduino, Freeduino, and so on. Typically these boards are fully compatible with their official Arduino counterpart, and they may include additional features not found on the original Arduino board they are derived from.

The Arduino also supports several industry-standard I/O buses and protocols. These include the One-Wire, Inter-Integrated Circuit (I^2C), and Serial Peripheral Interface (SPI) bus protocols, allowing simple connections to the Arduino using only a few wires. With some Arduino-compatible processor boards and add-on shields or modules, the Arduino can also support Wi-Fi, RS-232, USB, Bluetooth Serial, Ethernet, and other methods of communication as well.

And of course, it really wouldn't be a microcontroller without some form of individual I/O pins for discrete inputs and outputs. The basic Arduino Uno has 14 digital I/O pins and six 10-bit analog-to-digital (A/D) input pins. Six of the digital pins can also be used to generate pulse-width modulation (PWM) outputs. The analog pins can also be used as digital I/O pins if your projects calls for a few extra digital pins.

The Arduino is not just a couple of processor boards — it's an entire ecosystem. A sketch created for one Arduino processor can be easily recompiled and run on a different Arduino processor, simply by changing the board type in the IDE and recompiling the sketch. It really is just that simple.

It's this simplicity and ease of use that has made the Arduino the tool of choice for microcontroller enthusiasts worldwide. Now you can create complex projects without having to dig through datasheets and solder for months like you did in days gone by. Many complex Arduino projects can be built in a few hours and using only a handful of wires and parts.

Programming the Arduino is just as easy, many of the modules and devices for the Arduino have supporting software "libraries," which are similar to the device drivers used to add new hardware to your PC. Using these libraries, the complexity of communicating with these various modules and devices is greatly simplified, allowing you to focus on the project itself, without having to get down to the bits and bytes level of every little piece of your Arduino project.

The Arduino Processor Family

The original Arduino started out using the 8-bit Atmel AVR series of microcontrollers, the ATmega328 and the ATmega2560 in particular. Since then the family has been expanded to other processors, including the ARM Cortex series, Intel Quark, and others. Processors have increased in speed and capability, now sporting 32-bit CPUs and substantially increased amounts of flash and RAM memory.

Although there are now numerous variations on the Arduino, the most common Arduino, the Uno, consists of an Atmel ATmega328 8-bit microcontroller with a clock speed of 16 MHz. The ATmega328 has 32 KB of flash memory, 2 KB of static RAM (SRAM), and 1 KB of EEPROM onboard. The Uno has 14 digital I/O pins, and six of these pins can also do pulse width modulation (PWM). It has six 10-bit analog inputs that can also be used for digital I/O pins. Two of the digital pins also directly support external hardware interrupts, and all of the I/O pins support pin state change interrupts, allowing external hardware control of program execution. In 2015, there were an estimated 700,000 official Arduino boards in the hands of users, along with an equal estimated number of Arduino-compatible "clones" and variants.

Typically powered via the USB programming port or through the onboard dc power jack, with its low current drain and onboard 5 V power regulator, the Arduino is ideally suited for battery-powered projects. Recent low-power Arduino boards now run on just 3.3 V, further reducing power demands. For example, the Arduino boards used in amateur high-altitude balloon tracking systems can run exclusively from small solar cells that provide all the power needed to drive a 10 mW WSPR (Weak Signal Propagation Reporter) or APRS (Automatic Packet Reporting System) transmitter.

Designed for expandability, the Arduino Uno I/O and power connections are brought out to a series of headers on the main board. The header layout is standard among the majority of the Uno-type boards. Many of the Arduino add-on boards, also known as *shields*, can be plugged directly into these headers and stacked one on top of the other, providing power and I/O directly to the shield with no additional wiring needed.

Many types of shields are available, including all manner of displays, Ethernet, Wi-Fi, motor driver, MP3, and a wide array of other devices. My personal favorite is the prototyping shield, which allows you to build your own interfaces to an even wider array of Arduino-compatible components, modules, and breakout boards for applications such as Global Positioning System (GPS), real time clock (RTC), text-to-speech, and direct digital frequency synthesis (DDS). A number of display modules are available, ranging from 2-line monochrome LCDs to high-resolution TFT color display modules. You can also find an endless array of sensors such accelerometers, magnetometers, pressure, humidity, proximity, motion, vibration, temperature, and many more.

Arduino team leader Massimo Banzi's statement about the Arduino project, "You don't need anyone's permission to make something great," and this quote from Arduino team member David Cuartielles, "The philosophy behind Arduino is that if you want to learn electronics, you should be able to learn as you go from day one, instead of starting by learning algebra" sum up what has made the Arduino so popular among hobbyists and builders. The collective knowledge base of Arduino sketches and program libraries is immense and constantly growing, allowing the average hobbyist to quickly and easily develop complex projects that once took mountains of datasheets and components to build.

The Arduino phenomenon has sparked the establishment of a number of suppliers for add-on boards, modules, and sensors adapted for the Arduino. With these new boards and features, the Arduino now has the power needed to support processing-intensive applications and high speed communications.

What is Open Source?

Generally speaking, Open Source refers to software in which the source code is freely available to the general public for use and/or modification. Probably the best example of Open Source software is the Linux operating system created by Linus Torvalds. Linux has evolved into a very powerful operating system, and the vast majority of applications that run on Linux are Open Source. A large percentage of the web servers on the internet are Linux-based, running the Open Source Apache Web Server. The popular Firefox web browser is also Open Source, and the list goes on. Even the Android mobile device operating system is based on Linux, and is itself Open Source. This ability to modify and adapt existing software is one of the cornerstones of the Open Source movement, and is what has led to its popularity and success.

The Arduino team took the concept of Open Source to a new level. Everything about the Arduino — hardware and software — is released under the Creative Commons Open Source License. This means that not only is the Integrated Development Environment (IDE) software for the Arduino Open Source, but the Arduino hardware itself is also Open Source. All of the board design files and schematics are Open Source, meaning that anyone can use these files to create their own Arduino board. In fact, everything on the Arduino website, **www.arduino.cc**, is released as Open Source.

As the Arduino developer community grows, so does the number of applications and add-on products also released as Open Source. While it may be easier to buy an Arduino board, shield or module, in the vast majority of cases, everything you need to etch and build your own board is freely available for you to do as you wish. The only real restriction is that you have to give your work back to the Open Source Community under the same Open Source Licensing.

What more could a hobbyist ask for? Everything about the Arduino is either free or low cost. You have a whole community of developers at your back, creating code and projects that you can use in your own projects, saving you weeks and months of development. As you will see in some of the projects in this book, it takes longer to wire and solder things together than it does to actually get a project working. That is the true power of Open Source, everyone working together as a collective, freely sharing their work, so that others can join in on the fun.

Open Source Licensing and How it Works

There are several main variations on the Open Source licensing model, but all are intended to allow the general public to freely use, modify, and distribute their work. This is by no means a complete list of all of the various Open Source licensing models. If you want more detailed information than provided here, I recommend visiting **en.wikipedia.org/wiki/Comparison_of_free_and_open-source_software_licences** as a good starting point.

The most common Open Sources license models you will encounter in the Arduino world include the GNU General Public License (GPL), Lesser GPL (LGPL), MIT, and Creative Commons licenses. As a general rule, for the average hobbyist, this means you are free to do as you wish. However, there will always be those of us that come up with that really cool project we want to package up and sell to finance our Arduino and ham radio hobby. It is important for that group to review and understand the various license models you may encounter in the Open Source world.

The GNU GPL

As with all Open Source licensing models, the GNU General Public License (GPL) is intended to guarantee your freedom to share, modify, and distribute software freely. Developers who release software under the GPL desire their work to be free and remain free, no matter who changes or distributes the program. The GPL allows you to distribute and publish the software as long as you provide the copyright notice and disclaimer of warranty, and keep intact all notices that refer to the license. Any modified files must carry prominent notices stating that you changed the files, along with the date of any changes.

Any work that you distribute and publish must be licensed as a whole under the same license. You must also accompany the software with either a machine-readable copy of the source code or a written offer to provide a complete machine readable copy of the software. Recipients of your software will be automatically granted the same license to copy, distribute, and modify the software. One major restriction to the GPL is that it does not permit incorporating GPL software into proprietary programs.

The copyright usage in the GPL is commonly referred to as "copyleft," meaning that rather than using the copyright process to restrict users as with proprietary software, the GPL copyright is used to ensure that every user has the same freedoms as the creator of the software.

There are two major versions of the GPL: Version 2, and the more recent Version 3. There are no radical differences between the two versions. The changes are primarily to make the license easier for everyone to use and understand. Version 3 also addresses laws that prohibit bypassing Digital Rights Management (DRM). This is primarily for codecs and other software that deals with DRM content.

Additional changes were made to protect your right to "tinker" and to prevent hardware restrictions that don't allow modified GPL programs to run. In an effort to prevent this form of restriction, also known as "Tivoization," Version 3 of the GPL has language that specifically prevents such restrictions and restores your right to make changes to the software that works on the hardware it was originally intended to run on. Finally, Version 3 of the GPL also includes stronger protections against patent threats.

The Lesser GNU General Public License (LGPL)

The Lesser GNU General Public License (LGPL) is very similar to the GPL, except that it permits the usage of program libraries in proprietary programs. This form of licensing is generally to encourage more widespread usage of a program library in an effort for the library to become a *de-facto* standard, or as a substitute for a proprietary library. As with the GPL, you must make your library modifications available under the same licensing model, but under the LGPL, you do not have to release your proprietary code. In most cases, it is preferable to use the standard GPL licensing model.

The MIT License

Originating at the Massachusetts Institute of Technology, the MIT license is a permissive free software license. This license permits reuse of the software within proprietary software, provided all copies of the software include the MIT license terms. The proprietary software will retain its proprietary nature even though it incorporates software licensed under the MIT license. This license is considered to be GPL-compatible, meaning that the GPL permits combination and redistribution with software that uses the MIT License. The MIT license also states more explicitly the rights granted to the end user, including the right to use, copy, modify, merge, publish, distribute, sublicense, and/or sell the software.

The Creative Commons License

There are multiple versions of the Creative Commons License, each with different terms and conditions:

1) **Attribution (CC BY)** — This license allows others to distribute, remix, tweak, and build upon a work, even commercially, as long as they credit the creator for the original creation.

2) **Attribution-NonCommercial (CC BY-NC)** — This license allows others to remix, tweak, and build upon a work non-commercially. While any new works must also acknowledge the creator and be non-commercial, any derivative works are not required to be licensed on the same terms.

3) **Attribution-ShareAlike (CC BY-SA)** — This is the most common form of the Creative Commons License. As with the Attribution license, it allows others to distribute, remix, tweak, and build upon a work, even commercially, as long as they credit the creator for the original creation, and license their new creation under the same license terms. All new works based on yours convey the same license, so any derivatives will also allow commercial use.

4) **Attribution-NonCommercial-ShareAlike (CC BY-NC-SA)** — This license allows others to distribute, remix, tweak, and build upon a work non-commercially, as long as they credit the creator and license their new creations under the identical licensing terms.

5) **Attribution-No Derivs (CC BY-ND)** — This license allows for redistribution, both commercial and non-commercial, as long as it is passed along unchanged and in its entirety, with credit given to the original creator.

6) **Attribution-NonCommercial-NoDerivs (CC BY-NC-ND)** — This is the most restrictive of the Creative Commons licenses, only allowing others

to download a work and share it with others as long as they credit the creator. Works released under this license cannot be modified in any way, nor can they be used commercially.

The Arduino is released under the Creative Commons Attribution-ShareAlike (CC BY-SA) license. You can freely use the original design files and content from the Arduino website, **www.arduino.cc**, both commercially and non-commercially, as long as credit is given to Arduino and any derivative work is released under the same licensing. So, if by chance you do create something that you would like to sell, you are free to do so, as long as you give the appropriate credit to Arduino and follow the requirements outlined in the FAQ on the Arduino website, as you may not be required to release your source code if you follow specific guidelines. If you include libraries in your work, be sure you use them within their licensing guidelines. The core Arduino libraries are released under the LGPL and the Arduino IDE is released under the GPL.

Benefits of Open Source

Before I wrote this book, I knew very little about USB human interface device (HID) communication with the Arduino. But, lucky for me, there was a fully functional USB HID library out there just waiting for me to come up with a way to use it. Now, having built the USB CW Keyboard described later in this book and working with that USB HID library, a whole new world of USB and even Bluetooth HID projects has been opened up to me.

That's the beauty of the Arduino and Open Source. You don't have to be a programming genius to create fully functional projects as long as you have the entire Open Source developer community at your back. Every time I branch out into a new, relatively unexplored area such as the USB and Bluetooth HID devices, not only do I learn something new, but my mind just goes bonkers thinking up all kinds of new projects I can build using these new pieces of my Arduino programming toolbox.

Just one final thing. When you do start creating wonderful new things, remember to share them back to the community, so that others following in your footsteps can benefit from your work and create wonderful new things of their own.

Arduino Processor Boards — An Update

Deciding which Arduino board to use in your project can be a daunting task. In my case, I'll usually try starting out with an Arduino Uno or Nano, simply because of their versatility, low cost, and ease of use. If an Uno or Nano isn't right for the project in mind, then I'll move up the chain, looking for the Arduino-variant that has the power and features that the project will need. Amazingly, the Uno and Nano can handle about 90% of the projects I come up with, which just goes to show how versatile they really are.

Because the Arduino was first introduced in 2005, dozens of new Arduino and Arduino-variant boards have been developed, each with its own set of features and enhancements. Some of the older boards have gone by the wayside, fallen out of use, or even been discontinued. But due to the Open Source nature of the Arduino, many compatible versions of these older boards can still be found online.

The list of Arduino and Arduino-compatible processor boards discussed in this chapter is by no means complete. Many of the boards not listed here have already been extensively covered in other books, including my own *More Arduino Projects for Ham Radio*. Rather than repeat that information, in this book we'll focus primarily on the boards that are used in this book's projects and those that are best suited for amateur radio projects and projects in general.

The ATmega328 Series

The Atmel ATmega328 series of processors has formed the core of the Arduino family for a number of years, and only recently have newer, faster, and more powerful processors joined in on the fun. To avoid confusion, we'll try to group the various Arduino and Arduino variant boards by processor, starting with the ATmega328 processor. Boards in this category include the Uno, Nano,

Lilypad, and Mini, among others. Most are virtually identical in function and capability, with the board size and footprint being the major difference between them. The ATmega328 has 32 KB of flash memory, 2 KB of SRAM, and 1 KB of EEPROM. We'll discuss these three different memory types in the next section.

Moving forward, the ATmega328 has been complemented with the ATmega32U4, which is identical in functionality to the ATmega328, with the addition of a built-in USB controller and 512 bytes more of SRAM in the ATmega32U4 chip itself. We'll cover the ATmega32U4 boards in the next section.

The Arduino Uno

When you talk about Arduino boards, the Arduino Uno (**Figure 2.1**) is the first thing to come to mind. The Uno is considered to be the current mainstream Arduino board and is the board that most people start out with. The majority of the existing libraries and sketches are designed to work with the Uno, so you have a wide variety of existing code to work with as you start your Arduino adventures. I still prefer to use the Uno in many of my projects, primarily for its low cost, simplicity, and ease of use.

The Uno, while not the true first Arduino processor board, is often considered to be the most commonly used of the original Arduino boards. The Uno uses an 8-bit ATmega328 series microcontroller running at a clock speed of 16 MHz. It has 14 digital I/O pins and six 10-bit analog-to-digital pins that can also be used for digital I/O, if desired. Six of the digital I/O pins support pulse width modulation (PWM) that allows you to control the pulse width of a square wave on the output pin to do things such as dim an LED or play an audio tone. Two of the digital I/O pins can be configured to support external interrupts for hardware program control, and all 24 I/O pins can be configured to provide

Figure 2.1 — Arduino Uno R3.

a program interrupt when the I/O pin changes state. Interrupts can be pretty handy, and we'll talk more about them in the USB CW Keyboard project chapter later in this book. That project could never have been made to work properly without the use of interrupts. Seeing how that project came together will you help to understand how interrupts work, and the discussion demonstrates how they can be used to do some pretty interesting things in your own projects.

The Arduino Uno has three types of onboard memory: flash, static random access memory (SRAM), and electrically-erasable programmable read-only memory (EEPROM). Flash memory is rewritable memory that is primarily used to store your Arduino programs, known as sketches.

Flash memory is semi-permanent and retains its contents even when power to the Arduino is turned off. You can rewrite the contents of flash memory approximately 100,000 times, meaning that you can use the same Arduino in dozens, and even hundreds of projects, simply by loading a different sketch. Flash memory can also be used to hold data that doesn't change, such as lookup tables, text, and other constants, in order to save valuable SRAM space through the use of the PROGMEM keyword.

SRAM is used to hold variables and other dynamic data, and its contents are volatile, meaning that the contents are lost when the Arduino is reset or powered off.

The onboard EEPROM can be used to retain data such as calibration values and similar settings between reset or power cycles. As with flash memory, the Arduino EEPROM has a lifetime of approximately 100,000 write cycles.

The Arduino Uno can be powered through the USB port or by 7 – 20 V dc on either its on-board dc power jack or though the Vin pin. The ATmega-series microcontrollers support the industry-standard Serial Peripheral Interface (SPI), and Inter-Integrated (I^2C) bus communication protocols. The Uno, as with many other Arduino boards, has a standard 2.7 × 2.1-inch footprint. Female headers on the edges of the board allow for the stacking of add-on interface boards, also known as shields, without the need for additional wiring.

Good Things Come in Smaller Packages

Not all Arduinos come in the standard footprint that allows the use of shields. Some, such as the Arduino Nano (**Figure 2.2**), and Solarbotics Ardweeny (**Figure 2.3**) are much smaller, yet have the same functionality and

Figure 2.2 — Arduino Nano.

Figure 2.3 — Solarbotics Ardweeny.

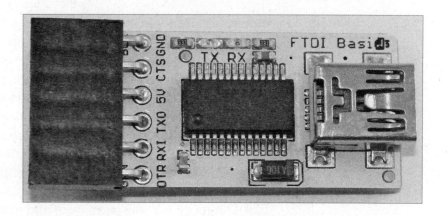

Figure 2.4 — FTDI interface.

features of their larger brothers. These smaller boards allow you to build smaller, more compact projects, yet still have access to the code base and all of the standard libraries available for the Uno. These baby Arduinos, some not much larger than a postage stamp or two, are often based on the ATmega328 and have the same features and functionality of the Arduino Uno. While many have a smaller mini-USB or micro-USB connector for programming, some, such as the Ardweeny from Solarbotics, require the use of a serial-to-USB interface module (also known as an FTDI interface — see **Figure 2.4**) for programming and interfacing with the Arduino IDE.

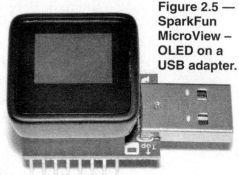

Figure 2.5 — SparkFun MicroView – OLED on a USB adapter.

Taking the concept of tiny Arduino to the next level, SparkFun has combined the ATmega328 with a 64 × 48 pixel organic LED (OLED) display. The SparkFun MicroView – OLED (**Figure 2.5**) is not much larger than the digital watches of old. Once programmed using the external USB programming adapter (**Figure 2.6**), all the MicroView needs is 3.3 – 16 V dc power and you're off and running.

Wearable Arduinos

One of the newer applications for the Arduino is in the area of wearable electronics. While you may not immediately think that wearable electronics are useful for ham radio projects, they might be of interest for electronic name/call sign badges and other ham accessories. Through the use of conductive thread, these wearable Arduinos, such as the Lilypad Arduino (**Figure 2.7**), can even be sewn into clothing and powered by small rechargeable Lithium-polymer (LiPo) batteries.

Figure 2.6 — SparkFun MicroView USB adapter.

While some of the original "wearable" Arduino boards have been officially discontinued, these boards are still available from various online sources, including SparkFun, Adafruit, and eBay, among others. Many of the boards in this category are based on the ATmega328 or ATmega 32U4 processors, with the major differences between them being size, available I/O pins, and type of power/battery and programming connections. These boards are programmed using the FTDI interface adapter or via an onboard micro-USB connector.

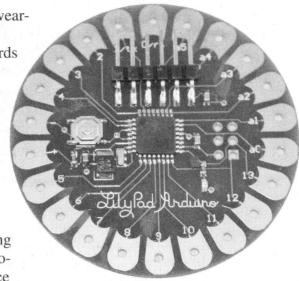

Figure 2.7 — Lilypad Arduino.

The Lilypad Arduino SimpleSnap (**Figure 2.8**) is unique in that it has an onboard LiPo battery and charging circuit, and is designed using conductive snap connectors, allowing you to remove the board from the project itself for washing or use in other projects. The Adafruit FLORA also has matching Bluetooth and GPS modules available.

The ATmega32U4 Series

Beginning with the Arduino Leonardo (**Figure 2.9**), the ATmega328 processor was replaced with the ATmega32U4. While nearly identical in functionality with the Uno and ATmega328, the 32U4 eliminates the need for the Uno's 16U2 external USB interface chip, and also allows the Leonardo to appear to a connected computer as a USB human interface device (HID) mouse and/or keyboard. The Leonardo and others in this class also feature 20 digital I/O pins instead

Figure 2.8 — LilyPad Arduino Simple Snap. (Photo courtesy Arduino.cc)

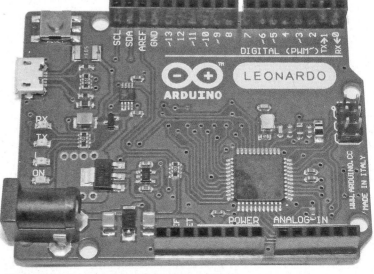

Figure 2.9 — Arduino Leonardo.

Arduino Processor Boards — An Update   2-5

of the Uno's 14, and they have 12 analog input pins versus the six on the Uno. Seven of the I/O pins support PWM instead of the Uno's six. The 32U4 also includes 2.5 KB of SRAM, 512 bytes more than that of the ATmega328-series.

The Leonardo does have the idiosyncrasy of the USB port number changing when the board is reset, which requires some getting used to during programming and Serial Monitor operations. Unless your project needs to emulate a keyboard or mouse, it would probably be best to choose an ATmega328-based board.

The Mega Series

With the advent of the larger touchscreen color graphic TFT (thin-film transistor) displays and the retirement of the Arduino Due, other Arduino processors with larger numbers of digital I/O pins, such as the Mega2560 are experiencing a resurgence. Still part of the mainstream Arduino line after more than a decade, the Arduino Mega series offers substantially more memory and I/O, along with other features.

Based on the 16-MHz ATmega2560 processor, the Mega2560 (**Figure 2.10**) packs a punch with 54 digital I/O pins, 16 10-bit analog inputs, and four hard-

Figure 2.10 — Arduino Mega2560.

ware serial ports on a larger 4 × 2.1-inch footprint. Six of the digital I/O pins can be configured for external hardware interrupts, and 15 of the digital I/O pins can provide PWM output. The I/O headers are compatible with most of the shields designed for the Uno and similar Arduinos, although care must be taken to ensure that the desired shield matches the Mega I/O pin layout. The Mega2560 ups the program flash memory to 256 KB, the SRAM to 8 KB, and the EEPROM to 4 KB. The Mega2560 ADK adds a USB Host port to enable communication with Android phones for use with Google's Android Accessory Development kit.

Recently, the Mega2560 has been downsized into an inexpensive, smaller package, roughly 75% of the Uno's footprint. Designated the Mega2560 Mini (**Figure 2.11**), it has the same ATmega2560 processor and all of the features in the full-size Mega2560. This board could easily become my "Go To" Arduino

Figure 2.11 — Mega2560 Mini.

board for when I need more memory or I/O than a standard Uno or Nano provides.

A New Batch of Choices

Moving beyond the somewhat limited processing and memory capability of the ATmega328, the latest Arduinos are using faster, more powerful processors with more memory and other added features. One of the nice things with these newer boards is that the Arduino IDE now has the capability of installing software "plug-ins" in order to support these processors. This means that you can create your sketch for one type of processor, and with a simple change in your IDE settings, you can compile the same sketch for another processor type without having to modify your sketch, other than to take advantage of features inherent in the new processor type.

Care must be taken when working with these processors, as many of the new boards are powered by 3.3 V and their I/O pins are generally not 5 V tolerant. That means you can burn out your board if you apply 5 V on any of the I/O pins. Be sure to check the power and voltage specifications when using these boards in your projects.

The Arduino Every (**Figure 2.12**) is a Nano-sized board with a beefier 20 MHz, 8-bit ATmega4809 processor onboard. The Every is 5 V compatible

Figure 2.12 — Arduino Every. (Photo courtesy of arduino.cc)

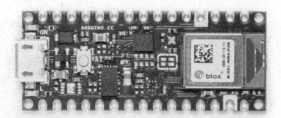

Figure 2.13 — Arduino Nano 33 BLE. (Photo courtesy of arduino.cc)

and has 48 KB of flash memory, 6 KB of SRAM, and 256 bytes of EEPROM. The Every has 14 digital I/O pins (five PWM) and eight analog pins, and can be used as a drop-in physical replacement for the Nano. It's an ideal swap, with slightly higher processor speed and significantly more memory.

The Arduino Nano 33 BLE boards (**Figure 2.13**) are 3.3 V boards using a 32-bit Nordic Semiconductors nRF52840, containing an ARM Cortex-M4 processor running at 64 MHz. The Nano 33 BLE has 1 MB of flash memory and 256 KB of SRAM. The Nano 33 BLE has 14 digital I/O pins (all can be PWM) and eight analog pins, making it an ideal replacement for the Nano, with significantly higher processor speed and more memory. Physically, the Nano 33 BLE is a drop-in replacement for the original Nano.

The standard BLE comes with a 9-axis inertial measurement unit (IMU) containing an accelerometer, gyroscope, and a magnetometer, along with Bluetooth BLE that allows it to be both a BLE and Bluetooth client and host device.

The Arduino Nano 33 BLE Sense version has all of the features of the standard Nano 33 BLE, and adds a series of embedded sensors, including humidity, temperature, barometric pressure, gesture, proximity, light color and intensity sensors, along with a built-in microphone.

The PJRC Teensy 4.0 and 4.1

The Teensy 4.0 and 4.1 (**Figure 2.14**) boards from PJRC are small-footprint boards that use an iMXRT1062, containing a 32-bit ARM Cortex-M7 processor running at a whopping 600 MHz, with 2 MB of flash (8 MB for the 4.1) and 1 MB of SRAM. These Teensy boards also include two USB ports, three CAN Bus ports, two I2S digital audio ports and one S/PDIF digital audio port. It also has SD card I/O capability, along with three SPI and three I²C bus interfaces. The 4.0 supports seven serial interfaces, 40 digital I/O pins, and 14 analog input pins. It also includes cryptographic acceleration, random number generation, and a real-time clock.

The Teensy 4.1 has all of the features of the 4.0, but also increases the flash

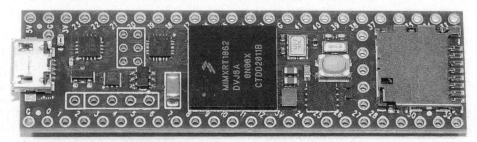

Figure 2.14 — Teensy 4.1.

memory to 8 MB. There are locations on the 4.1 board to expand the flash and SRAM memories with additional chips. The Teensy 4.1 increases the number of I/O pins to 55, and includes a 10/100 Ethernet interface onboard. One thing to be aware of with the Teensy 4.x boards is that not all of the pins are brought out to thru-hole solder connections. Some of the signals are brought out to small solder pads on the board, which does not lend it very well to socket mounting.

The SparkFun ESP8266

Traditionally, you will find the Espressif ESP8266 Wi-Fi-enabled controller as an added feature to a board, not as the microcontroller itself. However, the ESP8266 chip is powerful enough to handle both the microcontroller and the Wi-Fi function of an Arduino-compatible board.

The SparkFun ESP8266 Thing (**Figure 2.15**) has a 32-bit ESP8266 running at 80 MHz, with 512 KB flash and 96 KB SRAM, which also provides the built-in 2.4 GHz IEEE 802.11 b/g/n Wi-Fi. The SparkFun ESP8266 Thing has 11 digital I/O lines, all of which support PWM, and one 10-bit analog input pin.

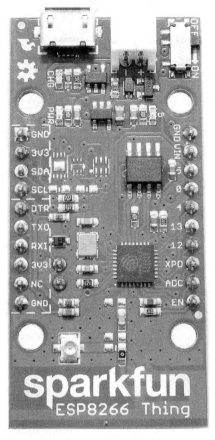

Figure 2.15 — SparkFun ESP8266 Thing.

It is important to note that the analog input pin on the ESP8266 can only accept a maximum of 1 V. The ESP8266 itself is a 3.3 V device and the I/O pins are not 5 V tolerant. SparkFun has created a plug-in for the Arduino IDE to allow programming the ESP8266 Thing using the standard Arduino IDE.

The SparkFun ESP32

The SparkFun ESP32 Thing uses the Espressif ESP32 Wi-Fi-enabled controller, which is a more powerful version of the ESP8266. The 32-bit ESP32 processor contains a dual-core Tensilica LX6 microcontroller running up to 240 MHz. In addition to the onboard 2.4 GHz IEEE 802.11 b/g/n Wi-Fi controller, the ESP32 also comes with integrated dual-mode Bluetooth, supporting both classic and BLE Bluetooth devices. The ESP32 Thing has 4 MB of flash and 520 KB of SRAM, along with 28 digital I/O pins (16 pins support PWM), and 18 analog input pins. The ESP32 Thing also has an onboard Hall-effect sensor, temperature sensor, and a microSD card slot, along with support for the SPI, I²C, and automotive CAN buses.

The ESP32 is a 3.3 V device and the I/O pins are not 5 V tolerant. As with the ESP8266, SparkFun has created a plug-in for the Arduino IDE to allow programming the ESP32 Thing using the standard Arduino IDE.

Figure 2.16 — Adafruit Feather32U4 LoRa.

The Adafruit Feathers

Adafruit has long been a leader in developing Arduino boards and modules. Adafruit's founder, Limor "Ladyada" Fried, AC2SN, passed all three of her amateur radio license tests in a single day. It therefore comes as no surprise that the Adafruit Feather series (**Figure 2.16**) of Arduino-compatible boards support a wide array of Wi-Fi, Bluetooth, and other methods of RF communication. The Adafruit Feathers are available with the Arduino 32U4 that is used in the Arduino Leonardo and other boards, including the ARM Cortex-M0, M0+, M3, and M4, the nRF52, the ESP-32, and the ESP8266, all on a smaller Arduino-Nano style footprint.

Depending on the version, the Adafruit Feather supports Bluetooth LE and 2.4 GHz IEEE 802.11 b/g/n Wi-Fi. More recently, the Feather supports the license-free 433 and 868/915 MHz Industrial/Scientific/Medical (ISM) bands, including 100-mW Long Range (LoRa), capable of communication over distances up to 20 km.

The memory and I/O specifications are different for each variant of the Feather. The best way to determine which Feather is right for you would be to visit the Adafruit website at **www.adafruit.com**. The Feather also has a wide array of add-on boards available. These boards, known as "FeatherWings," are similar to the shields used on the larger Arduinos. All of the Adafruit Feathers are supported in the Arduino IDE simply by downloading the Adafruit plug-ins for the Arduino IDE.

The Adafruit Grand Central M4 Express

Based on the Microchip ATSAMD51, the Adafruit Grand Central M4 (**Figure 2.17**) is a 3.3 V board that uses a 120 MHz Cortex M4 with floating point support for its processor, and uses the Arduino Mega form factor. The M4 Express has an impressive 54 digital I/O pins and 16 analog input pins. Two of the analog pins can be analog outputs using the onboard 12-bit digital-to-analog converter (DAC). The Express has 8 MB of flash storage and 256 KB of RAM, along with a microSD card slot.

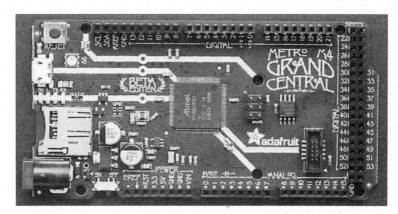

Figure 2.17 — Adafruit Grand Central M4 Express. (Photo courtesy Adafruit.com)

The Express also has eight ports of serial communication, including the onboard USB port, to provide any combination of I²C, SPI, or UART ports. As if that wasn't enough, it also supports stereo I2S audio I/O, and incorporates a 12-bit parallel capture controller for camera/video in. While the primary target for the M4 Express is CircuitPython development, it does have an Arduino board support package.

Arduino Board Summary

You have a wide array of Arduino boards to choose from for your Arduino projects, and we only covered a small portion of them. Newer, more powerful Arduino boards are being developed all of the time, and with the smaller footprint of these boards, creating smaller, portable Arduino projects has been made much easier. In this chapter you have seen most of the different Arduino boards currently available that you may want to use in your own projects. In the next chapter, we will discuss the add-on boards, modules, and components you are most likely to use in your ham-related Arduino projects.

3 Arduino Shields, Modules, and Devices

It seems like new shields, modules, and components for the Arduino are coming out daily. While researching the ideas for some of the projects planned for this book, I discovered even more new devices just begging to be used in a project. Even now, as I put the finishing touches on this book, I've already got a bunch of projects in mind that use some of these new components.

As with the Arduino processor boards, rather than discuss every possible shield, module, and component out there that can be used in Arduino projects, we'll be primarily focusing on those that would commonly be used in ham radio-related Arduino projects. At the end of this chapter, you'll find a list of some of the places I get my Arduino parts from. eBay has become the place I go to for the majority of my Arduino parts. I've bought tons of parts from eBay, where they are usually less expensive than anywhere else, and I have never had bad parts or other quality issues. When it comes to the Arduino, more often than not, "Parts is Parts," no matter where you choose to buy them.

The Arduino was designed to interface with all manner of shields, modules, and devices. The standard layout of I/O header pins on the Uno and Mega allow for a wide array of shields to simply be plugged in on top of the Arduino. Without any wiring at all, you can add display, networking, and other features to your Arduino project.

With its built-in support for the industry standard Serial Peripheral Interface (SPI), Inter-Integrated Circuit (I^2C), and One-Wire bus protocols, you can add any number of modules and components, including displays, temperature sensors, GPS modules, lightning detection, Ethernet modules, and all kinds of interesting sensors and control capabilities to the Arduino. The list of things you can add to your Arduino is nearly endless and growing daily.

You may be familiar with some of these modules from my previous books. I promise we'll get to the new stuff as quickly as we can.

Shields

As you start creating projects with the Arduino, you may find it easier to use preassembled shields rather than figuring out how to work with and wire up individual modules and devices. Shields are boards that are designed to plug into the headers on top of the Arduino processor board, allowing instant access to the features on the shield without having to do any additional wiring.

You can even stack shields one on top of the other, adding functionality with each additional shield. However, the shield stacking method can get rather cumbersome in a hurry, and you always run the risk of I/O pin usage conflicts between the various shields. As a general rule, you'll only want to stack one or two shields on top of the Arduino before deciding that it may be best to use a protoshield (prototyping shield) or some other method to move your project off the unwieldy shield stack and onto something more functional and aesthetically pleasing.

Shields are usually supported by program libraries and example code, which allow you to quickly and easily develop working sketches. A thorough list of Arduino shields available can be found at **www.shieldlist.org**.

HobbyPCB RS-UV3 Radio Shield

The HobbyPCB RS-UV3 (**Figure 3.1**) is a complete 144/222/450 MHz FM transceiver shield that allows you to create 1200- and 9600-baud packet radio repeaters, EchoLink stations, and other VHF/UHF radio projects. The RS-UV3 can also be used as a standalone radio, without the need for an Arduino.

While it is essentially a standalone board, the RS-UV3 can be used as an Arduino shield, and has pin headers to match the Arduino Uno and Mega board shield footprint. The RS-UV3 has built-in level shifters allowing the use of either 3.3 or 5 V Arduino or Raspberry Pi boards, with an onboard 7.2 V Lithium-ion (Li-ion) battery charger, and a power output of 200 mW. The RS-UV3 also includes software to allow you to control the Radio Shield from a PC, with a virtual front panel application to give access to all of the functions of the RS-UV3.

Figure 3.1 — HobbyPCB RS-UV3 shield.

The RS-UV3 is controlled via the Arduino Serial port, and because it is designed as a standalone transceiver, there are no specific Arduino libraries needed to use it. The RS-UV3 has an optional aluminum case available, but if you plan to use the RS-UV3 with this case, there is no room to stack it on top of an Arduino and fit everything inside the case. HobbyPCB also has an optional 5 W VHF/UHF amplifier that works with both the RS-UV3 and the HamShield. A complete review of the HobbyPCB Radio Shield can be found in the April 2016 issue of *QST*.

HamShield

The HamShield started out as a Kickstarter project that achieved more than 468% of their funding goal. The HamShield (**Figure 3.2**) is a complete VHF and UHF FM transceiver on an Arduino Uno-style shield. Based on the Auctus 1846S radio transceiver chip, the HamShield covers 134 – 174, 200 – 260, and 400 – 520 MHz with a power output of 500 mW. Offering such features as CTCSS/CDCSS subaudible tones and DTMF tone encoder, the HamShield also offers 12.5 and 25 kHz channel bandwidths.

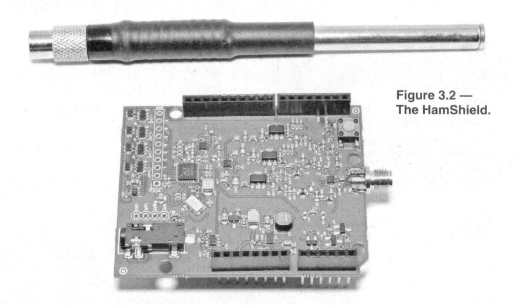

Figure 3.2 — The HamShield.

The HamShield Arduino Library allows you to control every aspect of the radio, and already has built-in functionality for CW, packet radio, SSTV, scanning, white channel seeking, empty channel detection, and more. I had originally considered using the HamShield for one of the projects in my books, but it is so complete as it is, the project would have ended up being something similar to just attach the shield, load a sketch and go — the HamShield and its library are just that powerful.

Argent Data Radio Shield

The Argent Data Radio shield (**Figure 3.3**) can be used to provide AX.25 packet radio send and receive capability with the Arduino. Packets are sent

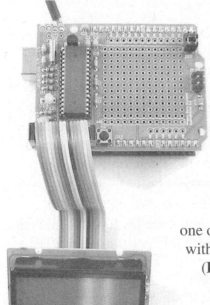

and received in AX.25 UI frames at 1200 baud, allowing operation on the VHF APRS network. There are a number of projects that use this board in ARRL's *Ham Radio for Arduino and PICAXE* by Leigh Klotz, WA5ZNU (**www.arrl.org/shop**, ISBN: 978-0-87259-324-4). The Argent Data Radio shield also includes a prototyping area, and an HD44780-compatible LCD interface.

Liquid Crystal Display (LCD) Shield

While it has largely been replaced by the newer display options, one of the first displays that just about everyone uses when starting out with the Arduino is the 16-character by 2-line (16×2) LCD display (**Figure 3.4**). Also available as a standalone module, the 16×2 LCD shield is a quick and easy way to add a display to your Arduino project. Many LCD shields also include several pushbutton switches, allowing you to add control functions to your Arduino sketches.

Figure 3.3 — Argent Data Radio shield.

Color TFT Display Shield

To give your Arduino project a really nice look, you can use one of the many color graphic thin-film-transistor (TFT) displays, such as the one shown in **Figure 3.5**. These high-resolution color graphic displays come in various sizes, from 1.8 inches (128×160 pixels) all the way up to (and beyond) a monster 7-inch, 800×480 pixel display.

The Adafruit Color TFT shield features a 1.8-inch TFT display, along with an onboard microSD card slot you can use to store images for display.

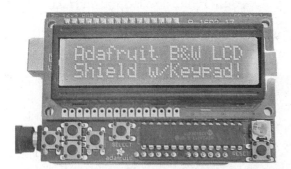

Figure 3.4 — Adafruit LCD shield.

Figure 3.5 — Adafruit Color TFT shield.

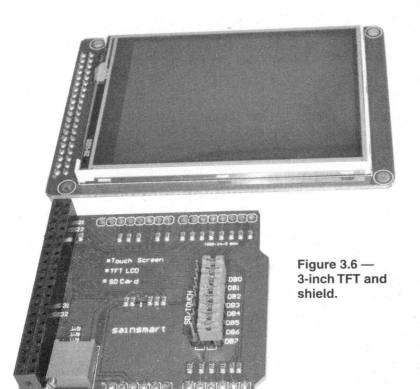

Figure 3.6 — 3-inch TFT and shield.

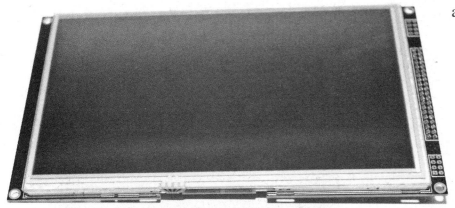

Figure 3.7 — 7-inch TFT and shield.

The 1.8-inch version of this shield also includes a five-way joystick, while the 2.8-inch version incorporates a resistive touch screen. Using the powerful Adafruit GFX library, these displays allow you to rotate the entire screen via software, draw pixels, lines, rectangles, circles, triangles, and other shapes, along with the ability to display normal text.

Recently, some of the larger and inexpensive color TFT displays have become available from places such as eBay. These displays range in size from 3 inches to 7 inches (and even larger). Due to the number of I/O pins needed to interface with these displays, you will need a display shield similar to the one in **Figure 3.6** and **Figure 3.7**. This shield is available for both the Arduino Uno and Arduino Mega footprint boards and will allow you to interface with many of the larger TFT displays now available. While these larger TFT displays can be used with the Arduino Uno, due to the number of I/O pins these larger displays require, you can't use the touchscreen features of the display with the Uno-type boards. For that, you would need a board with more available I/O pins, such as the ATmega2560 and the matching LCD shield.

Arduino Shields, Modules, and Devices   3-5

Relay and Motor Shields

Due to the fact that controlling external devices is one of the things that the Arduino does best, a relay shield such as the one shown in **Figure 3.8** or a motor shield (**Figure 3.9**) is often the quickest, easiest, and safest way to interface your Arduino to high-current and high-voltage devices such as motor and power control systems. The DFRobot relay shield, for example, allows you to control four onboard relays, and includes test buttons and indicator LEDs.

Ethernet Shield

The Arduino Ethernet shield (**Figure 3.10**) gives your Arduino the ability to communicate over your home network and the internet. With its powerful

◀Figure 3.8 — DFRobot 4 Channel Relay shield. (Photo courtesy DFRobot.com)

▶Figure 3.9 — DFRobot Motor Driver shield. (Photo courtesy DFRobot.com)

Figure 3.10 — Ethernet shield.

libraries and example sketches, you can turn your Arduino into a web-enabled simple chat server, web server, FTP server, or telnet client. It can be used to allow your Arduino to make http requests from websites and much more. The Ethernet shield also has an onboard microSD card slot that can be used to store files and other data. Some Ethernet shields also include an option to install a Power-over-Ethernet (PoE) module, allowing your Arduino to be powered via the Ethernet cable using a PoE switch or power injector.

Figure 3.11 — SparkFun USB Host shield.

USB Host Shield

The USB Host shield (**Figure 3.11**) adds USB host capability to your Arduino project, allowing you to communicate with USB devices such as keyboards, mice, joysticks, game controllers, and many other USB devices.

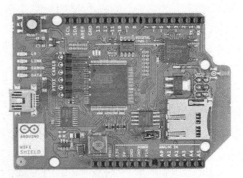

Figure 3.12 —Wi-Fi shield. (Photo courtesy Arduino.cc)

Wi-Fi Shield

While many of the newer Arduino processor boards have Wi-Fi built in, you may want to experiment with Wi-Fi by using something like an Uno and a Wi-Fi shield. The Arduino Wi-Fi shield (**Figure 3.12**) allows the Arduino to connect to IEEE 802.11 b/g networks, and supports the WEP and WPA encryption protocols.

With the proliferation of Arduino boards and variants that include Wi-Fi support onboard, the Wi-Fi shield is not as popular as it once was, but it is still a quick and easy way to add Wi-Fi capability to your Uno-based projects. Some of the original Wi-Fi shields are no longer being offered by places such as **arduino.cc**, but they are still available from suppliers such as eBay.

Audio Shields

Some of your Arduino projects may involve making sounds or playing audio, as in the case of a repeater identifier or contest voice keyer. Using one of the audio shields, you can quickly and easily add high-quality sound and even music to your Arduino projects.

The Adafruit Wave shield shown in **Figure 3.13** can play up to 22-kHz, 12-bit WAV audio files that are stored in the onboard microSD card. The

Figure 3.13 — Adafruit Wave Shield. (Photo courtesy Adafruit.com)

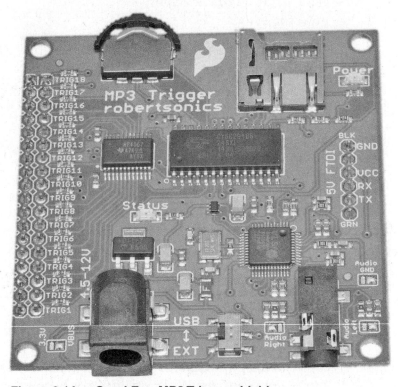

Figure 3.14 — SparkFun MP3 Trigger shield.

SparkFun MP3 Trigger (**Figure 3.14**) has 18 external trigger pins that allow you to directly play a selected MP3 file from the onboard microSD memory card. The MP3 Trigger shield has the capability for remote triggering of up to 256 separate MP3 files with up to a 192-kHz data rate.

XBee Shield

The XBee shield (**Figure 3.15**) allows an Arduino board to communicate wirelessly using the ZigBee protocol. ZigBee is based on the IEEE 802.15 standard and is used to create a small "personal area network" allowing you to

Figure 3.15 — Solarbotics XBee shield. (Photo courtesy Solarbotics.com)

communicate wirelessly with remote sensors and devices. The XBee can communicate at speeds up to 250 Kbit/s over distances up to 100 feet indoors or 300 feet outdoors, and can be used as a wireless serial/USB replacement or in a broadcast/mesh-style network.

CAN Bus Shield

Controller Area Network (CAN) is a bus protocol standard commonly used by the automotive industry for communication between the various onboard systems, as well as providing information for the on-board diagnostics system (OBD-II) mechanics use to monitor and troubleshoot your vehicle. While not specifically related to ham radio projects, the CAN Bus shield (**Figure 3.16**) offers some interesting possibilities for mobile operations, with the ability to extract data from a vehicle's computer and insert that data into telemetry information such as Automatic Packet Reporting System (APRS).

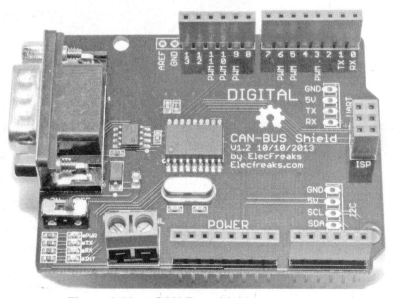

Figure 3.16 — CAN Bus shield.

Power Boost Shield

With its low power consumption, the Arduino is ideal for portable, battery-powered projects. While using 9 V batteries or AA battery packs does work, and I use them in a number of my projects, sometimes a self-contained rechargeable power source would do the job much better and a lot more efficiently. For that,

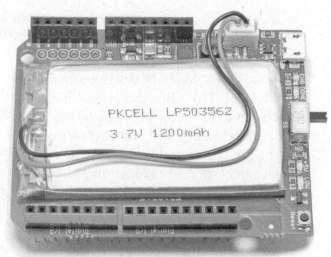

Figure 3.17 — Adafruit Power Boost shield.

Figure 3.18 — Easy VR Voice Recognition shield.

there is the Adafruit PowerBoost shield shown in **Figure 3.17**. This shield allows you to attach a high-capacity rechargeable Lithium Polymer (LiPo) battery directly to the Arduino. This shield contains a built-in battery charger as well as a dc-to-dc converter to provide the 5 V needed by the Arduino Uno-style boards. Capable of providing at least 500 mA of output current, with a peak output up to 1 A, the shield can use any 3.7/4.7 V Lithium-ion or LiPo battery. A 1200 or 2000 mAh capacity LiPo battery is recommended.

Voice Recognition Shield

I have had some wonderful conversations with people who contacted me about some of the projects in my first Arduino book. I probably should have expected it, but I was surprised by the interest from various organizations for the vision-impaired, asking for more voice-related Arduino projects. Also, just as I was finishing that book, I discovered the EasyVR Voice Recognition shield, but time and space constraints did not allow me to include any voice recognition projects in my first two books. Unfortunately, this is also the case with this book, there just wasn't enough time and room to do any voice recognition projects. They're definitely at the top of the list for any new projects.

The EasyVR Voice Recognition shield (**Figure 3.18**) has some very interesting potential for use in ham radio projects. The newer 3.0 version of this shield has up to 32 user-defined Speaker Dependent commands (available in US English, Italian, Japanese, German, Spanish, and French), along with 26 built-in Speaker Independent commands you can use in your Arduino projects. With the optional Quick T2SI Lite license, you can add up to 28 Speaker-Independent (SI) vocabularies, each with up to 12 SI commands, allowing a total of up to 336 additional SI commands. You can even generate DTMF tones with the EasyVR shield. This is a shield that I definitely plan to integrate into some of my future Arduino projects.

I/O Sensor Shields

Using an I/O Sensor shield (**Figure 3.19**) is great way to begin creating your own Arduino projects. This shield brings all of the Arduino pins out to groups of 3-pin headers, along with power and ground on each header group. This particular shield also has header sockets that can be used with either the Uno or Nano board footprint. This shield allows you to easily connect the I/O pins to your breadboard and prototyping boards using jumper wires while you are developing your project. Once your development is complete, all you have to do is remove the connecting wires and your Arduino is ready for the next project.

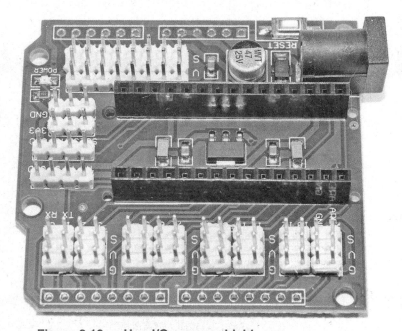

Figure 3.19 — Uno I/O sensor shield.

Figure 3.20 — Breadboard shield.

Breadboard Shield

For quick designing and prototyping for projects that only involve a few discrete components, a chip or two, and a couple of wires, there is the breadboard shield as shown in **Figure 3.20**. Smaller than a full-size breadboard, the breadboard shield allows you to create Arduino projects without the need for soldering, and you can quickly set up and rewire your project as needed. When you're finished, just as in a full-size breadboard, you can simply remove all the components and wires, and your breadboard shield is ready for the next project.

The DrDuino Shields

The DrDuino Shields are a great way to work with the Arduino. Several of the projects in this book started out as proof-of-concept and prototypes using the DrDuino Explorer Shield (**Figure 3.21**). The DrDuino shields have several

Figure 3.21 — DrDuino Explorer shield.

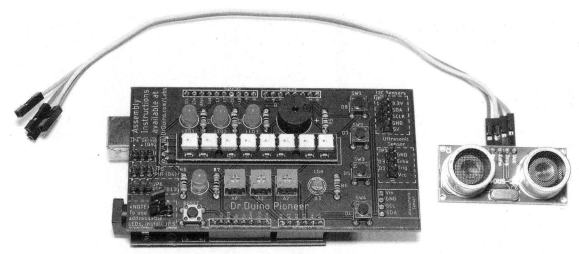

Figure 3.22 — DrDuino Pioneer shield.

of the components most commonly used when learning and working with the Arduino. For example, the DrDuino Pioneer (**Figure 3.22**), has three LED outputs, three potentiometers for analog inputs, four pushbutton switches, a piezo buzzer, and an addressable LED strip with eight RGB LEDs, along with connections for other sensors and devices such as a servo, an ultrasonic sensor, a light-dependent resistor, and a passive IR sensor, all on a shield that fits the Uno-type boards. The Pioneer is an excellent starter shield to learn how to use the Arduino. For more information, please see the DrDuino Pioneer product review in the November 2020 issue of *QST*.

The DrDuino Explorer Shield is targeted to be a mini development platform. The Explorer has two breadboard areas — the standard temporary wire jumper type, and a second area with solder pads for a more permanent use. The Explorer features four LED outputs, four pushbutton inputs, three potentiometers for analog input, a light-dependent resistor, an HC-05 Bluetooth module, a 128×32 organic LED display, a 5 V, 3 A dc-to-dc converter, and an addressable LED strip with eight RGB LEDs. The Explorer shield also has additional headers for external connections for an atmospheric sensor, an ultrasonic proximity sensor, a servo, and other external device connectors.

The Explorer also has all of the Arduino I/O pins brought out to jumpers on the board, allowing you to re-route one or more I/O pins to the Explorer on-board components or pass these signals on to the an Uno-type shield attached to the Explorer board. The Explorer has header sockets for either an Uno or Nano processor. The Explorer is an ideal tool to build and test your project prototypes. You can read a full review of the DrDuino Explorer in the January 2021 issue of *QST*.

Prototyping Shield

Once you have finished breadboarding and testing your project, you may want to create a more permanent version that involves wiring and soldering components. I have found the easiest method to interface my projects to the

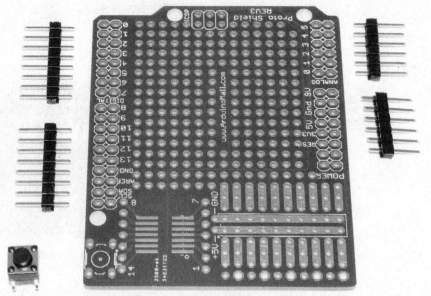

Figure 3.23 — A prototyping shield.

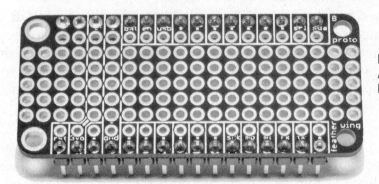

Figure 3.24 — Adafruit FeatherWing Proto.

Arduino is with a prototyping shield ("protoshield") such as the one shown in **Figure 3.23**. For their Feather line, Adafruit has the FeatherWing Proto (**Figure 3.24**) that can be stacked on top of a Feather board in a similar manner to a standard Arduino shield.

A protoshield brings all of the Arduino pins and voltages to solder pads, with the center of the prototyping shield laid out like a standard prototyping board. I like to solder DuPont-style 2.54 mm header pins in the prototyping area so that I can wire the Arduino pins to the header, and then use a connector to interface to my off-board modules and components. This also allows for easy troubleshooting, reconfiguration, and replacement of the external parts as needed. An added benefit to using prototyping shields is that you can quickly swap shields and sketches as needed and use the same Uno-style processor board for many different projects.

The Audeme MOVI Speech Recognition and Voice Synthesizer Shield

One of the latest Arduino shields is the Audeme MOVI speech recognition and voice synthesizer shield. The functions of speech and voice recognition are usually handled by two separate units, so this looks to be a handy shield to have around. The MOVI is programmed directly from the Arduino IDE, requires no voice samples for training, does not need an internet connection, and is speaker-independent. It has a built-in microphone with AGC, so it can detect speech up to 10 feet away. The MOVI shield is designed to work with the just about any Arduino board, including the Uno and Mega boards. In addition to English, the MOVI directly supports Spanish and German.

Arduino Modules

While shields provide a quick and easy way to interface with the Arduino, the real fun is when you begin interfacing your Arduino to external modules and components. With just a handful of modules and parts, you can wire up some pretty amazing projects. Most of the modules for the Arduino include software libraries, which greatly simplify working with these modules. You can create fully functional and powerful Arduino projects in just a few short hours.

Using individual modules and components also helps lessen the pain if (and in my case when) you do horrible things to your project and challenge Ohm's Law (Hint: Ohms Law usually wins, and signifies its victory by sending out smoke signals.) Shields are usually more expensive than individual modules and components. With many shields, when something goes horribly wrong, the entire shield no longer works as intended, and you have to buy a new one. With the lower cost of individual modules and components, fewer tears are shed, and if you're like me, you probably have more than one willing victim in your now rapidly growing supply of parts. Many of the various modules and components can be mounted to your project using header sockets, allowing for quick and easy replacement if bad things should happen.

In reality, I have found that the Arduino, shields, and modules are very forgiving when you mis-wire them. As long as you pay attention to the 3.3 and 5 V differences between boards and components, you have to really work to damage them. I'm not saying it can't be done, and it's amazing just how much smoke a tiny surface-mount component has inside of it, but you really have to do something wrong to get to that point.

Another big advantage of using modules instead of shields is that projects using shields tend to be bulky and not very aesthetically pleasing. Modules give you control over the size and shape of your project, allowing you to determine what the finished project will look like. To this end, there are a number of enclosures designed for the Arduino and Arduino projects that can really spice up the look of your completed project. Also, you are not just limited to enclosures designed for the Arduino, I've found a number of suitable enclosures at places such as Hobby Lobby and other craft stores that are ideal for Arduino projects. We'll discuss enclosures later in this chapter.

The following sections will introduce and briefly discuss the various modules and components that I have found to be most useful for creating Arduino-powered ham radio projects. Again, this is by no means an exhaustive list as new modules and components for the Arduino are being released every day. The modules and devices shown here are the ones most suited for Arduino ham radio projects.

Direct Digital Frequency Synthesis (DDS) Module

Prior to the development of the direct digital frequency synthesis module, about the only way to generate a radio frequency (RF) signal was to use a crystal or an unstable (relatively speaking) resistive-capacitive (RC) oscillator. Using a DDS module such as the AD9850 DDS module shown in **Figure 3.25**, you can programmatically generate highly stable waveforms up to 62.5 MHz, with an accuracy of 0.0291 Hz. Other DDS modules support higher frequencies than the AD9850, but I have found the inexpensive AD9850 to be more than adequate for most Arduino ham radio projects.

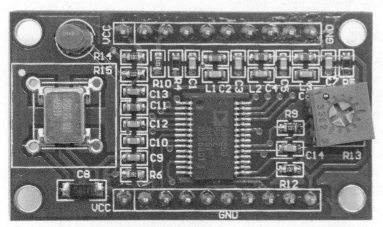

Figure 3.25 — AD9850 DDS module.

A DDS generates a smooth sine or square wave, while some DDS modules, such as the AD9833 also allow you to generate triangle waves. One of the major advantages of using a DDS module over processor-generated waveforms is that the DDS chip offloads all of the frequency and waveform generation from the processor. In the case of the AD9850, the module uses a 10-bit digital-to-analog converter on the chip to produce clean, stable waveforms across the entire operating range.

The AD9850 DDS also has the capability of shifting the phase of the output waveform from $0 - 2\pi$ ($0 - 720°$) and it interfaces to the Arduino using the SPI bus. There are a number of libraries supporting the DDS modules, including a very simple and easy to use library for the AD9850 written by Paul Darlington, MØXPD.

Si5351 High Frequency Programmable Clock Generator Module

The Si5351 High Frequency Programmable Clock Generator Module (**Figure 3.26**) is similar to a DDS in that both generate highly precise frequencies. In the case of the Si5351, It can generate up to eight different frequencies from 8 kHz to 160 MHz simultaneously, although many of the inexpensive Si5351 modules currently available use a three-output version of the chip.

The Adafruit Si5351A module communicates with the Arduino using the I²C bus and has three independent outputs, each capable of providing 3 V peak-to-peak square waves. The Si5351 is a 3.3 V device, so you will need level shifters on the Si5351 I/O pins if you are using a 5 V Arduino.

One major difference between a typical DDS and the Si5351 is that the

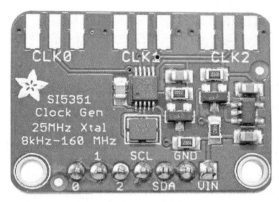

Figure 3.26 — Si5351 programmable clock generator module.

Si5351 only has a square wave output, as compared to the sine and square wave output of the AD9850 DDS. This means that an unfiltered signal generated by the Si5351 will also generate unwanted harmonics at multiples of the original frequency. For this reason, if you are going to use the Si5351 as part of a transmitter, for example, you will need to add a filtering section to smooth the square wave into a sine wave to reduce harmonic output. Adafruit and others have excellent libraries that support the Si5351, and allow you access to all of the features of the Si5351 chip.

Mini USB Host Module

While technically a mini-shield designed for use with the Arduino Pro Mini, the Mini USB Host shield/module shown in **Figure 3.27** is ideal for adding USB host capability to your Arduino project. The Arduino USB Host library designed for the full-sized USB host shield works with this module as well, allowing you to easily incorporate USB devices, including USB human interface devices (HID) to your Arduino projects. The Mini USB Host module also has eight general purpose input and eight general-purpose output pins that are accessible through the USB Host library.

Figure 3.27 — Mini USB Host module.

Ferroelectric RAM (FRAM) Module

Back when computers were just getting started and took up whole rooms (or racks as they got smaller), they predominately used ferrous core memory, which under magnification looks like a whole bunch of miniature toroids all woven together. With the advent of dynamic and static RAM chips, the good old core memory pretty much disappeared from use. Now it appears there is a new version of ferroelectric RAM (FRAM) (**Figure 3.28**) that is non-volatile, meaning that it retains the stored information even after power is turned off, similar to the flash memory used by the Arduino to store your sketches.

Unlike flash and EEPROM however, FRAM is much faster to write to. And, it has a much longer lifespan than flash memory. Flash and EEPROM memory can typically only be rewritten about 100,000 times, whereas FRAM can be rewritten 10,000,000,000,000 (ten trillion) times, and can retain data for up to 95 years at room temperature. FRAM comes in multiple sizes, typically from 64 Kbit (8 KB) to 256 Kbit (32 KB).

Figure 3.28 — Adafruit Ferroelectric RAM (FRAM) module.

Arduino Shields, Modules, and Devices   3-17

Relay Modules

There are many choices in relay modules for the Arduino. SainSmart has several interesting options for adding relay control to your project. Their iMatic series of relay control boards offers wired-Ethernet or Wi-Fi-based control for up to 16 relays. They also offer a Network Web Server controller version that allows you to remotely operate the relays through an onboard web server.

SparkFun also offers a high-power 40 A, 28 – 380 V ac four-channel solid-state relay board with a pre-programmed Atmel ATtiny84 microcontroller onboard. You can daisy-chain over 100 of these boards together for some serious ac power control configurations.

Stepper Motor Driver Module

There are a number of stepper motor drivers available for the Arduino. One of these is the SparkFun ProDriver, based on the TC78H670FTG bipolar stepper motor driver chip. The ProDriver can be controlled either by clock-input stepping or serial commands, and it has high-resolution stepping control — down to 1/128th of a step.

SparkFun I²S Module

The I²S bus interface is a new method of communication used for digitally connecting audio devices together. The SparkFun I2S Audio Breakout board/module uses the MAX98357A digital to analog converter (DAC) chip to drive an onboard Class-D amplifier to produce up to 3.2 W of audio power.

RobotDyn nRF24L01 2.4 GHz ISM Band Transceiver Module

The RobotDyn 2.4 GHz ISM Band Transceiver module is an ultra-low-power 2.4 GHz transceiver that operates in the Industrial, Scientific, and Medical (ISM) band and is capable of 2 megabits per second data rates.

LED Driver Modules

With the advent of the addressable RGB LEDs, you may find that you don't have enough pulse-width modulation (PWM) pins on your Arduino processor board to build the project you have in mind. To solve that problem, rather than using a different Arduino board with more PWM pins, you can use a PWM LED driver board. Adafruit has 12-channel (16-bit PWM duty cycle control) and 24-channel (12-bit PWM duty cycle control) PWM LED driver boards (**Figure 3.29** and **Figure 3.30**). These boards require just two or three SPI pins to operate, based on the board you select.

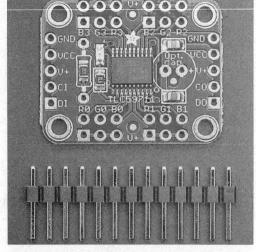

Figure 3.29 — Adafruit 12-channel LED driver module. (Photo courtesy adafruit.com)

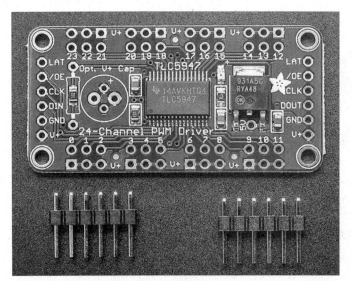

Figure 3.30 — Adafruit 24-channel LED driver module. (Photo courtesy adafruit.com)

You can even drive LEDs in series using these boards.

SparkFun also has a 16-channel LED driver module with a 12-bit PWM duty cycle control, with a 6-bit current limit control. It can be daisy chained with other boards.

Keypad Modules

In addition to the thin membrane keypads that have been associated with the Arduino from the early days, other keypad options are now available. Adafruit has several choices, including the original thin membrane keypad. They also offer the NeoTrellis 4×4 key matrix board with a silicone elastomer keypad cover and enclosure as shown in **Figure 3.31**. This board is available in several styles. One version uses monochrome 3 mm LEDs to backlight the keys, while the other uses addressable RGB LEDs. The NeoTrellis boards can also be connected together to use up to 32 boards (512 keys!) in a single configuration.

Figure 3.31 — Adafruit NeoTrellis 4×4 keypad and enclosure.

Figure 3.32 — Generic 4×4 keypad and enclosure.

Banggood and other suppliers also offer a 4×4 keypad (**Figure 3.32**) with removable keys and key covers that allow you to use your own key labels. It comes with blank keys, so you will need to create your own labels. I used laser-printed decal paper to create mine and transferred them to the keycap just like you would apply a decal to a model car or airplane. Unlike other keypads, this keypad outputs an analog voltage unique for each keypress, and the voltage can be read using an analog input pin on the Arduino. The keypad only needs power, ground, and an analog input pin to operate. It's available from many of the usual Arduino parts sources, including Banggood and eBay.

Arduino Shields, Modules, and Devices

Figure 3.33 — INA169 current sensor module.

Figure 3.34 — 30 A Hall-Effect current sensor.

Current Sensors

There are a number of current and voltage sensing modules, both invasive and non-invasive, for the Arduino. Starting out small, there is the INA169 (**Figure 3.33**) that can measure the dc current in a circuit. A "high-side" current sensor, the Texas Instruments INA169 chip converts the dc current flow across a shunt resistor into an output voltage that corresponds to 1 V/A (volt per amp), which can then be read using an analog input pin on the Arduino. Capable of monitoring up to 5 A at a maximum of 60 V continuously, the shunt resistor can be changed to increase or decrease the sensitivity of the sensor.

Another current sensor you can use is the Hall-Effect current sensor (**Figure 3.34**). Available in multiple ranges from ±5 A all the way up to ±75 A, a Hall-Effect current sensor uses the electromagnetic field flowing throw a conductor to measure current, thereby isolating the current source from the sensing electronics itself and providing a measure of protection for your measurement device.

Next, there is the non-invasive ac current sensor shown in **Figure 3.35**. The non-invasive current sensor clamps around one leg of an ac circuit, and through the use of a precision transformer, converts the current flowing in the ac circuit into a low voltage that

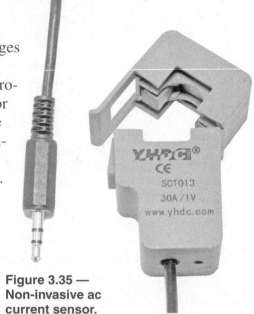

Figure 3.35 — Non-invasive ac current sensor.

Figure 3.36 — CJMCU-116 9-Axis IMU module.

can be rectified and read using an analog input pin on the Arduino. One of the benefits of a non-invasive current sensor is that you do not need to insert the sensor into the circuit path, providing a measure of isolation and protection for your measuring device.

9-Axis Motion/Position Sensor

The latest in motion and position sensing modules is the 9-axis, or 9 Degree of Freedom (9-DoF) sensor. Based on the InvenSense MPU-9250, the CJMCU-116 module (**Figure 3.36**) incorporates a 3-axis gyroscope, a 3-axis accelerometer, and a 3-axis magnetometer in a single small module, allowing you to create projects that can include a built-in compass, as well as azimuth and elevation sensing, and even motion sensing. This would an ideal module for a hand-held satellite tracking antenna, or for high-altitude balloon telemetry projects, among others.

ENC28J60 Ethernet Module

The ENC28J60 Ethernet Module (**Figure 3.37**) allows you to quickly add wired 10/100 Mbit/s Ethernet networking capability to your Arduino projects. Easily interfaced to the Arduino via the SPI bus, the Ethernet module is supported by a number of Arduino libraries and allows you to turn your Arduino into a web server, FTP file server, and even a Network Time Protocol (NTP) server.

Figure 3.37 — ENC28J60 Ethernet module.

GPS Module

With a GPS module such as the one shown in **Figure 3.38**, you can add all of the power of a standard GPS receiver to your Arduino, providing such information as time accurate to 1/100 of a second, latitude, longitude, altitude, speed, course, and other features of a standard GPS system. These GPS modules communicate with the Arduino via a standard TTL serial I/O port, typically with a default speed of 9600 baud, and they output standard National Marine Electronics Association (NMEA) NMEA-0183 messages. The TinyGPS and TinyGPS+++ Arduino libraries by Mikal Hart at **arduiniana.org** parse the NMEA messages sent by the GPS into data that can be accessed using simple function calls from within your Arduino sketch.

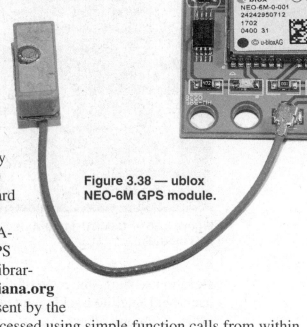

Figure 3.38 — ublox NEO-6M GPS module.

ISD1820 Voice Recorder and Playback Module

The ISD1820 (**Figure 3.39**) is capable of recording and playing back an 8 to 20 second message. The record duration and sampling frequency is selected by an external resistor. The module even includes an onboard microphone and LM386-driven speaker output, along with pushbuttons to active the recording and playback functions.

Emic 2 Text-to-Speech Module

Designed by Parallax in conjunction with Grand Idea Studios, the Emic 2 text-to-speech module (**Figure 3.40**) allows you to add natural sounding speech to your Arduino projects. Capable of speaking in English and Spanish, the Emic 2

Figure 3.39 — ISD1820 voice recorder and playback module.

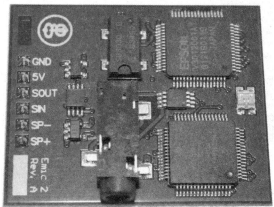

Figure 3.40 — Emic 2 text-to-speech module.

features nine preprogrammed voice styles in addition to program control of speech characteristics such as pitch, speaking rate, and word emphasis.

The Emic 2 communicates with the Arduino using a standard TTL serial I/O port at a default rate of 9600 baud, and includes an onboard audio amplifier with an audio jack. No program libraries are needed to use the Emic 2. All you have to do is send text to the serial port and the Emic 2 will convert it, along with any speech characteristic commands, directly into speech.

WT588D Speech Module

In addition to the Emic 2 Text-to-Speech module, the WT588D series of speech modules allow you to record and play back your own voice messages. The WT588D-U shown in **Figure 3.41** is a very inexpensive digital recording and playback device. The -U designation indicates the module has an onboard USB port to allow you to upload sound files to the board. The typical WT588D-U module has 32 MB of onboard flash memory for the WAV format sound files. The WT588D-U supports WAV files with a 6 to 22 kHz sampling rate for high-quality audio playback, and files can be uploaded while the WT588D-U is powered up and in-circuit.

Similar to the Arduino, the WT588D-U is programmed via its onboard USB port, using a software application that runs on a PC. The WT588D-U can have up to 220 different audio "segments." Each segment has the ability to combine up to 500 pre-stored phrases saved in WAV format, with programmed delays within each phrase/segment group.

The WT588D-U uses a 13-bit D/A or 12-bit PWM to generate 0.5 W of audio output into an 8 Ω speaker. The WT588D-U has 15 different key control modes, providing a number of different ways to trigger and control the audio output. Be sure to get the WT588D-U with the USB programming port — it makes programming and using this module a whole lot easier.

Figure 3.41 — WT588D-U speech module.

Figure 3.42 — HC-05 Bluetooth module.

Bluetooth Module

There are two basic modes for Bluetooth modules, slave and master. The HC-05-type Bluetooth module shown in **Figure 3.42** supports both modes. In slave mode, you can pair your Arduino to a device such as a workstation or cell phone and have the Arduino act as a mouse, keyboard, or other human interface device (HID). In master mode you can pair and connect multiple Bluetooth slave modules to your Arduino. Be careful if you want to connect a Bluetooth keyboard or mouse to your Arduino, as it will need to be an HID-compliant unit and the vast majority of the standard Arduino Bluetooth modules such as the HC-05 do not support HID devices. You might consider using a USB host shield and Bluetooth dongle for Arduino Bluetooth HID projects.

Some of the currently available Bluetooth modules can be switched between master and slave mode from within your Arduino sketch. The Arduino communicates with the Bluetooth module using standard TTL serial communications, usually at a speed of 9600 baud.

Be careful and verify the specifications for your particular Bluetooth module, as some accept power in the range of 3.6 to 6 V, but only allow a maximum of 3.3 V on the data pins. You can use a simple resistor divider network, or a level converter module to protect the data pins from overvoltage.

Figure 3.43 — Level converter module.

Level Converter Module

With the increasing number of 3.3 V Arduino boards and modules, you may find yourself needing to convert signal levels from 3.3 V to 5 V and vice-versa. Connecting 3.3 V pins to a 5 V source can often cause damage to the 3.3 V device, which is never a good thing. Fortunately, you can use a level converter module to convert between the two signal voltage levels. The SparkFun Logic Level Converter shown in **Figure 3.43** can convert between 3.3 V and 5 V signal levels on four pins (two input and two output). It can also be used to adapt 1.8 V and 2.7 V devices to 5 V.

Figure 3.44 — DS3231 real-time clock module.

Real-Time Clock Module

The DS3231 shown in **Figure 3.44** is an inexpensive but extremely accurate real-time clock (RTC) module for the Arduino. A vast improvement on earlier real-time clock modules, the DS3231 chip has an internal temperature-compensated crystal oscillator (TCXO). Because the Arduino does not have a real-time clock on board, the DS3231 is ideal for those projects where you need to keep track of date and time. The DS3231 module communicates with the Arduino via the I²C bus, and has an onboard coin-cell battery to maintain the clock when power is off. As with earlier RTC modules, the DS3231 can keep track of time, day, month, and year, with built-in leap-year compensation up to the year 2100. The DS3231 also has a built-in temperature sensor accurate to ±3 °C. As with most Arduino modules, the DS3231 is supported by excellent libraries, including my personal favorite, the DS3231 library from Henning Karlsen at Rinky-Dink Electronics (**www.rinkydinkelectronics.com**).

Lightning Sensor Module

The lightning sensor module shown in **Figure 3.45** is based on the Austria-microsystems Franklin AS3935 lightning sensor chip and can detect lightning at a distance up to 40 km. The AS3935 is capable of detecting both cloud-to-ground and cloud-to-cloud lightning. The AS3935 includes a proprietary embedded algorithm to reject man-made electrical noise, and has software-selectable threshold settings. The AS3935 will also statistically calculate the distance to the leading edge of the thunderstorm, along with the estimated strength of the lightning strike. The AS3935 can communicate with the Arduino via either the SPI or I²C bus.

Figure 3.45 — Lightning sensor module.

Analog-to-Digital Converter Modules

The analog-to-digital (A/D) converter used for the analog input pins on the Arduino is a single-ended 10-bit A/D, providing a count of 0 to 1023 over a 5 V range with respect to ground. For many Arduino projects, this resolution is more than adequate. However, there will be occasions, as in some of the antenna rotator control projects, where you will want higher resolution in your analog voltage measurements than the A/D onboard the Arduino allows. In the case of an antenna rotator position sensor, you will want to read a position in degrees from 0 to 360° (or 450° in the case of some rotators). Using a 10-bit A/D, this leaves you with a resolution of about 1°. A 12-bit A/D would provide a resolution of about 1/10°, and a 16-bit A/D would provide you with a resolution of about 1/100°. As you can see, the more bits your A/D has, the higher the resolution.

The Texas Instruments ADS1015 12-bit A/D module can provide a count of 0 to 4095 on four single-ended inputs or two differential inputs, with a program-selectable sampling speed up to 3300 samples per second. The TI ADS1115 16-bit A/D module (**Figure 3.46**) can provide a count range of 0 – 65535 on four single-ended inputs or two differential inputs, with a program-selectable sampling speed up to 860 samples per second. Both communicate with the Arduino using the I²C bus. They feature six program-selectable gain settings with a full-scale reading of 6.144 V (*warning*: do not ever exceed V_{DD} + 0.3 V on any input pin, so you will never be able to read more than the supply voltage at the ⅔ gain setting), all the way down to a full-scale reading of 0.256 V at the 16× gain setting. The Arduino libraries for these modules allow access to and control of all the internal parameters of these modules, and make using these modules in your Arduino projects a breeze.

Figure 3.46 — ADS1115 16-bit A/D module.

Digital-to-Analog Converters

Performing the reverse function of an A/D module, a digital-to-analog module (D/A) converts digital data into an analog voltage representation of the value of the digital data. The one thing the Arduino is missing is an onboard D/A converter, although some of the newer variants do have this feature built in.

In trying to add this functionality to my Arduino projects, I have built several types of D/A converters, from a simple 8-bit resistive ladder all the way up to using an I²C D/A module. The results of these tests showed the Microchip MCP4725 12-bit I²C D/A converter (**Figure 3.47**) to be my D/A of choice, both for its simplicity and ease of use. Using the MCP4725 Arduino library and sample sketches, you can quickly have your Arduino outputting analog voltages, sine waves, and even triangle waves, among other waveforms.

Figure 3.47 — Microchip MCP4725 12-bit I²C D/A converter.

I²C Digital I/O Expanders

Sometimes, the 14 pins of digital I/O available on the standard Arduino just aren't enough, and you don't want to splurge for a Mega or use a variant with more I/O pins. Using an 8- or 16-pin I²C I/O expander chip can get the job done, and cost you a whole lot less. The Microchip MCP23008 I²C serial I/O expander chip (**Figure 3.48**) will add eight pins, and the Microchip MCP23017 (**Figure 3.49**) will add 16 pins of digital I/O to your Arduino.

Fully supported by Arduino program libraries, the serial I/O expander chips communicate with the Arduino via the I²C bus, and have programmable address settings, allowing for up to eight expander chips on a single I²C bus. This means you can add up to 64 or 128 pins of digital I/O to your Arduino project to handle those bigger projects. The MCP23S08 and MCP23S17 provide similar capability, but communicate with the Arduino using the SPI bus instead of the I²C bus.

Figure 3.48 — Microchip MCP23008 I²C serial I/O expander.

Figure 3.49 — Microchip MCP23017 digital I/O expander.

I²C Bus Extender

While the average usable length for an I²C Bus connection is typically 9 to 12 feet, the Texas Instruments P82B715 I²C Bus Extender chip can be used to buffer and extend the bus to a length of 50 meters (164 feet).

DTMF Generator

Dual-tone multi-frequency (DTMF) is a mainstay in amateur radio. Just about every modern mobile rig has a DTMF tone keypad, and many repeaters utilize DTMF for control and other functions. DTMF is a simple way to control many of your Arduino projects remotely, without a lot of complex circuitry. The Holtek HT9200 DTMF generator shown in **Figure 3.50** allows your Arduino to generate the standard 16 DTMF tones, as well as 8 single tones. The HT9200B allows for both a serial and parallel mode interface, and it is easily interfaced to the Arduino without the need for a separate library.

Figure 3.50 — Holtek HT9200 DTMF generator.

DTMF Receiver

Of course, if you're going to use a DTMF tone generator, you most likely will also need a DTMF receiver. The Zarlink MT8870D shown in **Figure 3.51** is a DTMF receiver chip that converts the 16 DTMF tones into a 4-bit digital output. The only major external component required is a 3.79545 MHz crystal for the oscillator circuit portion of the chip. This is a very inexpensive crystal that was originally designed for use in extracting the color-burst portion of an analog television signal, and is readily available online from eBay and most electronics parts suppliers.

Figure 3.51 — Zarlink MT8870D DTMF receiver chip.

Addressable LEDs

Individually addressable LEDs are becoming very popular in Arduino projects. Adafruit refers to their version of these LEDs as NeoPixel LEDs. These LEDs can be wired together as individual LEDs, or purchased in sticks or strips that can be up to 10 meters long, and they can be daisy-chained together to create even longer strips. These LEDs are often powered by the WS2812B intelligent controller chip that is actually embedded inside each 5050-type surface-mount RGB LED as you can see in **Figure 3.52**. While they do come with plain white LEDs, the RGB color versions are far more popular. These LEDs are available in sticks (**Figure 3.53**), strips, rings (**Figure 3.54**), and even quarter-circle arcs for the larger LED rings.

The individual LEDs are usually pulse-width modulated at

Figure 3.52 — Close-up showing the WS2812B chip embedded in the LED itself.

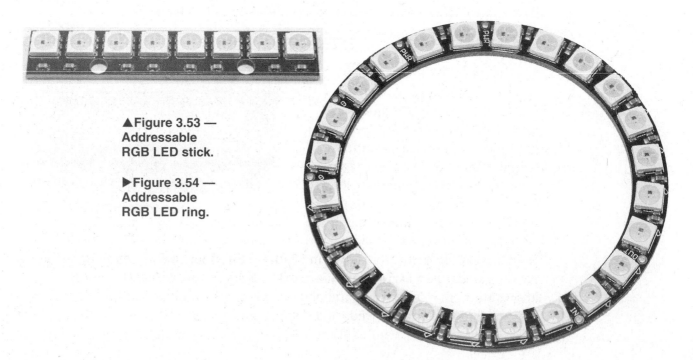

▲Figure 3.53 — Addressable RGB LED stick.

▶Figure 3.54 — Addressable RGB LED ring.

400 Hz or higher, which keeps the power utilization much lower than you would expect. For example, a 10 meter (32.8 foot) flexible strip generally has 300 individual LEDs. The super-bright 5050-type LEDs used in most strips can draw up to 50 mA per LED. The math says if you have all 300 LEDs lit at full brightness, the current draw would be 15 A.

In reality, you rarely operate at full brightness, and in actual operation you'll see that the current consumption will be substantially lower. Still, you do need to be aware of the power requirements for larger numbers of LEDs in your projects, and should consider adding an external fused power source to power your LED strips.

Arduino Displays

There are many different types of displays that can be used with the Arduino, and they can be interfaced using a variety of methods. The most commonly used display for the Arduino is the 16-character by 2-line (16×2) LCD shown in **Figure 3.55**. A larger 16-character by 4-line version of this

Figure 3.55 —
16×2 LCD display.

display is also available. Both are interfaced to the Arduino using six digital I/O pins, two pins for control and four pins for data. Libraries and example sketches for these displays are built into the Arduino IDE. Newer versions of these displays are also available that support I²C and/or TTL serial communication interfaces. I generally prefer the I²C version of these displays, as they only require a simple four-wire connection of +5 V, ground, and the Arduino SDA and SCL I/O lines to function.

Nokia 5110 Graphic LCD display

Originally used in the older Nokia cell phones, the Nokia 5110 84×48 pixel graphic LCD shown in **Figure 3.56** is one of the more popular Arduino displays. Easy to interface to the Arduino using five digital I/O pins, the Nokia 5110 is small, backlit, easily readable, and capable of displaying graphics and up to six lines of 14-character-per-line text. While the Nokia 5110 is technically an SPI-bus capable display, more often than not it is connected to the Arduino directly to digital I/O pins and the SPI bus is not used. The Nokia 5110 is based on the Phillips PCD8544 LCD controller. While specified

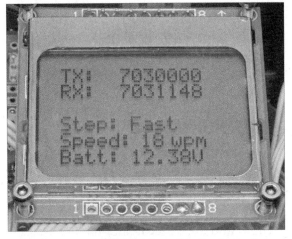

Figure 3.56 — Nokia 5110 LCD display.

for 3.3 V, I have used the Nokia 5110 with supply voltages from 2.7 to 5 V, without 3.3 to 5 V level shifters, with no damage to the display.

Contrast settings can vary widely from display to display, but most libraries allow you to adjust the contrast level via software. The Nokia 5110 also has a four LED backlight that can be controlled through the use of an additional digital I/O pin or a resistor from the backlight pin to ground. More recently, multiple variants of the Nokia 5110 have become available, with some minor electrical differences between them.

Figures 3.57 and **3.58** show three different versions of the Nokia 5110 display currently available. Note that the power pin is different on the one in the center, and that the pin designations may not exactly match up with the ones in the project schematics. The LCD on the far left is the one that I use most often for projects that call for using the Nokia display. While most of the Nokia 5110 displays use a dropping resistor for the backlight LEDs, the Adafruit version uses a transistor and requires a logic-high level to enable the backlight. When wiring up the Nokia 5110 module, always use the pin labels and not the pin numbers, as these designations differ among the various Nokia 5110 versions. When in doubt, build a simple test project on your breadboard to determine the correct wiring for the version of the Nokia display that you have on hand.

Figure 3.57 — Three types of the Nokia 5110 (front view).

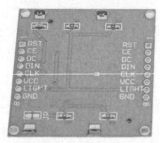 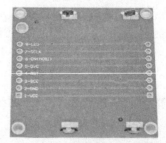 

Figure 3.58 — Three types of the Nokia 5110 (rear view).

Organic LED (OLED) Displays

The organic LED (OLED) displays are small graphic LED displays that can communicate with the Arduino using either the SPI or I²C bus. There are several versions of OLED displays available, primarily the 128×32 pixel and the 128×64 pixel (**Figure 3.59**) versions. These are small displays. The 128×32 module is about 1 inch wide by ⅜ inch high, and the 128×64 module is about 1 inch wide by ⅝ inch high. A newer OLED, the 64×48 pixel SparkFun

Figure 3.59 — Adafruit 128×64 OLED display. (Photo courtesy of Adafruit.com)

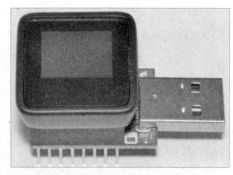

Figure 3.60 — SparkFun MicroView OLED display.

MicroView OLED (**Figure 3.60**), is a mere 0.66 inches across and about 0.5 inches high.

An OLED display is essentially composed of tiny individual LEDs, and because it uses LED technology, does not need a backlight. OLED displays are very bright and can be used in sunlit applications. The libraries supporting the OLED displays allow you to display both text and graphics. The OLED display is an ideal choice when a small, bright, and clearly readable display is desired.

Thin Film Transistor (TFT) Color Displays

If you really want to spice up your Arduino display, a color TFT graphic display as shown in **Figure 3.61** is the way to go. TFT displays for the Arduino come in a variety of sizes, from 1.8 inches (128×160 pixels) all the way up to a massive 7- inch (800×480) pixel display and offer up to 262,144 shades of color. The smaller 1.8, 2.2, and 2.8-inch versions of these displays communicate with the Arduino using either the SPI or I^{2}C bus. The 3.2-inch and larger displays will require an adapter shield because of the larger number of I/O pins to support these display sizes. Many of these display modules also include an onboard microSD card slot, allowing you to save bitmap images. Many of these displays also include a touchscreen, but you will more than likely need a board such as the Arduino Mega to support both the display and touchscreen simultaneously.

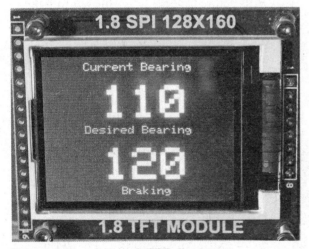

Figure 3.61 — 1.8 inch TFT display.

When purchasing the generic TFT displays from places such as eBay, there are a few things you need to be aware of, particularly the minor differences between the generic displays and the ones you get from vendors such as Adafruit and SparkFun. Don't be afraid to use the generic TFT displays — I use them all the time in my projects without any problems. The Adafruit TFT libraries work just fine with the generic TFT displays.

Arduino Shields, Modules, and Devices   3-31

Figure 3.62 — A generic 1.8-inch 128×160 pixel color TFT display module.

Figure 3.63 — Reverse side of a generic 1.8 inch TFT display showing the 3.3/5 V solder jumper.

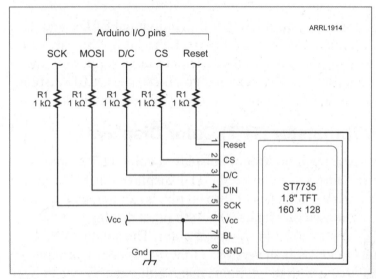

Figure 3.64 — Schematic for adding resistors to the signal lines for the 1.8-inch ST7735-type TFT display.

Figure 3.65 — A 2.2-inch ILI9341 generic TFT display.

The 1.8-inch generic displays (**Figure 3.62**) are usually based on the ST7735 controller. They are designed to operate from 3.3 or 5 V, depending on the presence of a solder jumper on the back of the module (see **Figure 3.63**). When using 5 V, the data lines still must be fed with 3.3 V. To achieve this, you can use a voltage divider, level shifter module, or do it the simplest way — just add a 1 kΩ resistor in series with each signal line as shown in **Figure 3.64**. Also, while the generic board may be designated as an I²C and SPI device, it actually does not support I²C and is strictly an SPI device. When wiring the display, the SCL label is actually the SPI SCK pin, and the SDA connection is actually the SPI MOSI pin. The Adafruit GFX and ST7735 libraries work fine with the generic ST7735 displays.

The 2.2-inch (**Figure 3.65**) and larger TFT displays also need a minor bit of extra consideration. These displays typically use an ILI9341 controller. Like the 1.8-inch displays, the signal lines must also be 3.3 V, but the 1 kΩ resistor trick doesn't work well with these displays. I have found that a voltage divider on the signal lines works just fine, and that is how I interfaced the 2.2-inch TFT displays in the projects in this book. As with the 1.8-inch displays, the Adafruit GFX and ILI9341 libraries work fine with these generic TFT units.

E-Ink Displays

If you've ever used a Kindle or Nook eReader, you've used an E-Ink display (**Figure 3.66**). E-Ink displays are unique in that the image stays on the display even after the power is completely disconnected. The image is high contrast and daylight-readable, and has been said to look just like printed paper, hence the term E-Ink.

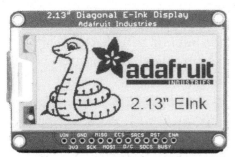

Figure 3.66 — Adafruit 2.13-inch E-Ink display. (Photo courtesy Adafruit.com)

Nextion TFT Touchscreen

The Nextion TFT touchscreen display (**Figure 3.67**) is part of the next generation of TFT displays. The Nextion displays combine an onboard processor and touchscreen in what they refer to as a Human Machine Interface (HMI). Unlike the standard TFT displays, the Nextion's screen display and touchscreen elements are created using the Nextion Editor software, similar to how you create and upload your sketch to the Arduino. The Nextion connects to your Arduino via a standard TTL serial connection to provide event notifications (screen presses) that the Arduino sketch can then process. This offloads a huge amount of the display and touchscreen processing load from the Arduino, making your sketches much easier to create and manage.

These displays are available in screen sizes from 2.4 inches all the way up to 10.1 inches. While creating displays is a bit more involved with the Nextion as compared to a standard TFT display, the results are well worth the effort and perform graphic display operations much faster than its standard TFT counterpart. While a bit on the pricey side, especially compared to the standard TFT displays, the Nextion display is rapidly gaining popularity among developers, and is the new graphic display for the μBITX V6 transceiver from HF Signals.

Figure 3.67 — 2.8 inch Nextion display.

Enclosures

Once you've built your project, you have a number of enclosure options available. To add to the fun, I've even found some rather unusual places you can get some nice enclosures that work very well for Arduino-based projects.

For those tiny projects that can fit in your pocket, there's the good old Altoids mint tin (**Figure 3.68**). The larger Altoids mint tin is a near perfect fit for an Arduino Uno, and a number of 3D designs are available to print a plastic insert to insulate the Arduino from the metal case.

Figure 3.68 — Arduino Uno in an Altoids mint tin.

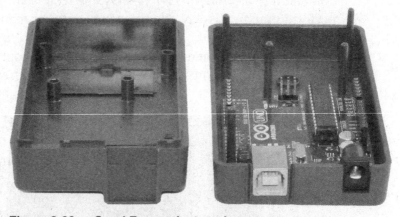

Figure 3.69 — SparkFun project enclosure.

The major Arduino suppliers are always a good place to find enclosures. Both SparkFun (**Figure 3.69**) and Adafruit have a variety of Arduino enclosures. Another good source for Arduino enclosures is on eBay. One of my personal favorites is the pcDuino/LinkSprite-style enclosures (**Figure 3.70**) that have an optional height expander to allow an Arduino and a shield to fit inside, with removable inserts and pre-cut holes for a 16×2 LCD. I also like the

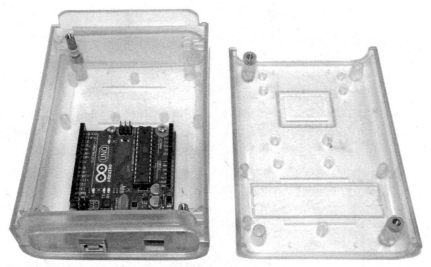

Figure 3.70 — pcDuino-LinkSprite enclosure.

Figure 3.71 — Solarbotics Mega SAFE.

Solarbotics SAFE and Mega SAFE enclosures (**Figure 3.71**). I particularly like the Mega SAFE enclosure for use with an Arduino Uno. There's enough room for the Uno, a prototyping shield stacked on top, and room left over for a 9 V battery and plenty of places to mount a display, switches, LEDs, rotary encoders, and other parts.

More recently, I have found that Hobby Lobby and other craft stores have clear acrylic golf ball, tennis, baseball, softball, and model car display cases (**Figures 3.72** and **3.73**) that are easily adapted into nice looking Arduino enclosures. I have started using these in a number of my Arduino projects and have been very pleased with the results. They're easy to work with and your finished Arduino project looks similar to one of those "ship in a bottle" models that we built as kids.

Figure 3.72 — Hobby Lobby baseball display case.

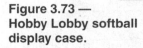

Figure 3.73 — Hobby Lobby softball display case.

Arduino Parts Suppliers

Adafruit — **www.adafruit.com**
Arduino — **www.arduino.cc**
Argent Data Systems — **www.argentdata.com**
Banggood — **www.banggood.com**
DFRobot — **www.dfrobot.com**
eBay — **www.ebay.com**
Embedded Adventures — **www.embeddedadventures.com**
Hamshield — **inductivetwig.com/products/hamshield**
HobbyPCB — **www.hobbypcb.com**
Nextion — **www.nextion.tech**
Solarbotics — **www.solarbotics.com**
SparkFun — **www.sparkfun.com**

Arduino I/O Methods

The Arduino has a variety of input/output (I/O) methods — ways that it can communicate with attached modules, devices and the outside world. For example, **Figure 4.1** shows a HyGain AR-40 antenna rotator simulator that requires the Arduino to communicate with a display, a PC, and several switches and potentiometers.

It is important to understand how each of these I/O methods can be used in your Arduino projects. Although some of this information was covered in my previous books, the goal is to provide information to help you understand what is going on inside your Arduino project, as well as understand how the various projects in this book function. This will help you to add your own features and enhancements.

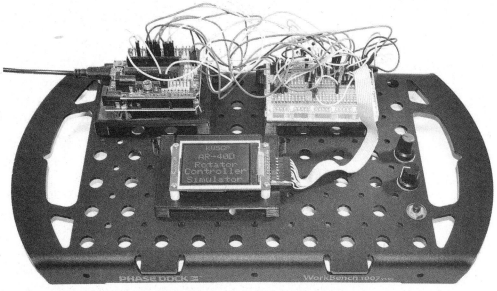

Figure 4.1 — A HyGain AR-40 Rotator Simulator.

The main purpose of a microcontroller such as the Arduino is to interface directly with sensors and devices that allow the Arduino to sense and control things. The primary methods of I/O on the Arduino are digital I/O, digital I/O with pulse width modulation (PWM), analog input, and TTL/USB serial communication. The Arduino also supports several bus-type protocols, including the Serial Peripheral Interface (SPI), Inter-Integrated Circuit (I^2C), and 1-Wire buses.

In addition to these methods, many of the newer Arduino variants support Bluetooth and Wi-Fi. Through the use of a shield or external module, your Arduino can also interface with and communicate with USB devices. The Arduino also supports hardware and software "interrupts," which allow program execution flow to be changed (interrupted) by an external event.

Digital I/O

The simplest and most common form of I/O on the Arduino is performed using the digital I/O pins. Digital I/O pins can be used to turn on LEDs, relays, and other external devices, as well as sense the position of a switch, pushbutton, or other form of on-off (digital) input. Through the use of several Arduino libraries, you can also use digital I/O pins to allow TTL serial, SPI, or I^2C communication using pins other than the pins normally designated for these communication protocols.

The Arduino digital I/O pins also have an internal pull-up resistor (normally disabled) that can be enabled via software, saving you from having to add an external pull-up resistor when sensing the position of a switch, for example. Some Arduino variants also have this internal pull-up resistor, while some do not. Be sure to verify that the Arduino variant you are using supports this feature when designing your project. On the Arduino Uno, Mega2560, and Leonardo, the value of the internal pull-up resistor is from 20 kΩ to 50 kΩ.

The Arduino digital I/O pins are defined as being tri-state, meaning that the I/O pin can either be a logic-level High, Low, or in a high-impedance (disabled) state, which effectively removes the I/O pin from the circuit.

Digital I/O with Pulse Width Modulation

Six of the digital I/O pins (pins 3, 5, 6, 9, 10, and 11) on the Arduino Uno can be configured to output data using pulse width modulation (PWM). PWM allows you to control the duty cycle of a square wave output on the digital I/O pin using an 8-bit value from 0 to 255. Using PWM, you can dim an LED, output an audio tone, control the speed of a dc motor, modulate a signal for an infrared LED remote, and perform other tasks where you need to modify the duty cycle of the output waveform. The Arduino Leonardo has seven PWM digital I/O pins, and the Arduino Mega2560 has 14 PWM digital I/O pins. The number of PWM digital I/O pins varies among Arduino variants, so be sure to check the digital I/O pin specifications if you plan to use an Arduino variant in your project.

Analog Input

The Arduino Uno has six analog input pins. Even though it uses the same Atmel ATmega328 processor, the Nano interestingly has eight analog input pins. These pins convert an analog input of 0 to 5 V (3.3 V on the 3.3 V Arduinos) into a 10-bit digital value from 0 to 1023. This range can be modified either by using an external reference voltage or by selecting the Uno's internal reference of 1.1 V. The Arduino Mega2560 also supports an internal reference voltage of 2.56 V. The Arduino Leonardo and Mega2560 each have 12 analog input pins. The Arduino analog input pins can also be used as digital I/O pins, and like the regular digital I/O pins, also have an internal pull-up resistor (normally disabled) that can be enabled via software. The value of the internal pull-up resistor on the analog pins is the same as for the standard digital I/O pins.

Analog Output

While the "standard" Arduinos such as the Uno, Nano, and Mega2560 do not have I/O pins to convert from a digital representation to an analog output voltage, some of the more recent Arduino boards and variants do have digital-to-analog (D/A) pins. Using the D/A pins, you can output an analog voltage, sine wave, and other waveforms directly from the Arduino itself, rather than having to use an external D/A module.

Serial I/O

The Arduino Integrated Development Environment (IDE) on your workstation communicates with the Arduino via the hardware serial port. Most Arduinos communicate with the IDE via the USB connector. However, some Arduinos, such as the Solarbotics Ardweeny, require a USB-to TTL serial converter also known as an FTDI module (**Figure 4.2**) to access the serial port on the Arduino. The USB port normally shares the I/O pins with the Arduino's hardware serial port. On the Arduino Uno and similar Arduinos, digital I/O pins 0 and 1 are used for serial communication. When designing your Arduino project, care must be taken if you use the dedicated hardware serial I/O pins in your project, as they can interfere with your ability to communicate with and upload sketches to the Arduino.

The IDE can be used to upload sketches and the IDE also has a Serial Monitor feature that allows you to view the serial port data, as well as sending serial data to, and from, the Arduino. If your project requires a separate serial I/O port, you can use the Software Serial library, which allows you to assign any

Figure 4.2 —
An FTDI module.

pair of digital I/O pins as a software-driven serial port. Additionally, some Arduinos and variants, such as the Mega2560 and others have additional hardware serial ports onboard.

1-Wire Bus

The 1-Wire bus interface was designed by Dallas Semiconductor Corp. (now part of Maxim Integrated Products) to provide low-speed data signaling and power over a single data line. Typically used to communicate with small devices such as temperature, voltage, and current sensors, along with external memory and other devices, the 1-Wire bus interface is a bus-type interface architecture that is implemented using a single wire (two if you count the ground wire). With a data rate of up to 16.3 Kbit/s, the 1-Wire bus can communicate reliably with devices over 100 meters away from the host.

Each 1-Wire device has its own unique 64-bit serial number embedded in the device, allowing many 1-Wire devices to be attached to the same digital I/O pin, without the need for any additional configuration or wiring. Using an advanced algorithm, the host, also known as the bus master, can quickly identify all of the devices attached to the 1-Wire bus, and determine their type based on the device type information that is embedded in the lower 8 bits of each device's serial number. This algorithm can scan the one the 1-Wire bus and identify up to 75 sensors per second.

A unique feature of the 1-Wire interface is that many 1-Wire devices can be powered entirely from a single digital I/O pin used as the data line for the 1-Wire bus. Using this method, known as "parasitic power," the device does not need a separate power supply. Instead, a small capacitor (typically 800 pF) is integrated inside the device to store enough of a charge to operate the device's interface to the bus.

MaxDetect also makes a series of relative humidity and temperature sensors that use a proprietary 1-Wire interface that is different than the 1-Wire bus described above. The MaxDetect 1-Wire interface however, is not a bus architecture, and the MaxDetect devices do not have embedded addresses, meaning that you can only have one MaxDetect 1-Wire device attached to the digital I/O pin at time. The MaxDetect 1-Wire interface does not support parasitic power mode, and must be supplied power on a pin separate from the device's data pin. The MaxDetect 1-Wire interface is not compatible with the Dallas Semiconductor/Maxim 1-Wire bus, so care must be taken when mixing these two types of 1-Wire devices in your projects.

Both types of 1-Wire devices and interfaces are supported by Arduino sketch libraries and example sketches, which makes interfacing and using 1-Wire devices in your Arduino projects simple and easy.

Serial Peripheral Interface (SPI) Bus

The Serial Peripheral Interface (SPI) bus protocol was developed by Motorola to be a high-speed, full-duplex communication, bus-type protocol between one master and multiple slave devices. SPI, along with I^{2}C (discussed in the next section) are the "workhorse" buses that you will use most often with your Arduino projects. The Arduino communicates with the SPI devices on a single shared bus using four signal lines. These lines are designated Clock (SCLK, SCK, or CLK), Slave Select or Chip Select (SS or CS), Master-Out Slave In (MOSI), and Master-In Slave Out (MISO or DI). Each device attached to the SPI bus requires a separate Slave Select line. Think of the Slave Select line as a device select line that tells the selected device that the information on the SPI bus is addressed to it.

SPI is a loosely defined standard and can be implemented in slightly differing ways between device manufacturers. Because SPI is considered to be a synchronous communications protocol, data is transferred using the Clock line, with no formally defined upper limit on speed. Some SPI implementations can run at over 100 Mbit/s. SPI has four defined modes (Modes 0, 1, 2, and 3) which define the clock edge on which the MOSI line clocks the data out, the clock edge on which the SPI master device samples the MISO line, and the clock signal polarity. Fortunately, the Arduino sketch libraries for SPI and the various SPI devices handle the proper signaling required to communicate with SPI devices you may attach to your Arduino project.

On the Arduino Uno, the SCLK, MISO, and MOSI pins are permanently defined at pins 13, 12, and 11 respectively. Digital I/O pin 10 is often used at the Slave Select pin for the first SPI device, but any available digital I/O pin may be used as a Slave Select pin. On the Arduino Mega2560, the SCLK, MOSI, and MISO pins are defined as digital I/O pins 52, 51, and 50 respectively, with the Slave Select pin for the first SPI device typically assigned to digital I/O pin 53. On some Arduinos, such as the Uno R3, Leonardo, and Mega2560 among others, the SPI signals are also brought out to the six pin In-Circuit Serial Programming (ICSP) header.

Some Arduinos, such as the Leonardo, do not have any digital I/O pins specifically assigned for SPI communication. Instead, the SPI signals are brought out to the ICSP header only. Because each SPI device attached to your Arduino requires a digital I/O pin assigned to the Slave Select of each SPI device, your projects are limited to the number of digital I/O pins available. Some Arduinos and variants have multiple SPI buses onboard.

There are also several Arduino sketch libraries available that allow you to software-define the SPI bus interface to use regular digital I/O pins and communicate with the SPI devices via these pins by using software-based timing of the SPI signals (also known as "bit-banging"). With this method, the digital I/O pins are turned on and off manually by the library to simulate the hardware timing needed to communicate with the SPI device. While this bit-banging method does work, it requires all of the bit timing and clocking to be performed in software, which is much less efficient than the hardware-based SPI method. It is usually better to design your projects to use hardware SPI, and only use software-based SPI communications when absolutely necessary.

Compared to the I²C bus, SPI is generally faster. Assuming you have the digital I/O pins available in your project, SPI is the preferred method for interfacing most modules and devices that support both SPI and I²C communications.

Inter-Integrated Circuit (I²C) Bus

The Inter-Integrated Circuit (I²C) bus was developed by Phillips (now NXP Semiconductors) for attaching low-speed peripherals to a host device. On the Arduino, the I²C bus is also known as the Two-Wire Interface (TWI) bus. I²C is a serial, bidirectional, 8-bit communications protocol used by many manufacturers that develop peripherals and devices for embedded systems and microcontrollers such as the Arduino.

The I²C bus requires only two communication lines, Serial Data (SDA) and Serial Clock (SCL). On the Arduino Uno, these are defined as analog pins A4 and A5 respectively. On the Arduino Leonardo and Mega2560, SDA and SCL are assigned to pins 20 and 21 respectively. Some Arduino variants have a second I²C interface, with the pins designated as SDA1 and SCL1.

The I²C standard defines the speed of the I²C bus as 100 Kbit/s (Standard Mode), 10 Kbit/s (Slow Mode), 400 Kbit/s (Fast Mode), 1 Mbit/s (Fast Mode plus), and 3.4 Mbit/s (High Speed mode). The Arduino I²C bus defaults to a bus speed of 100 Kbit/s, but the bus speed can be changed by modifying the internal Arduino Two Wire Bit Rate Register (TWBR), or by modifying the TWI speed definition in the Arduino Wire library. Unless your project requires it, and your devices can support it, it is best not to modify the Wire library, as it may cause issues when you compile other projects that use the same Wire library.

I²C devices have a unique 7- or 10-bit address, with some devices capable of having their I²C address reassigned using jumpers, switches, or some other hardware method. That allows you to have multiple devices of the same type co-exist on the I²C bus. On the Arduino, I²C addresses 0 – 7 and 120 – 127 are reserved, leaving 112 7-bit addresses available for devices. Every I²C device connects to the bus using open-drain (the same as open collector except for MOSFET devices), requiring the use of pull-up resistors on the SDA and SCL bus lines. Typically, the value of the pull-up resistors is 4.7 kΩ, so the Arduino's internal pull-up resistor will not suffice as a viable I²C bus pull-up resistor value.

As with software SPI, there are several Arduino sketch libraries available that allow you to use regular digital I/O pins for the I²C bus, and to communicate with I²C devices on those pins using software-based timing of the I²C signals (bit-banging, as mentioned in the previous section). With this method, the digital I/O pins are turned on and off manually by the library to simulate the hardware timing needed to communicate with the device. While this method does work, it requires that all of the bit timing and clocking be done in software, which is far less efficient than the hardware-based I²C method. It is far better to design your projects using hardware I²C, and use software I²C communication only when absolutely necessary. Rather than using software to create a second I²C bus if needed, choose an Arduino variant that has multiple I²C bus interfaces available onboard.

Interrupts

While not often seen as an Arduino I/O method, interrupts allow certain conditions and events to modify the way your Arduino sketches execute. The Arduino has two types of interrupts — hardware and timer. Hardware interrupts are triggered by an external event, such as a change in the logic level on a digital input pin. An interrupt will pause (interrupt) the current program execution and immediately execute a user-defined function, known as an interrupt handler or interrupt service routine (ISR). When the ISR function is complete, program execution resumes, normally at the command that was last executed prior to the interrupt. An interrupt can happen at any time, and allow your Arduino sketch to immediately respond to external events without having to constantly check (or poll) to see if the desired condition exists. Without using interrupts, the USB CW Keyboard project described in a later chapter in this book would not have been feasible.

There are four types of interrupt conditions that can be defined: Rising, Falling, Change, and Low. These interrupt conditions refer to the state of the digital I/O pin used to generate the interrupt. The Rising condition will generate an interrupt when the I/O pin goes from a logic-level low state to a high state; Falling will generate an interrupt when the I/O pin goes from high to low. The Change condition will generate an interrupt when the I/O pin changes from either low to high or high to low. The Low condition will generate an interrupt when the I/O pin is low. Some Arduinos and Arduino variants also have an additional interrupt condition, High, which generates an interrupt when the pin is high.

The Arduino Uno has two interrupts, assigned to digital I/O pins 2 and 3. The Arduino Leonardo has four interrupts, on Pins 0, 2, 3, and 7, the Arduino Mega2560 has six interrupts, on pins 2 and 3, and 18 to 21. Some Arduinos and Arduino variants allow you to configure interrupts on all available I/O pins. The Arduino Uno can also handle Change interrupts on all I/O pins, but unlike hardware interrupts on the defined interrupt pins, the ISR function must decode the interrupt and determine which I/O pin generated the interrupt.

The Arduino Uno has three internal timers, defined as `Timer0`, `Timer1`, and `Timer2`, which can be used to generate `Timer` interrupts. `Timer0` is an 8-bit timer used by the Arduino for internal timing functions such as `delay()` and `millis()`. Because it can affect these functions, modifying the `Timer0` settings is not recommended. `Timer1` is a 16-bit timer often used by some Arduino sketch libraries, such as the `Servo` library, and `Timer2` is an 8-bit timer used by the Arduino `tone()` function. As long as you are aware of any potential interaction with these functions, you can modify the settings on `Timer1` and `Timer2` for use in your sketches. The Mega2560 has three additional 16-bit timers — `Timer3`, `Timer4`, and `Timer5` — which are not used by any Arduino internal functions. Other Arduino and variant boards also have additional timers available, such as the Teensy 3.6, which has up to 58 digital I/O interrupts and up to 19 timers onboard.

You can configure the timers to generate a software interrupt on overflow, or when the timer count reaches a desired value. The timers are based on the Arduino CPU clock rate (such as 16 MHz on the Arduino Uno). You can use the timer counter/control register (TCCR) for each timer to control the timer clock setting. By modifying the Clock Select bits, you can control how fast the timer increments the counter. On the Arduino Uno, the available settings are Clk/1 (clock speed), Clk/8, Clk/64, Clk/256, and Clk/1024. At the maximum setting of Clk/1024, you can have 16-bit `Timer1` generate a software interrupt approximately every 4,194 seconds. Using the Clear Timer on Compare Match (CTC) setting, you can adjust the time to generate an interrupt when the timer reaches a preset value. Using 15624 as the preset value will cause the timer to generate an interrupt once per second. Using a 1-second interrupt in this manner, you can add precision timing to your sketches without having to manually keep track of time using the `millis()` function or other manual methods.

Implementing interrupts does add a level of complexity to your Arduino sketches, and you have to remind yourself that an interrupt can occur at any time during program execution. You will have to remember and plan for this as you write and troubleshoot your sketch. You can enable and disable the interrupts as needed from within your sketches, to allow for uninterrupted execution of critical or time-sensitive points in your sketch where you don't want the sketch execution to be interrupted.

Used properly, interrupts can be a powerful tool in developing your Arduino projects, and once you get comfortable using interrupts, you will find that they can greatly simplify and enhance your project development, because you no longer need to have your sketches running in timing loops waiting for an event to occur. Instead, you can have your sketch off doing other things, and only respond when the actual event occurs.

Bluetooth Communication

Using Bluetooth technology with the Arduino is an often misunderstood and frustrating exercise. Deciphering how Bluetooth technology can be integrated into your Arduino projects could easily take an entire book unto itself. I had hoped to include some Bluetooth projects in this book, but every step forward ended up being at least a half-step backward, and there was no proper way to come up with a viable project without some major additional research and time investment on my part, which unfortunately was not possible in time for this book. Rest assured, Bluetooth technology is at the top of the list for lab projects in the future.

Usually, when we think of Bluetooth, we typically envision the Bluetooth devices we use in conjunction with our smartphones, such as wireless headsets and microphones, keyboards, mice, and other accessories. These types of Bluetooth devices are often referred to as human interface devices (HID).

In reality, the vast majority of Arduino projects involving Bluetooth communication use a non-HID version of Bluetooth, known as Serial Port Profile (SPP). Some of the more recent Bluetooth modules can emulate the role of an HID device when communicating with a PC or smartphone, but you cannot connect a Bluetooth keyboard to your Arduino using one of these Bluetooth modules. So, when you think of Bluetooth and the Arduino, think along the lines of a wireless serial data connection from one Bluetooth module to another, or a Bluetooth module to a PC or smartphone. If you want to use a Bluetooth keyboard with an Arduino, more than likely the easiest way would be to use a USB host shield/module with a standard Bluetooth dongle and connect the Bluetooth keyboard that way.

To complicate matters even further, there are multiple versions of Bluetooth technology — Bluetooth and Bluetooth Low Energy (Bluetooth LE, or BLE),

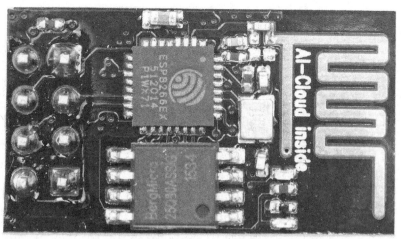

Figure 4.3 — The HC-05 Bluetooth module.

also known as Bluetooth 4.0. Both operate in the 2.4 GHz Industrial, Scientific, and Medical (ISM) band, but that's pretty much where the similarities end.

The older Bluetooth is often referred to as Classic Bluetooth, and incorporates the Bluetooth 1.x, 2.x, and 3.x standards. Often, you will hear the term "EDR" (Enhanced Data Rate) associated with this version of Bluetooth. EDR refers to enhancements within this version of the Bluetooth protocol allowing faster data rates, up to 3 Mbit/s. At 1 Mbit/s, Bluetooth LE has a slower data throughput than Classic Bluetooth, but operates at a much lower power level. Unfortunately, the two technologies don't often play well together. While a Bluetooth BLE smartphone can link and communicate with a Classic Bluetooth device, the reverse is not true. So for simplicity's sake, when working with Bluetooth devices, always use the same version on both ends of the connection to avoid any incompatibility issues.

When using the Bluetooth Serial Port Profile, one end of the connection is designated the Master and the other end is designated the Slave. The HC-05 Bluetooth module (**Figure 4.3**) is the Bluetooth module most often used with the Arduino. While the firmware varies widely among modules in the HC series, typically the HC-05 can be configured for either the master or slave role. The companion HC-06 module can only operate in the slave role. The HC-05 does not support the HID Bluetooth profile and therefore can only be used for serial data links. And the fun's not over yet. There is a lot of variance in the firmware for the various HC-05 modules used with the Arduino, so you really need to play with these modules to get a feel for how to integrate them into your projects.

Many of the HC-05-type modules can be powered by 5 V, but the signal lines operate at 3.3 V. To prevent damage to the module, you will need to use a voltage divider or level converter on the signal connections to the module.

The HC-05 has two operating modes, Data mode and AT Command mode. The AT Command mode is reminiscent of the old dial-up modem "AT" command prefixes such as ATA for answer or ATD for dial. The Command mode is used to perform any configuration required to communicate with the slave unit.

The HC-05 powers up into the default Data mode and there are differing methods of entering the AT Command mode based on which version of firmware you are running. The most recent and most common version of the HC-05 board has a small pushbutton switch that will put the HC-05 into AT Command mode if the switch is pressed and held while the unit is powered up. The LED on the HC-05 board should blink slowly on and off, about once every two seconds to indicate that it is in AT Command mode. When in Data mode, the LED blinks faster, about five times a second while waiting for a connection, and then a quick double flash about every two seconds when it is connected to another device.

There are other Bluetooth modules becoming popular for use with the Arduino, including the HC-10, HC-11, RM-42, and others. Again, it would take a whole book just to cover all of the various Arduino Bluetooth modules and their individual intricacies. But once you get the hang of working with the HC-05, working with the others should be just as easy, if not easier.

USB Communications

In many ways, USB technology is like an iceberg. At the surface, it appears to be just a simple, straightforward high-speed communication method. But it's that hidden 90% that will absolutely ruin your day, especially when it comes to the Arduino.

There are two basic components of a USB connection, the USB host and the device. This distinction is an important piece in understanding USB communication as it relates to the Arduino. In order to be programmed using the Arduino IDE, just about every Arduino communicates with your PC via the Arduino's onboard USB port (a handful of Arduino variants use only Bluetooth or Wi-Fi), or through an FTDI module to convert between the Arduino's serial port and the PC's USB port. In this type of connection, the PC is the USB host and the Arduino is the device. Most USB devices are human interface devices (HID) such as keyboards, mice, external hard drives, and so on. Here's where that bottom half of the iceberg comes into play.

The reason for this issue is the nature of the USB protocol. USB stands for Universal Serial Bus, meaning that it is designed to be the standard for interfacing all manner of devices to the host device. As such, USB plays a large role in determining (also known as enumerating) what device has been connected, as well as installing and/or configuring the device drivers needed to communicate with the device as necessary for the attached device to operate properly. Because of the huge variety of USB devices, you can see that setting up to communicate with a USB device is not really all that simple, but this all happens behind the scenes on your PC or smartphone. That can take a lot of processing overhead, horsepower, and memory to handle everything needed to communicate with a USB HID device.

Some Arduinos and Arduino variants such as the Leonardo can emulate a USB HID device to a host such as a PC or smartphone. But, if you want to connect a USB keyboard, mouse, or other HID device to your Arduino project, the Arduino has to have USB host functionality. There are only a few Arduino boards, such as the Arduino Due and Mega ADK that can function as a USB host without the use of a module or shield (as this was written, both of these boards are officially discontinued but still available through some suppliers). Some of the Teensy line of Arduino boards from PJRC can also function as a USB host.

A far simpler and less expensive method to add USB host capability to your Arduino project is to use a USB host shield (**Figure 4.4**) or module. Both are supported by the USBHost library and function identically. In actuality, the Mini USB Host Module (**Figure 4.5**) is a shield designed for the Arduino Pro Mini, but it works equally well as a module with other Arduino boards such as the Uno or Nano as you'll see in the USB CW Keyboard project later in this book.

Figure 4.4 — The SparkFun USB Host shield.

Figure 4.5 — The Mini USB Host module.

Amazingly, through the use of the USBHost library, you can connect all manner of USB devices simply and easily. While I have not personally tried it yet, it should be feasible to also add Bluetooth HID capability through the use of a Bluetooth USB dongle plugged into the Arduino's USB host shield/module port. This is definitely at the top of the list of things to test, as it opens up a whole new world of simple and inexpensive wireless Arduino projects.

5 Creating Arduino Sketches

In this chapter, we will discuss how to plan and design the program, or *sketch* as it's called in the Arduino world, for your Arduino project. We will also cover how to install Arduino libraries, and how to use the Arduino Integrated Development Environment (IDE) software shown in **Figure 5.1** to create your sketches.

Creating sketches is at the heart of every Arduino project. Using the Arduino IDE, you can create, edit, upload, and test your Arduino sketches. But before you even start to write the sketch for your Arduino project, it helps to plan out ahead of time exactly what you want your sketch to do, what I/O pins and methods it will use, and what shields or modules you'll be using. I refer to this form of program planning and design as the "Divide and Conquer" method. Each

Figure 5.1 — The Arduino IDE.

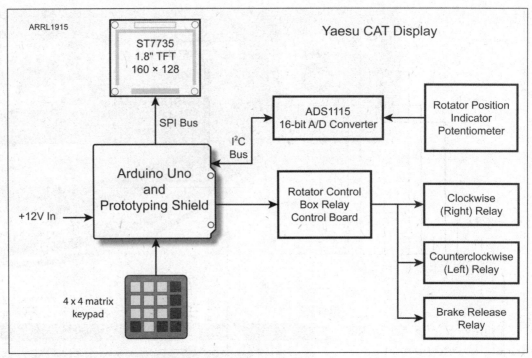

Figure 5.2 — An example of a block diagram.

major part of the sketch is broken into smaller chunks that can be coded and debugged, one piece at a time.

Basic Documentation

I recommend creating a *block diagram*, *schematic diagram*, and *flowchart* to help you stay on track with what you're trying to accomplish with your Arduino project and sketch.

Block Diagram

Generally, I like to start out with a block diagram of the project hardware. A block diagram similar to the one in **Figure 5.2** is a very high level overview of your project, giving you an idea of what shields, modules, and components you will need to assemble your project. A block diagram will also help identify what I/O pins and methods to use in your project. You can also use the block diagram to help determine which Arduino board or variant may be best suited for your project. This is also a good time to think about what type of enclosure you want to use.

Schematic Diagram

After I create the block diagram, I'll draw out a rough schematic diagram, similar to the one in **Figure 5.3**, only my drawing is a hand drawn version that I use to construct the project, using the block diagram as a rough guideline. As I construct the project prototype, I make any additions or changes to this hand-drawn version, and use this updated working drawing to create a nicer, easier to read version with a CAD program such as Autodesk's *Eagle*. A schematic

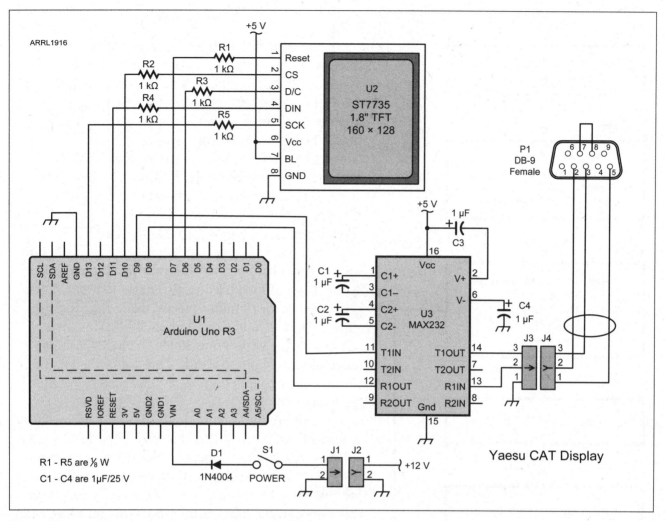

Figure 5.3 — An example of a schematic diagram.

diagram is a drawing that shows the physical connections for all of the components in your project in a standard format.

Assigning which I/O pins to use for the project is determined by the types of modules and components you plan to use. Because devices and modules that use the SPI or I²C bus will already be assigned to their respective bus pins, all that remains is figuring out which pins to use for the remaining components. A schematic diagram can be used as a step-by-step drawing to keep track of the project build and will also help prevent wiring or design errors that you may overlook otherwise. If you are unfamiliar with how to read a schematic diagram, the ARRL website has a section on circuit construction, including some basics on how to read circuit diagrams (schematics) at **www.arrl.org/circuit-construction**. Another excellent resource for learning how to read schematics is *Beginner's Guide to Reading Schematics* by Stan Gibilisco (ISBN 978-0-07-182778-2). Finally, if I am planning on using a block of code or library for a device that was created by someone else, I'll often use the same pins they used in their project, so that I can use their sketch as a way to test that piece of the project.

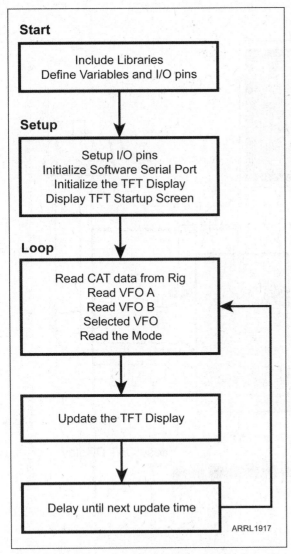

Figure 5.4 — Flowchart example.

Flowchart

Once I finish building the project, I'll use small blocks of code to test the various functions of each component on the board. After I have an assembled project that I am reasonably sure will work as intended, it's time to actually start on the sketch itself. Before I start writing code, first I'll create something that I've used since my dear dead mainframe days — a flowchart (**Figure 5.4**). Flowcharts are essentially a block diagram for your program. This helps you get your mind organized as to how the sketch needs to be written, what constants and variables and their type you may need, as well as helping determine what libraries you may need to include in your sketch. Flowcharting also helps to break down your sketch into smaller pieces, using the aforementioned "Divide and Conquer" method. This method allows you to write your sketch in blocks, testing each block as you go along. That way, when something goes wrong, you have a good idea of where the problem may be.

By taking the time to do this basic documentation ahead of time, you'll find that this preparation really helps when it comes time to build the project prototype and write the sketch. It gives you a chance to review your design and see if there is anything you may have overlooked during the initial building and testing phase. Because you did some basic testing when you completed the project build, you can be reasonably certain that any problems that crop up at this point are in the sketch itself and not with the hardware.

Final Documentation

When you finish your project, it's a good idea to take the time to document it, especially if you plan to share your project with others. Once each project in this book was finished and tested, I used Autodesk's *Eagle* software to create a finished schematic drawing, along with using the Open Source *Fritzing* program to document the finished project. Both packages allow you to create schematic diagrams, board layouts, and parts lists, and they can even generate a file you can use to have your project etched onto a circuit board. *Fritzing* also has a unique feature that allows you to create a breadboard image of your project, such as the one shown in **Figure 5.5**, which will create a pictorial view of your project that you can use to wire your project up on a breadboard. The projects in this book include a schematic diagram, with the colorful *Fritzing* diagram available as a download from **www.sunriseinnovators.com** or **www.kw5gp.com** as an aid to the construction process.

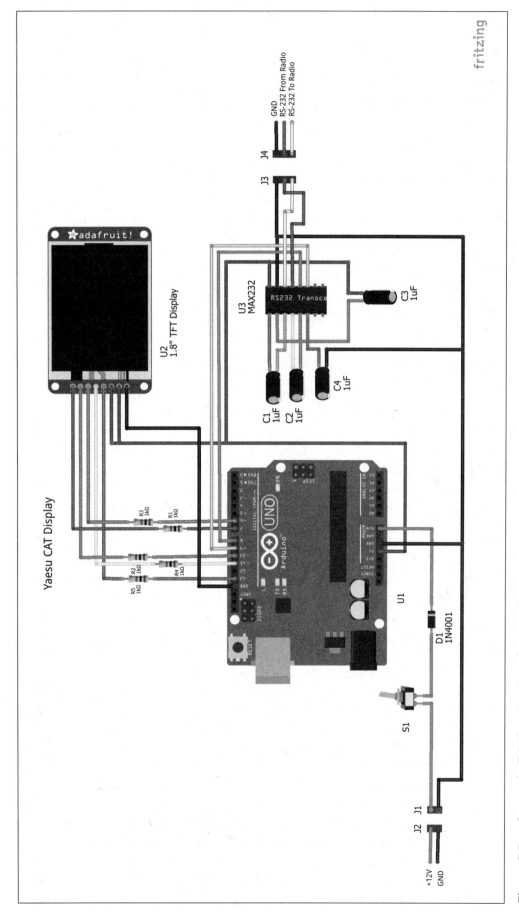

Figure 5.5 — A sample Fritzing drawing. Colorful Fritzing diagrams for all the projects in this book are available for download from www.sunriseinnovators.com or www.kw5gp.com.

The method you choose to plan and organize your projects is a matter of personal choice, so use whatever works best for you. The important thing is to take the time to plan as much as possible ahead of time and document your work as thoroughly as possible when the project is completed. A little bit of planning and documentation upfront can help keep you from soldering or coding yourself into a corner and getting frustrated. Remember, building things and working with the Arduino is supposed to be fun. Using block diagrams, schematics, and flowcharts can help keep things fun.

The Arduino IDE

Arduino sketches are created using the Arduino Integrated Development Environment (IDE). This book is not intended to teach you how to use the Arduino IDE, as there are already a number of good books on this topic available, including *The Arduino Cookbook* by Michael Margolis (ISBN 978-1449313876), and *Programming Arduino — Getting Started with Sketches* by Simon Monk (ISBN: 978-0-07-178422-1) that can devote far more pages to teaching you how to use the Arduino IDE than is possible here. There are also some excellent online tutorials for installing and learning how to use the Arduino IDE at **www.arduino.cc/en/Guide/Environment**, **learn.sparkfun.com/tutorials**, **learn.adafruit.com**, and other websites as well. In this book, we will focus on a few IDE basics, and the things you will need to know to use the Arduino IDE with the projects in this book, such as installing Arduino libraries, and how to add support for Arduino-variant boards to the Arduino IDE.

The Arduino IDE is used to create, edit, and upload your Arduino sketches. The Arduino IDE is very versatile and has the ability to support a wide range of different hardware and processors. Using the IDE's text editor, you can create an Arduino sketch. To help you, the IDE will highlight keywords, and match up your brackets and braces to help you identify the loop and if-then code blocks in your sketch. Under the **TOOLS** menu, the **AUTO-INDENT** feature will automatically indent your sketch text to make your sketch more readable. Preferences are set under the **FILES>PREFERENCES** menu option, allowing you to customize some important aspects of the IDE, including the location where the IDE will save your Arduino sketches. The location where the IDE saves your Arduino files is also known as the *sketchbook*, and is used by the IDE to create a menu listing of your available Arduino sketches. Please get familiar with the **PREFERENCES** menu as we will be using this option to add support for additional Arduino boards and variants in just a bit.

The Arduino IDE is used for more than simply creating sketches. The IDE has a message area that is used to provide feedback while saving, compiling, and uploading sketches, and is where any error messages or warnings are displayed. The toolbar icons near the top of the IDE window provide quick access to the most commonly used IDE functions — **VERIFY**, **UPLOAD**, **NEW**, **OPEN**, and **SAVE**. The **VERIFY** operation will compile and check your sketch for errors, but does not upload the sketch to the Arduino. The **UPLOAD** operation does the same as a **VERIFY**, except when the sketch compiles without any errors, it will upload it to the Arduino. The **NEW**, **OPEN**, and **SAVE** operations allow you to create a new sketch and open or save an existing sketch.

It is important to note that the Arduino IDE performs an auto-reset on the

Arduino board as part of the upload process. Some older Arduino boards do not support the auto-reset function. On these boards, you will have to manually press the Arduino's reset button immediately prior to selecting **UPLOAD** in the IDE. Also, some of the newer Arduino boards such as the Leonardo and some variants will actually lose connectivity with the workstation as part of the auto-reset process, which may cause the board to change the USB port it's connected to on the workstation. This is primarily due to a newer USB interface design that integrates the USB functionality on the Arduino (or variant) processor itself, which allows the Arduino to also act as a human interface device (HID) to a connected workstation.

As part of the reset process, your workstation attempts to identify the Arduino and its USB port, but may fail to do so in the allowed time frame. This is a case where you may have to go into the IDE and reset the **SERIAL PORT** selection back to the proper port. Sometimes, the Arduino's USB port doesn't show up at all after an upload or reset. The only solution I have found for this is to reset the Arduino, or disconnect and reconnect the Arduino USB port until the IDE finally figures things out. It can get annoying sometimes, and is the primary reason why I prefer the Arduino UNO over the Leonardo for many of my projects.

An interesting feature of the Arduino sketch editor is the ability to create your sketch using multiple tabs from with the editor. In theory, this would allow you to break a large sketch into smaller, more readable sections, as well as giving you the ability to incorporate other files usable by the Arduino IDE such as .h, .cpp, and .c files (such as the files used within a library). In reality, using multiple tabs to create your Arduino sketch can be confusing, because the IDE will compile your sketch in alphabetical order starting with the main sketch and add, or concatenate, the tabbed files so they effectively appear to be at the bottom of the main sketch. This can cause issues with any #define statements and variables defined in the tabbed files, but used in the main sketch, among other things. Also, some developers have used this as a quick way to add a library to their Arduino sketch. While this may work, it is not the ideal method to add a library to the Arduino IDE and your sketch, as we will see in a little bit. At the end of the day, creating a sketch using multiple tabbed files is often more trouble than it is worth, and it's best to just use a single tab for your Arduino sketch.

The **TOOLS** menu is where you select the type of Arduino board you are working with, along with the USB communications port the board is attached to. If you do not select the correct board and USB port, your sketch may verify and compile, but the upload process will fail.

You can also add support for third-party hardware, such as Arduino-variant boards, by placing their board definitions, core libraries, bootloaders, and programming definitions into a sub-folder in the hardware folder of your Arduino sketchbook. You can create the hardware folder if it does not exist, and the IDE will incorporate the new board into the IDE the next time the IDE is started. Fortunately, with the newer versions of the IDE, you can also add support for additional boards by adding a third-party URL for the new boards in the IDE preferences, and they will automatically be added to the IDE Board Manager, where you can select and download everything you need to support a new Arduino board or variant. We'll show you how to do this in just a bit.

Also on the **TOOLS** menu, you will find the **AUTO FORMAT** option. This will reformat the text in your sketch so that the curly braces all line up, and the text is properly indented, making your sketches easier to read. That feature is a great help when you get around to troubleshooting your sketch. You can also save your Arduino sketch in compressed .zip format by using the **ARCHIVE SKETCH** option. This will create a compressed .zip file of your Arduino sketch and save it in the sketchbook folder.

One of the most important options on the IDE **TOOLS** menu is the Serial Monitor. The Serial Monitor will display the serial data sent and received from the Arduino's USB/serial port. Typically, you will use the Serial Monitor to show basic program output and display any debugging information you have included in your sketch. You can also use the Serial Monitor to send characters and commands to the Arduino via the USB/serial port. The Arduino's USB/serial port can also be used to allow your Arduino to communicate with the Processing language installed on your PC, and to allow PC-side applications to interact with your Arduino sketch.

One of the most underutilized tools in the IDE is the Serial Plotter. Similar in operation to the Serial Monitor, the Serial Plotter allows you to generate a real-time graphical plot of a data string that your Arduino project sends to the serial port.

Adding New Boards

Starting with version 1.6.4, the Arduino IDE includes a Board Manager feature that allows you add support for new Arduino and Arduino-variant boards far more easily than in the past. The IDE Board Manager also keeps track of

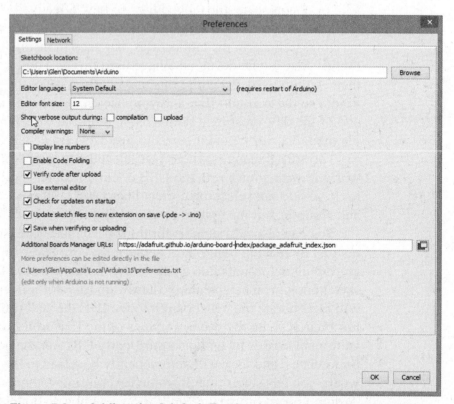

Figure 5.6 — Adding the Adafruit Feather boards to the Arduino IDE Board Manager.

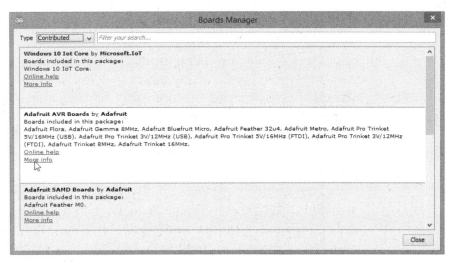

Figure 5.7 — Selecting the board types to add to the IDE Board Manager.

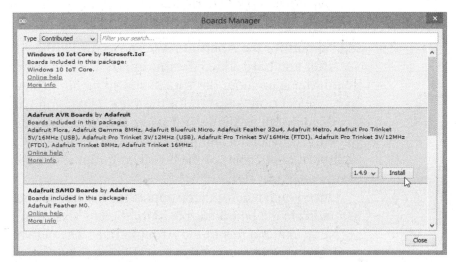

Figure 5.8 — The IDE Board Manager showing the board definition versions available to install.

any updates, and allows you to update existing board definitions and libraries as they are released.

Adding support for a new board has never been easier, and many Arduino board suppliers are adding support for their products in the IDE Board Manager. To add a new board to your Arduino IDE, all you have to do is add the support URL to the Additional Boards Manager URLs section of the IDE. You can edit this preference and have multiple entries, allowing you to add support for any number of new boards quickly and easily. An unofficial list of third-party URLs for the IDE is available at **github.com/arduino/Arduino/wiki/Unofficial-list-of-3rd-party-boards-support-urls**. Your board manufacturer may also provide you with a URL for adding support for their boards that you can add to the IDE Board Manager directly.

To begin adding support for a new board, in this case the Adafruit Feather boards, you first add the support URL to the Additional Boards Manager URLs list under the IDE Preferences as shown in **Figure 5.6**. In this example, we will

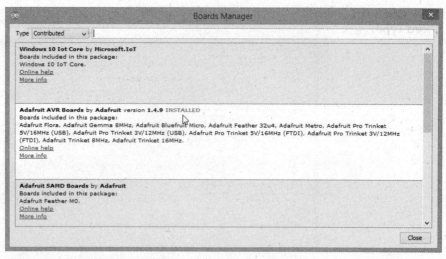

Figure 5.9 — The Board Manager showing that the new board type definitions have been successfully installed.

add support for the Adafruit AVR and SAMD (the ARM Cortex series).

After you have added the third-party URL to your preferences, select the Board Manager under the **TOOLS>BOARD** menu. When the Board Manager opens, select **CONTRIBUTED** for the **TYPE**. Next, you will see a list of available boards to install as shown in **Figure 5.7**. When you click on the **MORE INFO** link, you will see a screen similar to the one in **Figure 5.8**, where you can select the version (if applicable) of the board definitions, and can then install the new boards into your IDE.

Once you have installed support for the desired boards, the Board Manager will indicate the board has been installed as shown in **Figure 5.9**.

Next, exit and restart the Arduino IDE. The new boards should be listed and available for selection as shown in **Figure 5.10**. Select the desired board as shown in **Figure 5.11** and you can now create and upload sketches to the new board type. That's all there is to it for adding support for new boards to your Arduino IDE. Sometimes there are issues compiling sketches for new boards, but through the Board Manager and the Arduino IDE's **UPDATE** process, you can usually resolve these issues when updates are made available.

When you connect the new board to your Windows workstation for the first time, you may have to install a driver. In the case of the Adafruit Feather boards, you can download and run the Adafruit driver installer. Aside from clicking the default **I AGREE** and **INSTALL** boxes, the only screen to pay attention to is the screen where you are provided with checkbox options to select which board drivers to install (**Figure 5.12**). You can either select just the drivers for the boards you want, or take the default options and install the ones recommended. As a general rule, I take the default and get them all, saving me from having to go back and repeat this process when I get another type of board from the same supplier, in this case, Adafruit.

Figure 5.10 — Choosing the new Board Type.

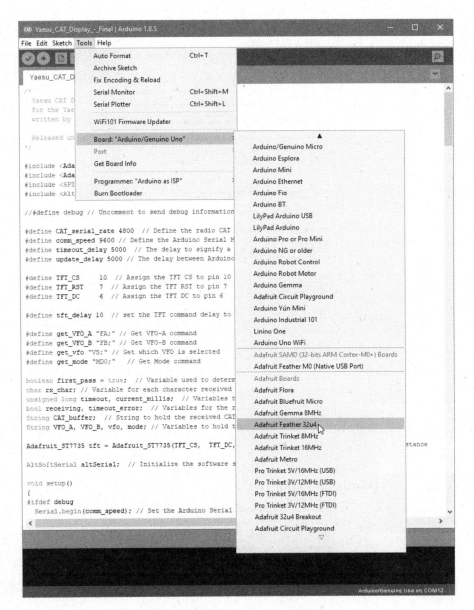

Arduino Libraries

Including libraries in your sketches is an integral part of programming the Arduino. An Arduino library is a prewritten set of functions that you can add to your Arduino sketch to support shields, modules, devices, and even add additional software functionality such as trigonometry functions. Through the use of libraries, a large percentage of the programming for a project may already be written for you. Another way to look at Arduino libraries is to think of them as the device drivers needed to interface to all of the various devices you connect to a workstation. By "including" a library in your sketch, you add the functionality of that library without any additional programming. Then, through the use of function calls from within your sketch, you can interface the shield or device to your sketch quickly and easily.

Quite often, your sketch is merely a little bit of programming "glue" used to tie everything together — the library does all the hard work for you. This is one

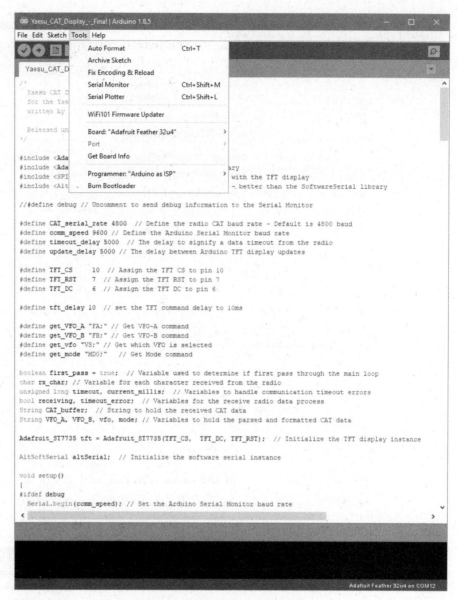

Figure 5.11 — The new Board Type has been selected.

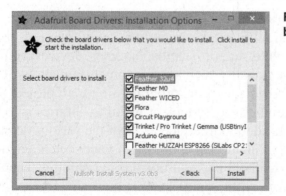

Figure 5.12 — Selecting the board drivers to install.

of the major advantages to the Arduino's Open Source model. There are literally thousands of Arduino enthusiasts like you out there, creating and sharing libraries that you can use in your Arduino projects. Using Arduino libraries can save you hours and even days of deciphering datasheets and writing test code just to interface to a new device. Many libraries even come with example code that you can integrate into your own sketches, thereby making the development and testing of your own project go much faster.

Many libraries are already integrated into the Arduino IDE, and a large number of other libraries available on the internet, written and shared by their creator. In many cases, you may find that a library already exists to suit the needs of your project. More recent versions of the IDE include the Library Manager feature, which has the built-in ability to search for updates to your existing libraries, as well as allowing you to install new libraries into your IDE.

Installing Libraries

One of the biggest issues I have discovered about using the Arduino is the confusion on what a library is and how to install it. Nearly 90% of the questions I receive about my Arduino books are library-related installation issues. If you have a sketch that uses a library, and that library is not installed, or installed incorrectly, the sketch will not compile. If you look at the message area on the IDE, you will see errors relating to the compiler being unable to locate functions contained within a library, along with the name of the .cpp or .h file associated with that library.

To make things even more confusing, starting with version 1.5.7 of the IDE, they "broke" some of the older libraries and you will need to either use an older version of the IDE or update the library that is failing in order to compile your sketch successfully. Among one of the older libraries that doesn't work with the newer IDE versions is the LCD5110_Basic library. You can download a newer version of this library from **www.rinkydinkelectronics.com**.

Additionally, some of the early, more popular libraries are now built into the IDE and no longer need to be installed. There have also been numerous continuing issues involving certain library functions, particularly scrolling text on the display with the LiquidCrystal_I2C library since version 1.5.7, so be sure to verify that you're not dealing with a broken library before throwing in the towel. Problems with these libraries is one of the reasons I now predominately use OLED or TFT displays in my Arduino projects.

Libraries are not that difficult to install, and the newer versions of the IDE now allow you to install libraries using several methods. The confusion regarding the installation of Arduino libraries may simply be the concept of "installing" a library. On the Arduino, technically, libraries are not installed in the classic sense of software installation. In a normal library installation, there is no executable file that will "install" a library for use with your IDE. The library "installation" process is actually far simpler.

The typical Arduino library consists of several files within a single top-level folder. In general, an Arduino library is written in C++ and contains a .h header file, and a .cpp file that contains the actual code for the library itself. These files are contained in the top level of a folder, or series of subfolders, with the top level folder having the same name as the .h and .cpp files it contains. A library may also contain additional files such as keywords.txt, and in the case of newer libraries that can be installed using the IDE Library Manager, a metadata file named library.properties, that contains information the IDE can use to install and update the library. A library often has additional subfolders containing example sketches and documentation on how to use the features of the library. So, essentially, at the end of the day, an Arduino library is just a group of files contained in a folder with the same name as the library's .h and .cpp files.

Library Manager

When you open the Library Manager located under the **SKETCH>INCLUDE LIBRARY>MANAGE LIBRARIES** menu option, you will see a list of libraries available for you to automatically download, install, and update using the Library Manager. All you have to do is select the library you wish to install and the Library Manager will take it from there. If available, the Library Manager will allow you to choose between the most recent version and older versions of the library. As with the installation of any new library, it's always best to exit and restart the IDE after installing the library to allow the IDE to properly add the new library to the list of available libraries.

A library can be also be installed manually or as a .zip file through the Library Manager. When you install a library, all you are doing is placing the library folder, and any subfolders that it contains, into the libraries folder in your Arduino sketchbook folder. In Windows, this is typically in your **DOCUMENTS>ARDUINO** folder. This is the same folder that will contain your Arduino sketches. To manually install a library, all you have to do is copy the folder of the library you wish to use into the "libraries" folder. If the libraries folder does not exist, simply create a folder named "libraries" in your Arduino sketchbook folder (be sure to use all lowercase for this folder name), and copy your libraries into this folder.

In order for the IDE to see the new library, you may have to exit and restart the IDE. At startup, the IDE will check the contents of the libraries folder, and any libraries in that folder will then be available for use in your sketches. You will see all of your available libraries under the **SKETCH>INCLUDE LIBRARY** menu option. Any available example sketches for your libraries should also be available under the **FILE>EXAMPLES** option of the IDE.

All of the libraries for the projects in this book that are on the **www.sunriseinnovators.com** and **www.kw5gp.com** websites were designed to be installed manually, and you will need to download and extract the project compressed .zip file, and copy each individual library folder into your sketchbook's libraries folder. **Figure 5.13** shows what the sketchbook libraries folder should look like when you have libraries properly installed.

Another way to install an Arduino file is using the Library Manager's .zip file method. While many libraries have been modified to use the new Library Manager, some are still available for download as a compressed .zip file. To install these libraries, all you need to do is download them and use the **ADD .ZIP LIBRARY** option under the **SKETCH>INCLUDE LIBRARY** menu. The Library Manager will extract the zipped library file and place it into the sketchbook libraries

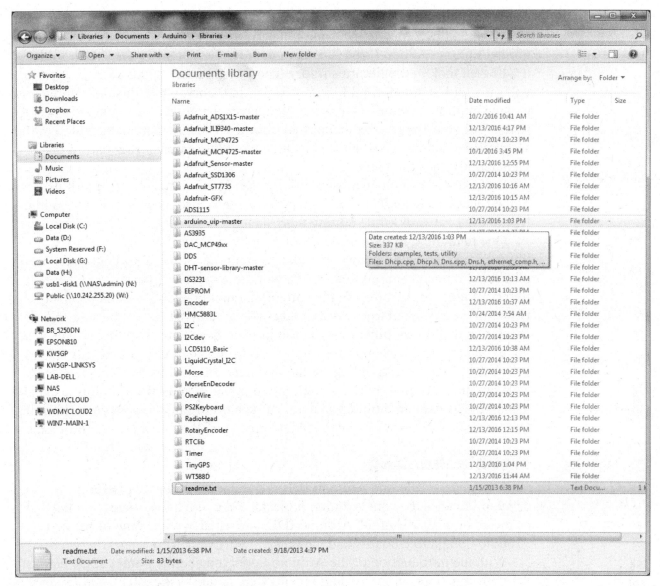

Figure 5.13 — Contents of the libraries folder.

folder for you. After you exit and restart the IDE, the library you just installed will then be available for use.

The Library Manager feature is very nice, and has really helped with problems related to installing libraries. Additional information on how to install Arduino libraries can be found at **www.arduino.cc/en/Guide/Libraries**.

Using Libraries

To use a library in your Arduino sketch, all you have to do is either select the library using the **SKETCH>INCLUDE LIBRARY** option in the IDE. You can also manually select the library by using a `#include <library.h>` preprocessor definition near the beginning of your sketch. Sometimes you will see the `include` statement using double quotes instead of the left and right carat surrounding the library name, for example `#include "library.h"` versus `#include <library.h>`. For all intents, these two forms for the include statement are interchangeable, but the double quote form allows you to have the library files in the same folder as your sketch .ino file instead of the normal libraries folder. If the IDE does not find the library in the sketch folder, it will then look in the libraries folder. The `#include <library.h>` form will only look in the normal libraries folder. In theory, while this does allow you to have multiple versions of a library you have customized, it's generally best to have all of your libraries in the normal libraries folder. When using a library, the include statement must always include the .h file extension as part of the included library's file name.

Some libraries include a keywords.txt file. This is a text file that will list all of the functions available in the library. This file is also used by the IDE to highlight these keywords while editing your sketch.

One final thing to remember about Arduino libraries. Many Arduino libraries are created by people just like you and me. Sometimes bugs and glitches slip in and cause things to not work as expected. I have found these library glitches to be rare, and they are often quickly addressed and resolved. However, you do need to be aware that if you are having issues, be sure to check that the library functions you use in your sketch are working as expected. For example, there was an issue with the built-in Liquid Crystal I2C library, where the library would only print the first letter of a string sent to the LCD. A work-around was quickly found and posted to the various Arduino forums, and the issue was fixed in a library update, but you do need to be aware that libraries can sometimes not work as intended and give you all sorts of strange problems.

Troubleshooting

As a general rule, the Arduino IDE is a stable and reliable platform to develop sketches for your Arduino projects. There are a few issues you need to be aware of when using the Arduino IDE, particularly with some of the newer Arduino-compatible boards and variants.

With the introduction of the Arduino Leonardo and the Atmel 32U4 chip, the USB functions on the Arduino board were added to the 32U4, and the 16U2 USB controller on the original Arduino boards was no longer needed. As mentioned earlier, the USB controller in the 32U4 chip functions a little bit differently than with the UNO. When you upload a sketch to a 32U4-based Arduino, the auto-reset that the IDE performs on the board as part of the upload process can cause your workstation to briefly lose connectivity with the Arduino and when it reconnects, the Arduino may be assigned to a different USB port than what you have set in the IDE. The simple answer here is to select the new USB serial port in the IDE and continue normally.

Sometimes, the board will not reconnect after uploading a sketch. This is primarily due to the USB driver on your workstation not seeing the Arduino before timing out while trying to re-establish communication. In these cases, you may have to disconnect the USB cable from your workstation to the Arduino a couple of times before everything gets figured out.

Some of the Arduino-compatible boards use the WCH CH340/341 series of USB converter chip in place of the Atmel 16U2. When you connect an Arduino-compatible board that uses the CH34x USB chip, your workstation will not be able to recognize the new board until you install the CH34x USB drivers on your workstation. Locating these drivers can be somewhat troublesome, but there are more and more websites now offering the CH340/341 drivers for download. I was able to locate the drivers for my CH34x boards on the CH340/341 manufacturer's website at **www.wch.cn/download/CH341SER_EXE.html**. The driver for the newer CH341 will also work with the CH340 chip, so all you need is the CH341 driver file for your workstation (Windows, macOS, or Linux). If you are unable to locate the drivers on the site listed above, you can find them simply by doing a web search for Arduino CH340 or CH341 drivers. Once the CH34x driver is installed, your workstation should be able to identify and connect to the Arduino-compatible board.

With some of the newer, more powerful Arduino-variant boards, as well as some of the inexpensive Arduino UNO and Nano-compatible boards, I have had issues with a workstation not being able to recognize and communicate with the board, even when the correct drivers had been installed. The board worked fine when I connected it to a different workstation. As it turned out, the board I was using was drawing too much power for the USB port I had it connected to. I ended up having to use a powered USB hub in order for the workstation to recognize the board.

The message area in the Arduino IDE can provide a great deal of information if you have issues compiling a sketch. In your **FILE>PREFERENCES** menu, you can select verbose compiler output to get more detail on any compiler or upload messages, as well as setting the type of compiler warnings shown. Compiler warnings are just that, warnings to alert you to a possible issue. Many compiler warnings can be ignored, but it is beneficial to pay attention to the information provided if, and when, you see a compiler warning. As with many compilers, a compile error may not show you the exact statement that failed and why, but if you carefully read the information, it will usually provide enough information to track down the issue.

Memory Issues

The Arduino memory architecture is not the same as in a workstation. The Arduino has three, and some cases, two separate types of memory onboard. The Arduino UNO has just 32 KB of flash memory, which is used to store the compiled sketch, and 2 KB of static RAM (SRAM) to store your program variables, data, and provide memory required by the Arduino sketch to operate properly. The UNO also has 1 KB of EEPROM that can be used to permanently store information, but few sketches actually use or need the EEPROM, and many of the Arduino-variant boards do not have EEPROM.

As such, memory in the Arduino UNO and similar boards is at a premium, to say the least. In general, you can fit the vast majority of your sketches in memory just fine. With the larger memory capabilities of the newer Arduino and variant boards, lack of memory is becoming less and less of an issue. It is important for you to keep the limited memory capacity of the Arduino in the back of your mind while creating your Arduino sketches. Unlike a regular workstation that will politely give you an "Insufficient Memory" error and refuse to run the program, the Arduino will do exactly as it is told, keep on running, and then start doing weird things, and failing to execute the simplest of commands properly. There are some programs and libraries, such as the Memory Free library, that you can incorporate into your sketches to keep track of available memory and let you know how much memory you have remaining.

If you find yourself needing to save precious SRAM, you can instruct the IDE to place static data and constants into flash memory by including the avr/pgmspace.h library and using the PROGMEM keyword in your sketch. The pgmspace library allows you to store static variables, constants, strings, and arrays in flash memory instead of SRAM. You can also use the FLASH library to store things like tables and string arrays in flash memory. Finally, you can also use the F() syntax for storing string constants in flash. Just remember, that if you put it in flash memory, it cannot be modified during execution of the sketch, so you can only use these memory saving features for data that doesn't change.

Simple Debugging Methods

You will see in many of the sketches for the projects in this book that I have included the preprocessor directive `#define debug_mode` or a similar definition to incorporate a debugging mode. When `debug_mode` is set to 1, or true, the IDE will compile additional debug code that has been added throughout the sketch. This debug code is used to output additional information to the Serial Monitor that can be used to monitor critical points in the sketch as an aid to troubleshooting. Once the sketch is tested and working, the debug mode is set to 0, or false, and the debug code is ignored and not compiled. This is a quick and easy way to add troubleshooting code to your sketch, and then turn it off once troubleshooting is complete. If you use the `#ifdef` and `#endif` preprocessor directives around the block of troubleshooting code, when debug mode is set to 0, or false, the IDE compiler will ignore the troubleshooting code and not compile it, making your finished compiled sketch smaller and more efficient.

Arduino Create

Arduino has an integrated online platform known as Arduino Create, which allows you to write code, save it to the cloud, and even upload your cloud-based sketches to your Arduino boards, all from within your web browser. You can even share this content with others. It has a free version, and all you need to do is create an account at **create.arduino.cc**. The Web IDE will automatically detect any Arduino board connected to your PC and configure it accordingly. All of the libraries available in the Arduino Library Manager are automatically detected, so you don't need to install any libraries to get your code to compile. The Web IDE even supports the Serial Monitor. The Arduino Web Editor is supported in the Windows, Linux, Mac, and Chrome operating systems.

Arduino Create also gives you access to the Arduino Project Hub, a tutorial platform dedicated to discovering the Arduino, electronics, and software. All of the content is categorized and searchable and covers a wide variety of project categories and projects.

Arduino IoT Cloud is yet another piece of Arduino Create. IoT Cloud is designed to be an easy to use Internet of Things applications platform. IoT Cloud allows you to access your internet-connected objects quickly, simply, and securely. You can connect multiple objects to each other and allow them to exchange real-time data. You can also monitor your objects from anywhere using a simple user interface. Arduino IoT Cloud supports a number of third party devices, such as the ESP8266. Arduino IoT Cloud has both free and paid options for all levels of Arduino enthusiasts. And here's a fun fact: you can even use Arduino Create to create Arduino sketches that run on the Raspberry Pi.

Multitasking and Real-Time Arduino Operating Systems

By nature of its design and intent, the Arduino in and of itself is not considered to be a multitasking system. This is changing now, with the advent of several libraries and multitasking methods designed for microcontrollers such as the Arduino. For simplicity's sake, these are often referred to as operating systems, while in fact, they are actually more along the lines of a multitasking kernel for embedded and microcontroller systems.

HeliOS is a very small multitasking kernel written in C that runs on most 8-bit microcontrollers such as the Arduino. HeliOS is installed as an Arduino library using the Library Manager and integrates easily into your Arduino project. It's easy to learn with an application programming interface (API) that consists of only 21 function calls. HeliOS allows you to implement cooperative and event-driven multitasking operations with task notification and messaging, timers, and memory management. HeliOS utilizes a run-time balancing feature that ensures tasks with shorter run-times are prioritized over tasks with longer run-times. HeliOS is definitely something to consider if your Arduino project needs to deal with multiple conditions and events simultaneously.

Another multitasking option to provide real-time operating system features to your Arduino project is FreeRTOS. While small enough to run on a microcontroller such as the Arduino, FreeRTOS is not limited to just microcontroller applications. It includes a kernel and a growing set of software libraries that are suitable for use across multiple industry sectors and applications. FreeRTOS is installed as a library to the Arduino IDE using the Library Manager. The FreeRTOS main kernel offers a number of features, including fixed priority pre-emptive time slice scheduling, inter-process communication, task management features such as task priority, task suspension, deletion and delay, along with software timer and interrupt management.

While the concept of multitasking with the Arduino is relatively new, the real-time multitasking libraries offer up some interesting possibilities for a variety of Arduino projects that would require the ability to handle multiple conditions, events, and operations simultaneously and are definitely something to consider experimenting with.

6 Tools, Construction Techniques, and Troubleshooting

Part of the fun of working with the Arduino is creating your projects. There's nothing quite like the satisfaction of building something yourself and then getting it to work as intended. As with any electronic project, you will need a few tools to properly construct your projects. You'll also need some basic test equipment to make sure things are working properly and to troubleshoot with when they're not. Developing a good troubleshooting method is also an important part of getting your Arduino projects to work when they don't do what they're supposed to do.

While creating projects on a breadboard is often a good way to start out, eventually you will want a more permanent solution, involving soldering your project's components to some form of prototyping or etched circuit board, and mounting the finished project in an attractive enclosure similar to the one shown in **Figure 6.1**. For this, you will need a soldering iron or soldering station. You'll also want to have a multimeter handy to read voltages, double-check component values, check continuity, look for shorts, and make other measurements. As your projects become more and more advanced, particularly if you're working with analog signals, you will probably want to get some more sophisticated test equipment such as a frequency counter or oscilloscope, and perhaps a logic analyzer for digital signals.

Breadboard

As I begin designing and building a new project, I will usually build a prototype version of the project on a breadboard. A breadboard is typically a rectangular plastic block, in my case 7 inches long by 3 inches wide, which has a series of interconnected holes where you can plug in components and jumpers.

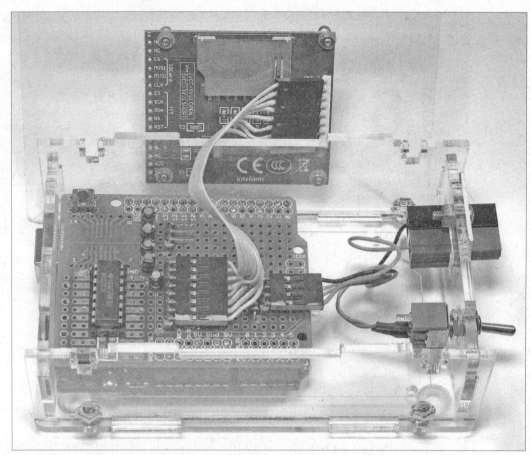

Figure 6.1 — A completed project in a Solarbotics enclosure.

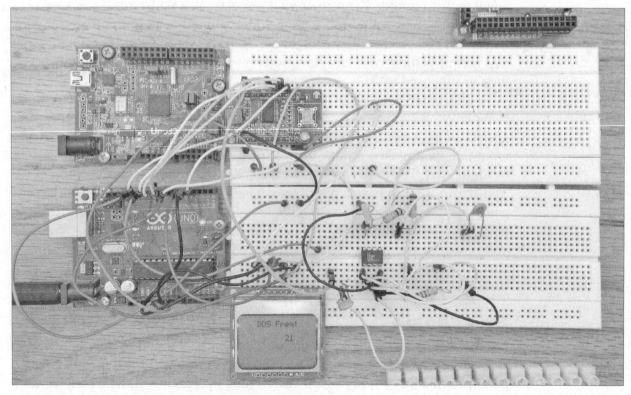

Figure 6.2 — My original homebrew breadboard setup.

On the breadboard, I can quickly build a project and make wiring changes simply by moving the jumpers around.

Figure 6.2 shows my original Arduino breadboard during the development of the DDS VFO project for my second book. Everything was mounted on a 12- by 15-inch piece of wood for easy storage without having to take apart the project when I needed my workbench for other purposes. In addition to having two breadboard units for those really big projects, it had an Arduino UNO, a Digilent Uno32, and an Arduino Due mounted around the edges of the board. Also mounted on my development breadboard was a 9 V battery clip, a servo, a 16×2 I²C LCD display, and a terminal block to connect to off-board devices such as antenna rotator motors and the like. This setup allowed me to design and develop many projects without having everything spread out all over the workbench.

However, as times changed and new Arduino processor boards become popular, others such as the Due and the Digilent Uno32 faded from popularity, and in the Due's case, discontinued altogether. To modernize my development board, I would have to unscrew the Due and Uno32 and mount different, more commonly used processor boards. As this progressed, eventually the wooden board would be nothing but a mass of screw holes and would need to be replaced.

The good news is, that as the Arduino becomes more popular, so do the commercially available development systems. Rather than upgrade my original breadboard system, I made the switch over to the PhaseDock WorkBench (**Figure 6.3**) for my proof of concept and large prototype development projects. The PhaseDock WorkBench is unique in that each building block-style component is mounted on a "slide," which is then mounted to a "click" platform that then snaps into holes in the work area itself. There are a number of pre-built clicks and slides available, and you can also customize your own. One feature that comes in handy, particularly for me, is the plastic cover option, to keep a project

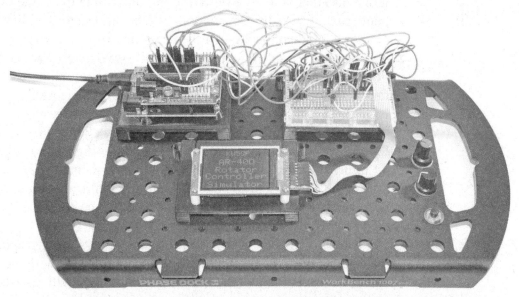

Figure 6.3 — The PhaseDock WorkBench with a prototype of the AR-40 rotator project described in a later chapter.

safe from cats and other hazards when I'm not working on when I'm not working on the project. A complete review of the PhaseDock WorkBench can be found in the April 2020 issue of *QST*.

Component Tester

When building electronic projects, sometimes it's difficult to determine the value of a component by its color code or the value stamped on the body of the component. When looking for a transistor, how can you easily tell if it's a PNP or NPN transistor, what the correct pinout is, or if it's even any good at all? Sure you can test it with a multimeter, or look up the datasheet online, but there is a far easier way.

Figure 6.4 — Component tester.

There are a number of inexpensive multifunction component testers (**Figure 6.4**) that you can use to test and determine the value of electronic components such as resistors, capacitors, diodes, transistors, inductors, MOSFETs, and other components with just the press of a button. Powered by a 9 V battery, the multifunction tester will try to determine the type of component, test it, and display the value of the component, along with some basic parameters such as pin configuration and gain in the case of a transistor, forward voltage drop in the case of a diode, and other characteristics.

I don't know about you, but ever since they changed over to the five-band color codes on resistors, and started using three-digit numbers on capacitors, I can't figure out what the value is without using a cheat sheet or multimeter. Now, I just put the component to be tested into the multifunction tester's onboard socket, verify that the part is good, and that it's the correct value that I'm looking for, all at the press of a button.

The newer versions of the multifunction testers use a graphic LCD display to actually draw the schematic symbol of the component with the pin configuration of the device overlaid on the schematic symbol and will also display some basic device parameters. I got my multifunction tester from eBay for about $10, and it is easily the most-used piece of test equipment on my workbench.

Multimeter

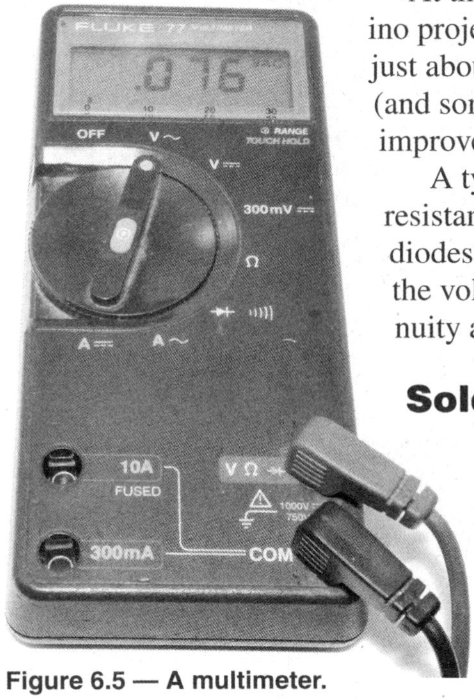

At the very least, you will also need a multimeter to test your Arduino projects. It doesn't have to be a fancy one like mine (**Figure 6.5**) — just about any multimeter will do. You can get one for as little as $8 (and sometimes even free with a coupon and purchase) from home improvement and hardware stores.

A typical multimeter will allow you to read voltage, current, and resistance. Some models will even measure capacitance and test diodes or transistors. Generally, you will use your multimeter to read the voltage on the Arduino power and I/O pins, and check for continuity and short circuits in your project.

Figure 6.5 — A multimeter.

Soldering Station

Once you have finished developing and testing your project on a breadboard, the next step usually is to solder everything together on either on a piece of copper-clad perfboard or an Arduino prototyping shield. Things sometimes get tight when working with small boards, so you'll want a good soldering iron or maybe even a soldering station similar to the one shown in **Figure 6.6**.

For years, I used a plain old soldering iron and was quite happy with how things turned out. Then, when I needed to try my hand at soldering some surface mount components, I discovered rather quickly that using a regular soldering iron just wasn't going to get the job done. I borrowed a soldering station from my friend, the late Tim Billingsley, KD5CKP, to take a crack at it. Needless to say, for a variety of reasons I failed miserably at my first surface mount soldering attempt, but I was hooked on the soldering station itself. I liked the fact that the soldering pencil had an adjustable temperature control and used standard interchangeable tips.

It also had a heat gun attachment, with adjustable airflow and temperature

Figure 6.6 — An adjustable temperature soldering station with hot air attachment.

Tools, Construction Techniques, and Troubleshooting

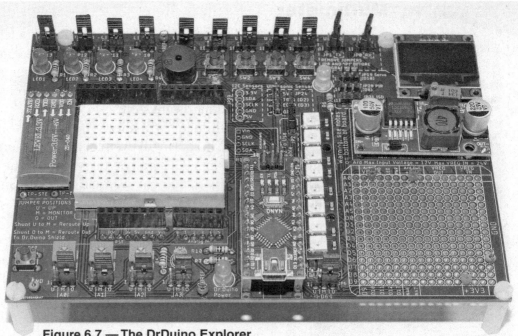

Figure 6.7 — The DrDuino Explorer.

controls. While I may have been less than successful in my dealings with surface mount components, the heat gun worked great with heat shrink tubing. I ended up getting a soldering station of my own, with all the bells and whistles, online for about $60.

DrDuino

As I mentioned earlier in Chapter 3, the DrDuino Explorer shown in **Figure 6.7** is a great way to quickly create a proof of concept or prototype project. The DrDuino Explorer can act as an I/O shield inserted between either an Arduino UNO or Nano and your prototyping shield or other type of shield. This allows you to use a multimeter or oscilloscope to view the status of the Arduino's power and I/O pins. You can also do some basic proof of concept and prototyping projects directly on the Explorer board itself, and use its onboard components to simulate inputs and outputs to help you with your project design. You can even do a hybrid of both methods and use the Explorer's onboard I/O jumpers to reroute the I/O pins to either your prototyping shield or the Explorer's onboard components.

The Explorer also has small breadboard and soldering areas for you to construct projects on the Explorer board itself. This is a great compact prototyping solution, especially if you're cramped for space in your work area. I relied heavily on the Explorer to create the working prototypes and sketches for several of the projects in this book. A complete review of the DrDuino Explorer can be found in the January 2021 issue of *QST*.

I/O Shields

An I/O Shield (**Figure 6.8**) was one of the first add-ons I ever bought for my Arduino. An I/O shield brings all of the Arduino I/O pins out to 3-pin headers,

Figure 6.8 — An Arduino UNO/Nano I/O shield.

with the I/O pin, power, and ground in each 3-pin group. You can then use jumper wires from there to a breadboard or access the pins with a multimeter or oscilloscope for testing and troubleshooting. Many of the modern I/O shields for the UNO also have a socket onboard that will allow you to use the shield with a Nano as well. You can also get an I/O shield that will support the Arduino Mega footprint type boards (**Figure 6.9**).

Frequency Counter

As you begin working with some of the modules capable of generating radio frequencies (RF), such as the AD9850 DDS module and the Si5351 clock generator module, you will most likely want to have a good frequency counter handy. I was able to pick mine up at a hamfest (**Figure 6.10**) for about $50. All it needed was a new display driver chip and it was as good as new. Mine goes up to 50 MHz, but that's generally all I need it for. At that same hamfest, for $50 I was able to pick up another perfectly functional frequency counter that goes up to 1.3 GHz, but I haven't had any real need to use it, so it usually sits on the shelf as a backup.

Having a frequency counter is extremely helpful for both audio and RF signals. You may not have a need for a frequency counter in your Arduino project development, but it never hurts to be on the lookout for an inexpensive one at a

Figure 6.9 — Arduino Mega I/O shield.

Figure 6.10 — A frequency counter.

hamfest. There are also some very inexpensive ones available online from eBay and Banggood that look like they'd work well for some applications where you need a basic frequency counter.

Oscilloscope

If I could only have one piece of test equipment in my lab, it would be an oscilloscope. With an oscilloscope, you can view waveforms, measure voltage, and with a little math, you can determine the frequency of the waveform. Some oscilloscopes have add-on units that can also display voltage and frequency, with some of the newer digital storage scopes incorporating that as part of the waveform display.

There are basically two types of oscilloscopes, the older traditional analog oscilloscopes that use a small cathode ray tube (CRT) with a high persistence phosphor to display the signal waveform similar to how an older CRT-based TV does, and the newer, digital storage scopes. Digital storage scopes actually sample the input signal at a high rate using an A/D converter, and graph the signal on either a CRT or TFT display. Because they use digital data, a digital storage scope can save and analyze the waveform, and add calculations such as voltage and frequency to the display. Many of the newer digital storage scopes can also interface with a PC over a USB port to allow capturing the waveform on the PC, as well as controlling the oscilloscope settings from the PC.

Depending on what kind of Arduino projects you plan on building, you can even build your own mini-oscilloscope using an Arduino. There are several Arduino oscilloscope projects on the Instructables website (**www.instructables.com**) that work quite well for basic troubleshooting on the Arduino. The main limitation with these Arduino-based oscilloscopes is the speed of the Arduino itself, and you can only use it to view waveforms up to about 200 kHz depending on the oscilloscope project you build. This type of oscilloscope would therefore only be usable at lower frequencies (in and just above the audio range). If you don't want to build one, there are a number of pre-built mini-digital oscilloscopes that use a TFT display available on eBay for around $10.

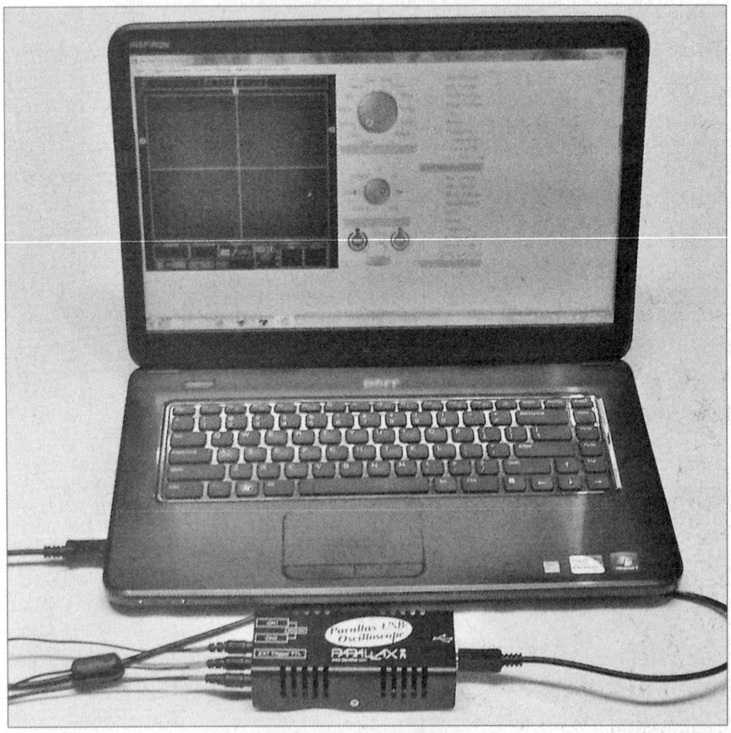

PC-based USB Digital Storage Scope

Moving up in functionality, PC-based USB digital storage scopes connect to your PC via a USB port, and are controlled by a software application running on the PC (**Figure 6.11**). This type of scope generally does not have a standalone display, and all waveforms

Figure 6.11 — A USB digital oscilloscope. (Photo courtesy Pat Vickers, KD5RCX)

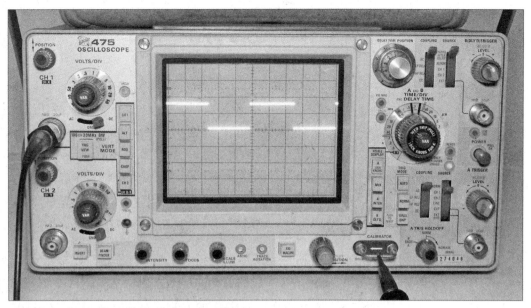

Figure 6.12 — Tektronix 475 oscilloscope.

are displayed on the PC itself. A good quality 100 MHz USB digital scope costs about $65 on eBay. The one downside of this type of oscilloscope is that you need a PC to use it, so while this type of scope is portable, they tend to take up workspace on your lab bench. I used one for quite a while before I decided to invest in an old analog scope, simply because that's what I'm used to using.

Traditional Analog Oscilloscope

When you think of an oscilloscope, typically you think of the larger analog scopes such as the Tektronix 475 shown in **Figure 6.12**. When I decided to move up from my USB-based scope hooked to a laptop PC, I picked up an old school Tektronix 465 100-MHz dual-channel oscilloscope with the voltage/frequency add-on at a hamfest for about $150.

At the same hamfest, I ran into an old friend, Henry Wingate, K4HAL, who gave me a 200-MHz Tektronix 475 scope that was the exact same scope I used while working with him at Control Data Corporation many, many years ago. He managed to get it when they closed the doors a number of years back, so there's a sentimental attachment to this particular scope, and it has been the mainstay in my lab for several years now.

A typical analog scope can display two channels worth of waveforms on a small cathode ray tube (CRT), with a high persistence phosphor coating inside the CRT that allows you to see the full screen trace, similar to the way the old-style CRT-based TV sets displayed the picture. While there were some Polaroid camera options available that you could use to take pictures of the waveforms on the CRT screen, most analog scopes can't be interfaced to a computer, and they only display the waveform in real-time, meaning a waveform can't be saved and analyzed at a later date.

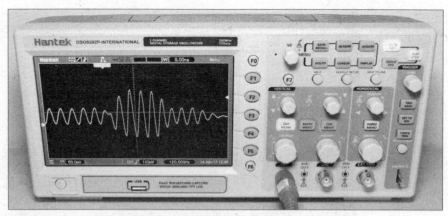

Figure 6.13 — A digital storage oscilloscope.

Digital Storage Oscilloscope

In the past, the best digital storage oscilloscopes were the big, expensive commercial units that really didn't fit in the average ham's budget, including mine. With the introduction of the larger high-resolution color TFT displays, good quality digital storage oscilloscopes have become more affordable. I recently treated myself to a Hantek DSO5202P 200-MHz dual-channel digital storage scope (**Figure 6.13**) for $300. While not top-of-the-line, it's about the size of a lunchbox, with a 7-inch high resolution color TFT display. It has most, if not all, of the features of the high-priced top-end digital storage scopes. It even has a USB port and a PC application where you can view and control the scope from a PC, allowing you to save and print waveforms like the one shown in **Figure 6.14**.

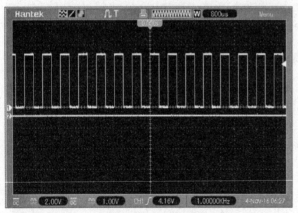

Figure 6.14 — A digital oscilloscope's waveform displayed on a workstation.

The Hantek scope easily fits on a shelf on my workbench, and I can hook up the USB cable to save any waveforms on the PC when I need to. While $300 for a piece of test equipment may seem a bit on the expensive side, if you do a lot of building and work with HF and VHF frequencies, it's hard to beat a scope of this type and price. I'd been worried what I would do if, and when, my 30+ year old analog scope died, but now I have a backup plan. I usually use the new digital storage scope for quick tests and troubleshooting and break out the old analog scope for the more involved troubleshooting, not for any reason other than the comfort and experience using the older scope. Similar scopes from Rigol and Siglent have been reviewed in *QST*.

Logic Analyzer

One of the most recent test equipment acquisitions for my lab is a USB-based 16-channel logic analyzer (**Figure 6.15**). In the past, logic analyzers tended to be a lot like analog oscilloscopes — bulky and unwieldy. They didn't have very good display quality for the multi-trace analyzer output unless you

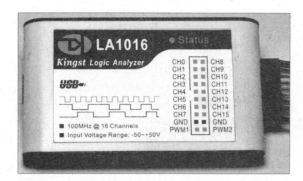

Figure 6.15 — The Kingst logic analyzer.

Figure 6.16 — An example of logic analyzer output.

bought one of the very expensive high-end units. As with just about everything else through the years, the logic analyzer has been vastly improved, and is now available as a USB device that attaches to a workstation with excellent functionality and display quality.

As a major oversimplification, think of a logic analyzer as an oscilloscope for digital-only signals, with a large number of individual traces, typically 16 or 32 signal traces as shown in **Figure 6.16**. A logic analyzer has a programmable triggering point based on individual or multiple conditions. This allows you to see multiple points of data simultaneously and is ideal for viewing complex timing conditions such as SPI or I²C bus transmissions. This particular logic analyzer can also decode and display SPI and I²C communications.

For my logic analyzer, I went a bit overboard and bought a 16-channel, 100-MHz Kingst LA1016 unit for about $80 from eBay. This particular device can safely allow input voltages from –50 to +50 V, so it would take a lot of effort on my part to damage it input-wise. Less expensive and lower bandwidth ones are available with as few as 8 channels and for less than $10. You may never need a logic analyzer, but in my case, there will always be a time when I would give just about anything to have one at hand in the lab.

Figure 6.17 — Adafruit's Bluefruit LE Sniffer. (Photo courtesy Adafruit.com)

Bluetooth LE Sniffer

As you start working with Bluetooth projects, having a Bluetooth 4.0 LE Sniffer such as the one in **Figure 6.17** can be a great benefit. Unlike regular Bluetooth devices, a sniffer allows you to see and decode all Bluetooth LE activity within range. You can also use software such as Wireshark to see the data inside each individual packet. This would be a great benefit when you're trying to figure out why your Bluetooth LE projects aren't working properly. Note that the sniffers I was able to locate are only for BLE 4.0. I was unable to locate a Classic Bluetooth sniffer, although they may be available as well.

Design Tools

Designing and documenting is a major part of every project. While many of us start from a hand-drawn circuit design, at some point in your project, you will want a more professional-looking method of documenting the project, especially if you plan to share it with others. If you plan to have a circuit board fabricated for your project, you will need to use design software that will generate a printed circuit board (PCB) layout in a format that your board fabricator can support. There are a number of design tools to choose from, but I'll only focus on a few of the most popular ones here.

I am a big fan of Autodesk's *Eagle*, which will allow you to create schematic diagrams as well as create the PCB layout based on the schematic diagram. The program has the capability to automatically route the PCB traces for you. *Eagle* has a huge library of components you can incorporate into your projects, and you can even create new parts as needed. *Eagle* has both a free and paid version. For most of us, the free version will do just about everything needed. It is free for hobbyists and allows you create two schematic sheets, two signal layers, and an 80 × 80 cm (12.4-inch square) board area.

Another popular PCB design tool is *KiCad*, which is free and Open Source. Unlike *Eagle*, there are no licensing restrictions with *KiCad*. You can even download the source code and modify it if you so choose. If you are an *Eagle* user like me, the transition to *KiCad* is not difficult at all. *KiCad* also has a number of keyboard shortcuts for doing standard tasks such as component rotation or drawing a wire, whereas *Eagle* does not. Compared to *Eagle*, creating components in *KiCad* is a bit more confusing at first, but as with all things new, it gets easier in time. *KiCad* has a very nice 3D Board Viewer to help you with component placement on the board. Both programs can generate board layouts in the industry-standard Gerber file formats as well as the NC (numeric controlled) file for drilling holes in the PCB.

KiCad and *Eagle* are both supported on Windows, macOS, and Linux workstations. Both have their pros and cons, and the choice as to which one to use is up to you. I would recommend working with both and see which one you prefer.

Circuit Simulators

There are a number of general-purpose electronic circuit simulators as well as Arduino/microcontroller-specific simulators. As with the design tools, we'll focus on just a couple of the more popular circuit simulators, and recommend that you do further research to determine which one may be best for you.

LTspice is a *SPICE*-based analog electronic simulator from Linear Technology. Based on the original Open Source *Simulation Program with Integrated Circuit Emphasis* (*SPICE*), *LTspice* is free and it easy to use. Although it comes with component libraries that contain only Linear Technology products, additional components can be added by the user. The good thing about using a *SPICE*-based simulator is that once you learn one version, it's relatively easy to switch to other versions that may better suit your needs.

Spectrum Software's *Micro-Cap 12* is a powerful full-featured, commercial *SPICE*-based integrated schematic editor and mixed analog/digital simulator that is now free from **www.spectrum-soft.com**.

Tinkercad is a free, easy-to-use web-based modeling program from Autodesk, the makers of the *Eagle* PCB design tool discussed earlier. *Tinkercad* can be used for 3D design, electronic simulation, and coding. *Tinkercad* allows you to design your circuit by placing and wiring components, programming your project using either Codeblocks or C++, and then simulating how the circuit performs in real-life. *Tinkercad* supports the Arduino UNO and ATtiny microcontrollers, but you are limited to the devices in the built-in components library, as *Tinkercad* does not support user-created virtual "hardware." However it does support a broad range of libraries and components.

Codeblocks is a very interesting part of *Tinkercad*. Rather than coding each individual line of an Arduino sketch, you can use a "codeblock" to incorporate a standardized block of code into your sketch that can then be expanded into normal Arduino code, and then edited as you would normally. Once the circuit and sketch are complete, you simply select the **START SIMULATION** button and *Tinkercad* will then virtually "run" your sketch. *Tinkercad* not only supports a virtual version of the Arduino IDE's Serial Monitor, it also incorporates a virtual multimeter and oscilloscope to allow you to view the voltages and signals in your project. Once you have everything working in *Tinkercad*, you can easily export the Arduino sketch, import it into your real Arduino IDE, and upload it to a physical Arduino. *Tinkercad* even includes a debugger that allows you to step through your virtual sketch and look at the variables while the sketch is running.

Considered to be one of the better Arduino simulators currently available is the *Virtronics Simulator for Arduino* v1.12 at **www.virtronics.com.au**. This simulator has a free version that incorporates a delay on loading sketches after the initial 45-day trial period expires or the 200 sketch limit is exceeded. It does remain full-featured after the trial period ends, except for the initial sketch loading delay. You can also purchase the Pro version for $19.99. *Simulator for Arduino* supports the Arduino UNO, Nano, Mega, Leonardo, and other common Arduino boards. It has a debugging mode that allows you to step through your sketch line by line, and can simulate all of the analog and digital I/O pins, as well as a number of devices such as LCD display, Ethernet, servos, and other add-ons.

Construction Tips

There are several tips and techniques I have found helpful when constructing my Arduino projects. First and foremost, is the use of connectors between my Arduino project and any external components such as switches, rotary encoders, displays, and so on. This allows me easily to remove the Arduino assembly from its enclosure for testing and modification.

DuPont Headers and Sockets

I have found that using the DuPont-style 2.54 mm headers and sockets (**Figure 6.18**) is ideal for working with Arduino prototyping shields and copper-clad perfboards. The prototyping boards also have 2.54 mm spacing between holes, so no modifications are needed to use these connectors. The DuPont-style headers and sockets have a male header assembly similar to the standard jumper pins that you find on many electronic devices. The male part of the header typically comes in 40-pin strips that are designed to be separated into shorter strips containing the number of pins that your project requires. The male header is also available in a right-angle configuration, allowing for a lower overall height of the connector assembly for use in tight quarters.

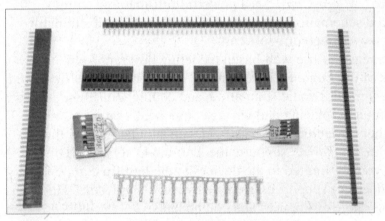

Figure 6.18 — DuPont-style headers, pins, and cables.

The female portion of the DuPont-style header, also known as a "header socket," consists of a plastic shell with holes for inserting a locking female crimped pin. The female connector is available in a wide variety of pin configurations, from a single pin up to 10 pins and higher.

I have found that using these female sockets is perfect for cabling between the Arduino projects and external components, such as switches and displays, using old PC ribbon cables. I was able to buy an entire box of old PC ribbon cables for $5 at a hamfest several years ago, and I still have about a quarter of the box left over. More than likely, you'll want the older cables that were used for floppy drives and the older IDE hard drives, as the conductors in these cables are typically stranded rather than solid. I've found that the stranded wire in these ribbon cables is much easier to work with, and more durable as well, as compared to the solid conductor versions.

To complete the cable, I just crimp and solder the female socket pin to the end of the ribbon cable, and insert the socket pin into the appropriate hole in the female header socket. I'll also use a permanent marker to identify pin 1 on the cable and on the board to help prevent wiring errors.

Sockets

I learned early on in my work with the Arduino that all ICs, including Arduino Nanos and module assemblies that will be mounted on a prototyping shield

Figure 6.19 — Example of two rows of header sockets to create a Nano socket.

or perfboard, should be in sockets. After having to desolder and replace a Nano a couple of times, I started putting everything I could into a socket. While as a general rule the Arduino is forgiving and pretty hard to burn up, it can be done. When soldering a new project, wiring errors can, and do, occur. Trust me, it's a royal pain when you've just spent five hours wiring up that new project, only to discover you were one pin off and just put 12 V into a digital I/O pin on a Nano. So, from here on out, if I can use a connector or a socket, I will.

You can get 40 pin strips of the female version of the 2.54 mm male headers than can be soldered to a board. For chips and modules such as the Nano, I'll cut these strips down to the number of pins I need, and use two rows to create a socket for the device as shown in **Figure 6.19**.

The DuPont 2.54 mm headers and sockets, along with the 40-pin SIP (single in-line package) header and socket strips are inexpensive. Typically, you can get them in bulk quantity on eBay for $1 per 100, with free shipping. I generally order each individual type of these connectors separately, so for roughly $10, I have enough of a quantity and variety of connectors to last me for quite a while.

Prototyping Shield

When building projects that use an Arduino UNO-style footprint, I like to use a prototyping shield, also known as a protoshield (**Figure 6.20**). The protoshield is designed with the same footprint as the UNO, so you can use header pins soldered to the board, and connect it to your UNO just as you would with any other shield. This allows me to easily remove the protoshield from the project so I can fix wiring errors or make additional modifications to the board. This technique also allows for the removal of the protoshield and reusing the Arduino UNO in another project simply by attaching a different protoshield and uploading a new sketch.

Anderson Powerpole Connectors

When I need a power connector on a project enclosure, I will usually use Anderson Powerpole connectors (**Figure 6.21**). A Powerpole connector is a plastic housing used with a latching crimped contact pin inserted into the housing. Powerpole connector housings are keyed in such a way that multi-conductor cables can be built simply by mating as many Powerpole housings together as necessary.

Figure 6.20 — An Arduino UNO prototyping shield.

Tools, Construction Techniques, and Troubleshooting

Figure 6.21 — Anderson Powerpole connectors.

While most often used for providing dc power to ham equipment, larger multi-conductor cables can also be built using Powerpoles. As an example, for portable events such as ARRL Field Day, our club's antenna rotator control cables have a block of Powerpole connectors on each end, with pigtails on the rotator and control unit used to adapt to the long rotator cable. Using a common wiring scheme, we can use just about any club member's rotator and controller without having to modify the cable. All we have to do is plug in the rotator cable to the rotator and controller pigtail cables and we're ready to go. No more messing with those screws on the back of the rotator or controller that always fall into the grass, never to be seen again, and no more wiring errors while attaching the rotator cable.

You can also get chassis mount adapters and connectors that allow you to easily integrate Powerpole connectors in your Arduino project enclosures as I have done with several of the projects presented in this book. Powerpole adapters come in a wide variety of current ratings, up to 180 A, and with different colored housings, allowing you to color code your connections. Most hams use the 15/30/45 A version, because these all use the same size connector shell.

Troubleshooting

One of the most frustrating things about building your own electronics projects is that they don't always work as they should. While there's no simple tried and true method to magically fix everything that doesn't go right, there are a few things you can do to help make the troubleshooting process go easier.

The first step in troubleshooting is to sit back and analyze what is happening — or not happening. Quite often, a simple analysis of the problem's symptoms will be enough to get you into the area where the problem lies. Troubleshooting basically comes down to logical, deductive reasoning, combined with the old "Divide and Conquer" technique we mentioned earlier. And of course, a little bit of luck never hurt anyone.

When something is not working correctly, trust nothing. Every component is suspect, every solder joint is suspect, every wire is suspect, your Arduino sketch is suspect, the new library you're using is suspect — everything is a suspect until you prove to yourself otherwise. Using deductive reasoning, narrow the problem down to the area where you believe the problem lies, and, starting with the most logical suspect, work your way down the list. Test for shorts, test the continuity of each wire, and verify that you're not soldered to the wrong pin.

If you're not making any headway, stop, get up, and walk away. Take a break and work things out in your head. When I was just starting out in my career, I was the assistant to the project's design specialist on a completely new prototype VHF and microwave data telemetry project. Occasionally, when he'd hit a snag in his design, the circuit didn't do what it was supposed to, or he had to figure out a completely new way of making things work, he'd just get up and take me along for a brief walk. Along the way, he'd throw some thoughts about the problem my way to get my insight and viewpoint. As a 19-year-old kid, that was a tremendous experience to be thought highly enough of that I could offer some insight to the problem and serve as his sounding board. More often than not, about halfway through the walk, he'd have that "eureka" moment and we'd go back at it and usually resolve the issue. It's those kinds of experiences through the years that have taught me some of the best troubleshooting techniques I have learned.

A key thing to start out with is to evaluate what is working, what isn't working, and come up with possible reasons that match the symptoms you're experiencing. Have a friend double-check your wiring. You would be amazed how many times you can look at something a hundred times, swear that it's right, and then a fresh set of eyes spots the problem in a matter of seconds.

If that fails, don't hesitate to ask for help or search online for possible answers. There's a good chance that someone, somewhere has experienced the same issue you have and can offer up a simple solution or troubleshooting idea.

Break your project down into sections and test each piece separately. Again, this will help you narrow the problem down to a specific area. As you can see, troubleshooting is more of a method than using a specific set of tools. But, with the Arduino, there are some additional considerations and tools you can use to help troubleshoot your projects.

With the Arduino, there is no built-in tool to monitor your memory usage, so it's easy to run out of available memory and not know it. When this happens, things just go totally bonkers and don't make sense at all. Fortunately, there are some small memory usage monitoring sketches you can integrate into your sketches if you feel that your sketch may be running out of memory. As mentioned earlier in Chapter 5, using the preprocessor `#define` and `#ifdef...#endif` statements to incorporate Serial Monitor debug code is another method to take a peek inside what your sketch is doing. Finally, while the Arduino does not have a debugger built in, there is a sketch debugging tool you can use to troubleshoot your Arduino sketches.

Debuggers

Debuggers are valuable tools for testing and debugging programs and have been around since the early days of computing. Most often used with the lower-level programming languages such as assembler and C, debuggers allow you to run the target program under controlled conditions that allow you to track the program's operation and monitor changes in variables and other components of your program. Unfortunately, the Arduino does not have a built-in debugger, but there is an Arduino library that can provide many of the features of a debugger.

The `serialDebug` library adds multi-level debugging capability to your sketch, displaying the debug messages on your Serial Monitor in different colors according to its debug level. You can also implement the serialDebug App which runs on your Windows, macOS, or Linux workstation to better format and display the debugger output. While not a full-featured debugger, the serialDebug library does offer several valuable debugging features, including the ability call a function, show and change global variables, and add or change "watch" conditions for global variables. A debugger "watch" condition means that the debugger will monitor and notify you if the watch variable changes value. The serialDebug library is compatible with any board supported by the Arduino IDE. There is even a converter that can convert an existing sketch and generate a new sketch with the debugger functions implemented.

At the present time, adding a debugger to the Arduino IDE is planned for version 2.0 of the IDE, but no release date has been announced. Until this, the serialDebug library can be used to fill the role of an Arduino debugger to help troubleshoot your sketches.

And finally, I can offer up this one last, but major piece of troubleshooting advice. I'm guilty of this more often that I care to admit, but I've found it to be true. Don't overthink the problem. Sometimes we get so wrapped up in the complexity of things that we don't stop and think things out at a simpler level. Instead, we just push on and get more frustrated as we go. The Arduino is a simple microcontroller, not a supercomputer. When installed correctly, the libraries work and can be thought of as "black boxes" that you don't need to see inside. More often than not, while trying to figure out some complex solution to a problem, backing up and "keeping it simple" has worked for me. So, if you find that your design isn't working out, stop and look for a simpler way. You'd be surprised how many times you'll come across a much simpler and more elegant solution to the complex corner you've painted yourself into.

Reminders

There are a couple of things to keep in mind when you are building your projects. These are some of the "gotcha's" that can catch you unaware when building your project and cause you all kinds of grief.

The I²C bus is an open-drain bus that requires pull-up resistors to function properly. The internal pull-up resistors on the Arduino SCL and SDA pins used by the I²C bus are not sufficient to serve as pull-up resistors, so you will need some form of external pull-up resistor for things to work right. A 4.7 kΩ resistor is a commonly used value for the pull-up resistor, but some modules have pull-up resistors built into the module. You only want one set of pull-up resistors on the I²C bus, so be sure to check whether any module on the bus already has the pull-up resistors onboard before adding external ones.

Some Arduino UNO-style boards have the circuit board copper traces run very closely to the mounting hole on the corner of the board near the A4 and A5 pins. It is very easy to chafe through insulating material on top of the board and short these pins when mounting the board. It's best to use a nylon washer between the screw head and the board to prevent this.

Finally, remember that there are now multiple versions of the Nokia 5110 LCD display, as mentioned in Chapter 3. Be sure that you verify the wiring for the type of Nokia LCD module that you are using in your projects. When in doubt, use your breadboard to test the wiring and pin configuration of your Nokia 5110 using one of the Example sketches so you'll know how you need to wire your Nokia display into your project.

Finishing Touches

As a finishing touch to your project, you'll probably want to have everything in a nice looking enclosure. As part of that, you'll also want some form of labeling as to what each switch and knob does. Trust me, I don't have the handwriting skills to even think about hand labeling my projects, so I need some form of printed labels.

In the past, I'd used a typical label maker or go back to my early days and try my hand at dry transfer labels (yes, you can still get those at a hobby shop or craft store). While at the local Hobby Lobby looking for enclosure ideas, I came across the latest in model decaling. They now have decal paper that you can use with an inkjet printer. Using this specially coated paper, you can create and print full-color decals for your projects, complete with text and graphics. All you do is print the design on the paper and spray it with a decal bonding spray to keep the ink from running when you apply the decal.

Once the bonding spray is dry, you can apply the decal as you would with a plastic model. Simply dip the decal in water for a few seconds, and the clear decal just slides off and adheres to the enclosure right where you want it. As with a regular decal, you can reposition the decal as needed. When dry, you have a nice-looking permanently mounted label without the mess. You can also get decal paper that works with laser printers and doesn't require the bonding spray.

Cheat Sheets

There are a number of "cheat sheets" and online calculators I have found that really help me when creating my Arduino projects. I have the cheat sheets printed out and next to my work area in my "lab" for quick reference. These are the resources that I have found to be most helpful when creating Arduino projects:

- Arduino Cheat Sheet — **blog.arduino.cc/2010/04/23/arduino-cheatsheet/**
- Arduino UNO Pinout Chart — **itp.nyu.edu/physcomp/wp-content/uploads/Uno_pinouts.png**
- Capacitor Codes — **www.dnatechindia.com/Capacitor-Colour-Code.html**
- Electronics Reference Sheet — **www.akafugu.jp/images/electronics_reference_sheet.pdf**
- Online LED Current Limiting Resistor Calculator — **www.ledcalc.com**
- Online Toroid Calculator — **www.toroids.info**
- SparkFun Arduino Cheat Sheet — **learn.sparkfun.com/resources/90**
- Teensy Pinout Chart — **www.pjrc.com/teensy/pinout.html**
- Wire Gauge and Current Limits — **www.powerstream.com/Wire_Size.htm**

Final Notes

All of the projects in this book were created and compiled using version 1.8.5 of the Arduino IDE. If you are having issues compiling any projects using a later version of the IDE, try using version 1.8.5 to compile and upload the sketch.

All of the sketches and libraries for the projects in this book are available for download at **www.sunriseinnovators.com/Arduino3** and **www.kw5gp.com/Arduino3**. The libraries on these sites will have to be installed using the manual library installation process discussed in Chapter 5, but many of them can be directly installed from its original source online using the IDE's built-in Library Manager.

My editor and I have worked hard to make this book as error-free as possible, but as with any project of this scope, sometimes errors do happen. Any updates, corrections, and other relevant information will be listed in the Errata link on the download websites, and corrected as soon as possible in later printings of this book.

Now it's time to have fun and build some projects. Enjoy!

Yaesu FH-2 Keypad

Figure 7.1 — The completed Yaesu FH-2 keypad in its enclosure.

Part of the fun of writing these Arduino books is thinking up and creating the projects that eventually end up in the book. People often ask where I come up with the ideas for some of these projects. Sometimes, I'll just look around my station thinking about things that could make a good Arduino project. Other times, it's sort of like the way I decide what to fix for dinner — going through all the cabinets to see what I have, and then ordering in pizza. In the case of Arduino projects, I'll look through the parts bins in the lab and look for something challenging that I haven't experimented with yet. Because my job requires that I spend a lot of time driving, I'll use that driving time to think up some wild and crazy projects that I'll jot down in the little notepad I always keep handy while on the road. Every now and then, someone will suggest a project idea to me and I'll try to make that into a project. And finally, as I go through the various ham radio catalogs and magazines, I'm always looking for something that would make a fun and interesting Arduino project.

Such is the case with the Arduino-powered Yaesu FH-2 Keypad project. The official Yaesu FH-2 shown in **Figure 7.2** is a multifunction remote control keypad for use with many Yaesu transceivers, including the FT-450, FT-891, FT-950, FT-991, FTDX1200, and FT-1000/2000. The FH-2 keypad comes standard with the FTDX3000, FTDX5000, and FTDX9000 transceivers. After doing some research on the internet, I found a couple of homebrew versions using a matrix keypad and

Figure 7.2 — The official Yaesu FH-2 keypad.

resistors. As it turns out, the FH-2 is little more than a keypad that selects a specific resistance value based on the key that is pressed. The functions of the Yaesu FH-2 keypad differ among the various models of compatible transceivers, so please consult your radio's manual to see what operations are supported by the FH-2.

Design Considerations

Simply building a version of the FH-2 with a membrane keypad and resistors didn't meet my "need to do it with an Arduino" project criteria, so I had to think up a new way to do it that was inexpensive, fun, and used new and interesting components and techniques (especially ones that could have other uses in ham radio projects). The FH-2 keypad project was not only a quick, inexpensive and fun build, it's now sitting in my station, just waiting to be used the next time I'm on the air.

Just about everyone starting out with the Arduino begins with a starter kit that includes a thin-membrane matrix keypad. At least a hundred times I've gone through the parts bins looking for ideas and thinking just what kind of cool project could I create with that keypad, but was never able to come up with a good one. When it was time to come up with projects for this book, suddenly, out of the blue, I had a whole handful of project ideas using keypads and the Arduino.

The MCP42010 Dual Digital Potentiometer Chip

What makes this project different from the standard Yaesu FH-2 keypad is that it introduces a component that has also been in my parts inventory for a while, with no ideas for a project to use it in. While I bought these parts with the knowledge of what they do and how I can use them, I just couldn't come up with a worthwhile project to use them in…until now.

This project uses two inexpensive MCP42010 dual digital potentiometer chips (**Figure 7.3**) to create the resistance values Yaesu transceivers expect from the real FH-2 keypad. The MCP42010 contains two 10 kΩ digital potentiometers within a single chip. Each potentiometer has 256 selectable positions that are digitally controlled via the SPI bus. The MCP42xxx series comes in 10 kΩ, 50 kΩ, and 100 kΩ versions. The MCP41xxx series is identical, except it only has a single potentiometer inside the chip. For this project, we will use the MCP42010, the 10 kΩ dual potentiometer version. Because the resistance value needed for the **DEC** key on the FH2 keypad exceeds 20 kΩ, we'll need to use two chips in series to give us a total resistance range of 40 kΩ.

Figure 7.3 — MCP42010 dual digital potentiometer chip.

Table 7.1 shows a chart of the resistances needed. The first column is the button on the actual FH-2, the second column shows the key for that button on the Arduino membrane keypad, and the third column shows the switch numbering

Table 7.1
Keypad Resistance Data

FH-2 Button	Membrane Keypad Button	FH-2 Switch Numbering	Specified Resistance (Ω)	FH-2 Measured Resistance (Ω)	Project Measured Resistance (Ω)
1	1	1	866	910	834
2	2	2	1,330	1,303	1,268
3	3	3	1,820	1,804	1,705
4	4	4	2,490	2,397	2,319
5	5	5	3,240	3,300	3,044
Mem	6	6	4,120	4,300	3,840
Left	7	7	5,360	5,100	5,000
Up	8	8	6,810	6,800	6,380
Right	9	9	8,870	9,100	8,260
Down	0	10	12,000	12,000	11,190
P/B	*	11	16,900	16,050	15,760
Dec	#	12	25,500	24,080	23,980

convention commonly used to designate the FH-2 keys. The next column shows what the Yaesu specified resistance is for each key, followed by a column showing the actual value read on the real FH-2 keypad. As you can see, the resistors are not high-precision resistors, and the measured tolerance appears to be 5%. The surface mount resistors inside the actual FH-2 keypad do not have a tolerance value printed on them, so there's no way to be sure, but we'll assume that the tolerance is 5%.

The final column shows the values from the Arduino FH-2 version. These values, even with a variance of 8 to 10% from the specified values on several keys, worked just fine with my Yaesu FT-950 and FT-991A. All of the digital resistance values are stored in an array within the sketch, and you can adjust them accordingly. Each step adds or subtracts a value of approximately 39.1 Ω, allowing you to adjust the resistance on a per-key basis if needed.

Figure 7.4 shows the block diagram for the Arduino-powered FH-2 keypad.

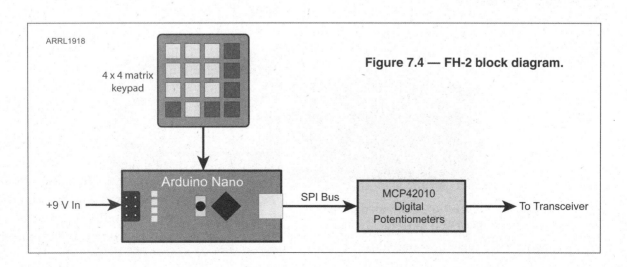

Figure 7.4 — FH-2 block diagram.

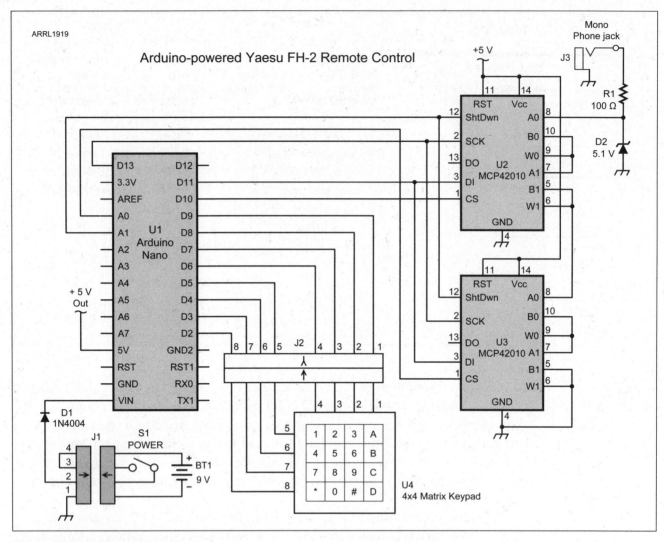

Figure 7.5 — FH-2 schematic diagram.
BT1 — 9 V battery
D1 — 1N4004 diode
D2 — 5.1 V Zener diode
J1 — 4-pin DuPont-style header
J2 — 8-pin DuPont-style header
J3 — Mini mono phone jack
R1 — 100 Ω, ⅛ W resistor

S1 — SPST switch
U1 — Arduino Nano
U2, U3 — MCP42010 10 kΩ dual digital potentiometer IC
U4 — 4×4 matrix membrane keypad
Enclosure — Solarbotics Mega SAFE

Figure 7.6 — Finished project board and keypad.

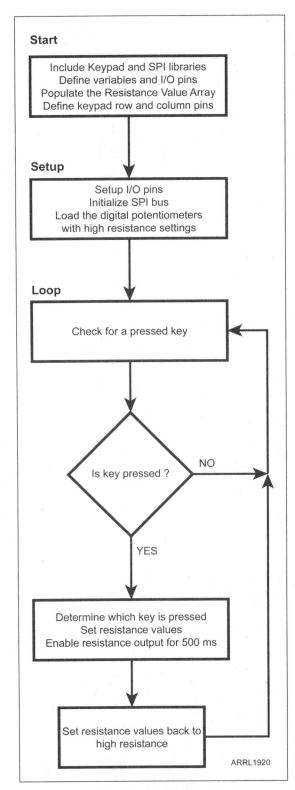

Figure 7.7 — The Yaesu FH-2 project flowchart.

The MCP42010 chips have both potentiometer sections wired as rheostats (the wiper connected to one leg of the potentiometer) and all of the potentiometer sections on both chips are wired in series, effectively creating a 40 kΩ programmable potentiometer.

You may notice that this book does not contain Fritzing diagrams for the projects. The color Fritzing diagrams do not reproduce well in black and white, so we are making full-color versions available online with the project sketches at **www.sunriseinnovators.com** and **www.kw5gp.com** instead.

Figure 7.5 shows the schematic diagram for the FH-2 Keypad. The project is designed to be powered by a 9 V battery and connects to the transceiver using the same type of mini-mono phone plug as the real FH-2. The membrane keypad is a 4×4 matrix keypad, and the row and column output for the pressed key will be read using digital input pins and the Arduino keypad library. The MCP42010 digital potentiometers connect to the Nano via the SPI bus and are cascaded together. I chose to use a Nano with this project, but it could easily be built using an UNO and a prototyping shield.

The Yaesu transceiver itself supplies a 5 V pullup voltage to the FH-2 plug. In order to protect the MCP42010 chips, a 5.1 V Zener diode is added to the transceiver connection, preventing any voltage higher than 5.1 V from reaching and damaging the potentiometer chips.

Figure 7.6 shows the finished project board and keypad ready for an enclosure.

Designing the Sketch

Figure 7.7 shows the flowchart for the FH-2 Keypad. This project is actually fairly simple and straightforward. Basically, the sketch will wait for a keypress on the 4×4 matrix keypad and then select the appropriate resistance value to supply to the output. A four-dimensional variable array is used to select the resistance settings to be applied to each of the digital potentiometers.

The sketch for this project is relatively simple, and it's been made even easier through the use of the keypad library. You can add this library to the Arduino IDE using the IDE's Library Manager. The sketch also uses the SPI library to communicate with the MCP42010 digital potentiometer chips, but you don't need to install the SPI library because it's already built into the IDE.

The sketch starts out by including the Keypad and SPI libraries. Next, we define the keypad object as having 4 columns and 4 rows. You could also choose to use a 4 row by 3 column matrix keypad simply by changing the keypad layout values. Next, we'll assign a character value to each key so that we can use a `Switch...Case` statement to determine which key was pressed. Finally, we define the row and column pins that connect to the 4×4 matrix keypad and instantiate (create) the keypad object. Note that if you change the `#define debug` statement to true, it will incorporate helpful debugging information into the sketch that you can see using the Serial Monitor.

```
/* Yaesu FH-2 Remote Control Emulator
   By Glen Popiel - KW5GP
   KW5GP@ARRL.NET

   Emulates the Yaesu FH-2 Remote Control
*/

#define debug false

#include <Keypad.h> // Include the Keypad library

#include <SPI.h>  // Include the SPI library

// Define the Keypad layout
const byte ROWS = 4; //four rows
const byte COLS = 4; //three columns

// Assign a character to each key
char keys[ROWS][COLS] =
{
  {'1', '2', '3', 'A'},
  {'4', '5', '6', 'B'},
  {'7', '8', '9', 'C'},
  {'*', '0', '#', 'D'}
};

byte rowPins[ROWS] = {5, 4, 3, 2}; //connect to the row pinouts of the keypad
byte colPins[COLS] = {9, 8, 7, 6}; //connect to the column pinouts of the keypad

Keypad keypad = Keypad( makeKeymap(keys), rowPins, colPins, ROWS, COLS );
// Instantiate the keypad object
```

Next, we define the `Chip Select` and `Shutdown` pins for the digital potentiometer chips and create the potentiometer value array. Because each digital potentiometer is controlled by an 8-bit value (0 – 255), we create a four-dimensional array element for each key pressed.

The value used to select the potentiometer resistance is relatively straightforward, with one hidden "gotcha." The formula to calculate the resistance of each potentiometer is the total potentiometer resistance multiplied by 256 minus the 8-bit potentiometer value, and that value is then divided by 256. Then you add the resistance of the potentiometer wiper. Repeat this calculation for all of the potentiometers in the circuit and you'll get the total resistance, sort of.

This is the "gotcha" part of digital potentiometers. When you look at the schematic, you'll see that a 100 Ω resistor is added to the output from the potentiometer chips, so the natural inclination would be to assume that this is the wiper value to plug into the equation, but it's not. You will need to add that value at the end of the calculations to get the actual resistance the transceiver sees, but that value is not what the formula is referring to.

Instead, each wiper connection inside the chip itself has a unique wiper resistance that can vary based on the usual things such as temperature, and it also varies based on the voltage applied to the potentiometer itself. At 5 V, this value typically averages 52 Ω, but there is no specified minimum, and a possible maximum wiper resistance of 100 Ω. So, you'll need to add that value into the equation, which is why when you work out the numbers in the sketch array, they don't match the resistance value that is actually set in the potentiometer chips. As a general rule, I just use the datasheet's specified typical value of 52 Ω and adjust from there as needed. For a 10 kΩ potentiometer, 52 Ω is relatively insignificant (0.52%) and well with the standard tolerance of 5% for resistors. But you do need to be aware of this "hidden" resistance value when using these digital potentiometers.

```
const int Chip_A    = 10; // Set the Chip Select for digital pot A
const int Chip_B = 14;   // Set the Chip Select for digital pot B
const int Shutdown_Pin = 15;// Set the Shutdown I/O pin for the digital pots

int Row = 0;   // Variable for the Row of Pot Arrays Values to use

// Define the Pot Array Values
int Value[13][4] =
{
  248, 254, 254, 254, // Key 1
  236, 254, 254, 254, // Key 2
  224, 254, 254, 254, // Key 3
  207, 254, 254, 254, // Key 4
  187, 254, 254, 254, // Key 5
  165, 254, 254, 254, // Key 6
  133, 254, 254, 254, // Key 7
  95, 254, 254, 254, // Key 18
  43, 254, 254, 254, // Key 9
  0, 90, 254, 254, // Key 10
  0, 216, 254, 254, // Key 11
  0, 0, 127, 254, // Key 12
  254, 254, 254, 254 // Maximum resistance value
};
```

There's not a whole lot of setup needed for the sketch. All we need to do is set the pin modes for the MCP42010 chips, make sure their output is turned off, and preload the chips with a high resistance value. To make things easier, we've created a `PotWrite()` function that will write the values from the selected row in the pot value array to the potentiometer chips.

```
void setup()
{
  // Setup digital pot I/O pins
  pinMode (Chip_A, OUTPUT);
  pinMode (Chip_B, OUTPUT);
  pinMode (Shutdown_Pin, OUTPUT);

  digitalWrite(Shutdown_Pin, LOW);     // Shutdown the Pots - Shows Output
                                       //as open circuit

#if debug
  Serial.begin(9600); // Start the serial port if debugging
#endif

  SPI.begin();    // initialize SPI

#if debug
  Serial.println("Ready");
#endif

  //Preload the Pots with all 254's
  Row = 12;
  PotWrite();;
}
```

In the main `loop()`, we use the keypad library to check for a keypress. We then use a `Switch...Case` statement to decode which switch has been pressed and assign the correct row of potentiometer values from the pot value array. Note that if the key pressed is an unused key, the high resistance value is loaded into the potentiometer chips and the chips remain in shutdown mode. Otherwise, the values for the selected key are written to the potentiometer chips and they are then enabled for 500 ms before they are shut down again, and wait for the next keypress.

```
void loop()
{
  // Check the keypad
  char key = keypad.getKey();

  if (key)   //If there is a key pressed
  {
#if debug
    Serial.print("Key pressed: ");
    Serial.println(key);
#endif
```

```c
// Determine which key is pressed and select the pot value array row number
    switch (key)
    {
      case '1':
        Row = 0;
        break;

      case '2':
        Row = 1;
        break;

      case '3':
        Row = 2;
        break;

      case '4':
        Row = 3;
        break;

      case '5':
        Row = 4;
        break;

      case '6':
        Row = 5;
        break;

      case '7':
        Row = 6;
        break;

      case '8':
        Row = 7;
        break;

      case '9':
        Row = 8;
        break;

      case '*':
        Row = 9;
        break;

      case '0':
        Row = 10;
        break;

      case '#':
        Row = 11;
        break;

      case 'A':
        Row = -1;
        break;

      case 'B':
        Row = -1;
        break;

      case 'C':
        Row = -1;
        break;

      case 'D':
        Row = -1;
        break;

      default:
        Row = -1;
    }
#if debug
    Serial.print("Array Index: ");
    Serial.println(Row);
#endif

    if (Row != -1)
    {
      PotWrite(); // Write the array
                  //  values to the pots
      digitalWrite(Shutdown_Pin, HIGH);
          // Enable the Pot chips for 500ms
      delay(500);

#if debug
  delay(5000);
#endif

      digitalWrite(Shutdown_Pin, LOW);
          // Disable the Pot chips
    }
  }
}
```

The sketch has only one function, the `PotWrite()` function. This function will read the required digital resistance settings and load them into the four digital potentiometers. First, we'll select digital potentiometer U2 by setting the CS pin low. Then we send a value of B00010001 over the SPI bus to the chip. This is the command to select Potentiometer A in the chip. After Pot A has been selected, we load it with the correct value from the pot value array. We then deselect and reselect the U2 pot chip, only this time we send a value of B00010010, which selects Pot B. Once Pot B has been selected, we load it with its value from the pot array. The process is then repeated for digital potentiometer U3 until all of the values have been loaded and the chips are ready to be released from the shutdown condition to apply the proper resistance to the transceiver.

Enhancement Ideas

How many times have you built something, or started to build something, and an idea on how to improve the project hits you, but you're already committed to doing it the way you started? I started creating the projects for this book a year or two ahead of the publication date. The FH-2 project was actually the first project built for this book. In the year or so since this project was designed and built, several new keypads have become available, some of them in only the past few months.

Adafruit has released their NeoTrellis 4×4 keypad (**Figure 7.8**) with monochrome or RGB LED key backlighting and an elastomer keypad area. Rather than use a 4×4 matrix that requires 8 digital I/O pins, the NeoTrellis interfaces with the Arduino via the I²C bus for both button management and LED driving.

Figure 7.8 — The Adafruit NeoTrellis 4×4 keypad.

Even more recently, and discovered just in time to make it into a project for this book, online sources such as Banggood offer an analog version of the 4×4 keypad (**Figure 7.9**) with removable keycaps that allow you to create your own key labels. These come in various sizes, including a 4×4 and a 4×3 key arrangement, allowing you to label your FH-2 identically to the real FH-2 keypad. That gives you two additional options to implement the Yaesu FH-2 keypad, allowing you to choose the look and feel that you prefer for your FH-2 project.

Figure 7.9 — The analog 4×4 keypad.

Peltier Cooler Controller

Another group of parts I've had in the parts bins for a while without a project to use them in is a handful of thermoelectric modules, also known as Peltier devices. Peltier devices are interesting in that they can be used for heating, cooling, or both, depending on the polarity of the dc power applied or which side of the module you choose to use. Essentially, heat is transferred from one side of the module to the other, making one side very cold, and the other side very warm. Reverse the polarity, and the heat transfer reverses.

Peltier devices can even be used to generate electricity thermally. The Peltier effect was first discovered by two 19th century scientists, Thomas Seebeck and Jean Peltier. Seebeck discovered that if you created a temperature difference across the junctions of two dissimilar conductors, electric current would flow. Peltier discovered that passing current through two dissimilar conductors caused heat to be emitted or absorbed at the junction of the two materials. Fast forward to the mid-20th century and the introduction of semiconductor technology, which allowed for the manufacture of practical thermoelectric modules.

These days, just about every electronics hobbyist "experimenter's" parts kit has a Peltier module. I had a few in my pile of parts, and then came across a whole bunch more at a local "returned goods" clearance store at a price I couldn't ignore. So I had a whole mess of these things with no project to use them in. Surely there had to be a viable Arduino project in there somewhere.

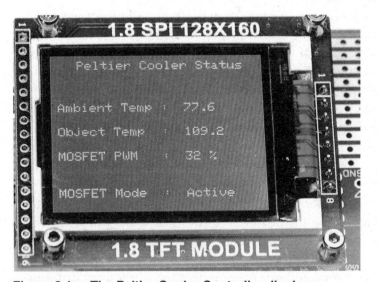

Figure 8.1 — The Peltier Cooler Controller display.

A Project Evolves

Then, along came the COVID-19 pandemic in 2020, and I started thinking of an Arduino project along the lines of creating a touchless infrared thermometer, like the ones that became popular during the pandemic. In my research, I came across the MLX90614 infrared temperature sensor (also known as a GY-906) that could measure temperatures from –94 to 719.96 °F (–70 to 382.2 °C) with a resolution of 0.9 °F (0.5°C) and a sensing distance of approximately 2 feet. Unbeknownst to me at the time, a high-accuracy medical grade version is also available that has a measurement accuracy of 0.036 °F (0.02 °C), but it turns out that the inexpensive modules available for the Arduino are of the standard, non-medical resolution.

I'm still wondering why I would need to read a temperature up to 720 °F, but it does raise some interesting possibilities in automotive and other applications. So, rather than experimenting with a medical-use Arduino project, I decided to switch gears and see how well the infrared sensor worked as the temperature sensing component in a temperature controller project. And, in the spirit of experimentation and doing things in a new and unique way, I decided it was time to put those Peltier modules to use and create an Arduino-powered Peltier temperature cooler controller.

These small Peltier modules are typically 40 mm square, 3 – 4 mm thick, and can operate up to 15.5 V dc. They can get a bit power hungry and draw up to 10 A, although the TEC1-12706 module used with this project is specified for 15.4 V dc maximum and a current draw of 6.4 A. The specifications of the TEC1-12706 module can be easily derived from the device nomenclature. The "TE" stands for Thermoelectric; "C" indicates the module size of 40 mm × 40 mm; and 1 indicates the number of stages, which is typically a single stage in most hobbyist versions of the module. The "127" portion of the label indicate the number of couplings. The higher the number of couplings, the more temperature conductive the device is. Finally, the "06" indicates the current rating for the device, in this case 6 A. Amazingly, these modules can generate 158 °F (70 °C) or more of a temperature difference between the two sides. Not bad for a little 1.6-inch-square module.

These modules work silently and do their thing without the need for a fan. You do have to be careful when using them, as the cold side can get very cold, and the hot side can get very hot, all very quickly. When you no longer need them to operate, it's best to turn off the device they are cooling and let the temperature of the module stabilize before shutting off its power. Otherwise, when you remove power to the module, all the heat you just pulled from whatever you're keeping cool tries to quickly stabilize by dissipating the heat to the entire module, which tends to defeat the purpose of keeping things safe and cool.

Because they are semiconductor devices, Peltier devices can be controlled using pulse-width modulation (PWM) for a more gradual control of the temperature. In my observations and testing, I have found that it is best to control the level of cooling rather than letting the device free-run. The temperature differential can get quite large, and the cold temperatures and the resulting condensation a Peltier module can generate can be just as bad for some devices as excessive heat can be. When powered off, the equalization of the heat differential can be equally damaging to whatever you're trying to keep cool. But, all in all, the Peltier module is a very cool (pun intended) device to work with.

Because the Peltier device draws a lot of current when in operation, you won't be able to power this project with a small battery. You'll need a source of 12 V dc that can provide at least 7 A of current for the Peltier module to function. Because we'll need to control the 6 A or more of current driving the Peltier module with pulse-width modulation, we'll us a 50N06 N-channel power MOSFET to do the power switching to the Peltier module.

Peltier modules work best when in contact with the object you need temperature control, so you'll need a way to dissipate the heat created by the hot side of the module. While you could use a fan, that would defeat the purpose of my concept of a fanless cooler, so instead, I chose to use a large workstation CPU heatsink.

Temperature Sensor

The MLX90614 infrared temperature sensor is an I²C device that uses infrared light to read the temperature of an object without the need to be in contact with the object itself. This allows the sensor to read temperatures far in excess of those from a standard temperature sensor. The MLX90614 actually produces two temperature measurements that can be read via the I²C bus. The MLX90614 can read the object temperature as you would expect it to, but it will also measure the ambient temperature on the module itself. The object temperature measurement ranges from –94 to 719.96 °F, while the ambient

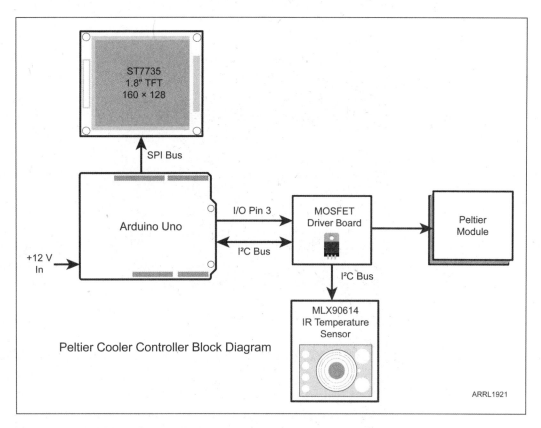

Figure 8.2 — Peltier Cooler Controller block diagram.

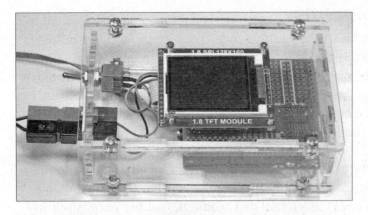

Figure 8.3 — Completed Peltier Cool Controller main unit, and with TFT display and top cover removed.

temperature measurement ranges from –40 to 257 °F. For this project, I mounted the temperature sensor about an inch away from the object being cooled to keep the sensor's field-of-view small. The sensing area will broaden about 2 inches for every inch of distance from the object being measured.

One final feature of the MLX90614 is the ability to output a pulse-width modulated signal that represents the measured temperature for direct PWM temperature controlled output to an external device. The minimum and maximum PWM value is configurable to allow you set a PWM range that matches the temperature range you would like to manage via the PWM output.

Figure 8.2 shows the block diagram of the Peltier Cooler Controller project. The Arduino and main part of the project will be constructed in a Solarbotics Mega SAFE enclosure (**Figure 8.3**), with the MOSFET driver on a separate piece of perfboard (**Figure 8.4**) mounted inside a separate standard Solarbotics SAFE enclosure. Short cables connect the Peltier module/heatsink assembly and the MLX90614 temperature sensor.

I left the CPU fan on the heatsink assembly but did not connect it, just in case it would be needed a later date. This will allow you to build the MOSFET driver in an enclosure close to the Peltier module mounted on the heatsink. The MLX90614 sensor also attaches to the

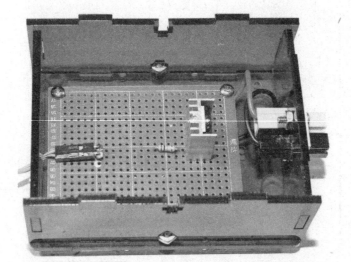

Figure 8.4 — Peltier Cooler Controller MOSFET board.

external MOSFET board and connects to the UNO using the MOSFET board's cable, so that you can mount the temperature sensor close to the device being cooled. You can even mount the MLX90614 module on a small metal strip attached to the heatsink and aim it at the area to monitor.

Building the Temperature Controller

Figure 8.5 shows the schematic diagram for the main portion of the Peltier Cooler Controller and **Figure 8.6** shows the MOSFET driver board schematic. A Fritzing diagram and sketch for this project can be found at **www.sunriseinnovators.com** and **www.kw5gp.com**. While the 50N06 MOSFET is a 50 A device and high temperatures on the MOSFET were not experienced during testing, as an extra measure of protection, I added a small TO-220 heat-

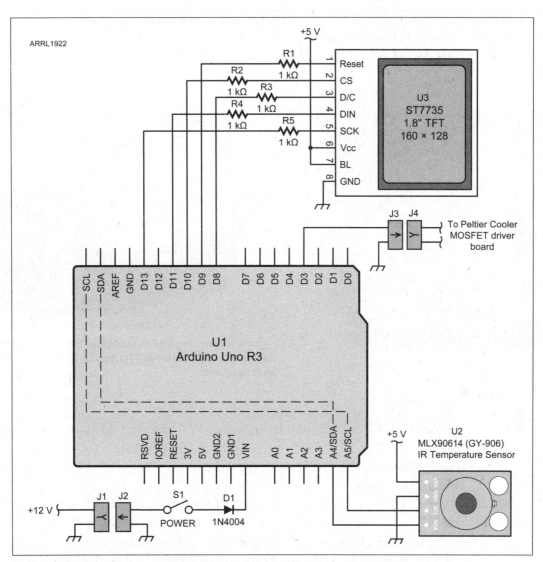

Figure 8.5 — Peltier Cooler Controller main unit schematic diagram.
D1 — 1N4004 diode
J1 — 2-pin DuPont-style female header socket
J2 — 2-pin DuPont-style male header
J3 — 2-pin DuPont-style male header
J4 — 2-pin DuPont-style female header socket
R1-R5 — 1 kΩ, ⅛ W resistor
S1 — SPST switch
U1 — Arduino Uno
U2 — MLX90614 (GY-906) infrared temperature sensor
U3 — ST-7735 1.8-inch color TFT display
Enclosure — Solarbotics Mega SAFE

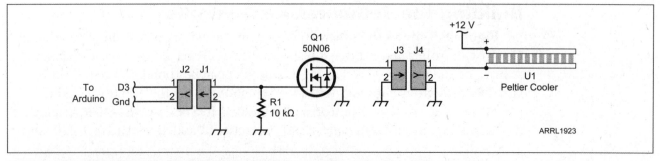

Figure 8.6 — Peltier Cooler Controller MOSFET driver board schematic diagram.
J1 — 2-pin DuPont-style male header
J2 — 2-pin DuPont-style female header socket
J3 — 2-pin DuPont-style male header
J4 — 2-pin DuPont-style female header socket
Q1 — 50N06 N-channel MOSFET
R1 — 10 kΩ, ⅛ W resistor
U1 — TEC1-12706 or similar Peltier thermoelectric module

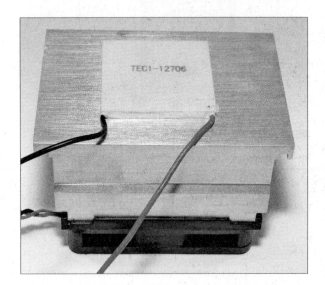
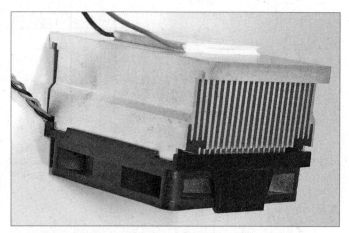

Figure 8.7 — The Peltier module mounted to the surplus CPU heatsink. The fan is not used, but could be if needed.

sink to the MOSFET. It will have 6 A at 12 V flowing through it, meaning that it will be handling 72 W of power. That could generate a lot of heat and even damage the MOSFET if something were to go wrong. It's always better to be safe than sorry.

The Peltier module itself was attached to a regular CPU heatsink (**Figure 8.7**) using a small amount of high-temperature adhesive known as Rocksett at each corner of the module. Rocksett is a gel-type adhesive. It sets and dries quickly, and can resist temperatures up to 2,000 °F. You could also use a high-temperature silicone sealant such as J-B Weld Red RTV Gasket Maker and Sealant that is available from automotive supply stores. A thin layer of regular heatsink thermal grease was spread on the Peltier module to allow for good heat transfer, and a small area on each corner of the Peltier module was cleaned to apply the adhesive.

It is important to know the hot and cold sides of the Peltier module before you glue it into place. As a general rule, especially with the TEC-12706 module, when you place the module on a flat surface with the positive (red) power wire to the left, the hot side will be facing up. If in doubt, apply power to the module *briefly* and you should be able to feel which side is which. Be sure not to touch it for too long or leave it powered on, as the module will heat quickly!

The Sketch

Figure 8.8 shows the flowchart for the Peltier Cooler Controller sketch. We'll start out by including all of the libraries we'll need. While the overall sketch is pretty straightforward, we'll be using a bunch of libraries to help do the heavy lifting. We'll need to include the Adafruit GFX and ST7735 libraries, along with the built-in SPI library for the TFT. Next, we'll need to add the Adafruit MLX90614 library and the built-in I2C Wire library. The Adafruit GFX, ST7735, and MLX90614 libraries can all be installed using the IDE's Library Manager. Next, we'll define the I/O pins and variables we'll need for the sketch, then we'll instantiate the ST7735 TFT display and MLX90614 IR temperature sensor objects.

In the `setup()` loop, we'll set the I/O pin modes, and initialize the TFT display and MLX90614 sensor. We'll make sure the MOSFET drive is off and display the startup screen on the TFT.

In the main `loop()`, we'll check to see if the update timer has expired. If the timer has expired, we'll read the ambient and object temperatures, set the MOSFET drive PWM level as needed, and update the TFT display.

```
/*
  Peltier Cooler with Infrared Temperature Sensor
  written by Glen Popiel - KW5GP

  Uses Adafruit ST7735 TFT Library
  Uses Adafruit GFX Library
  Uses Adafruit MLX90614 Library
  Uses SPI Library
  Uses Wire Library
*/

#define debug // Uncomment this to enable the display of debug information on the
Serial Monitor

#include <Adafruit_GFX.h>     // Core graphics library
#include <Adafruit_ST7735.h> // Hardware-specific library for the ST7735 1.8" TFT
#include <SPI.h>  // For SPI Communication
#include <Wire.h> // For I2C Communication
#include <Adafruit_MLX90614.h>  // For the MLX90614 IR Temperature Sensor
```

To start the sketch, we'll include the Adafruit_ST7735.h, Adafruit_GFX.h, SPI.h, Wire.h, and Adafruit_MLX90614.h libraries and enable debugging output. Then we'll define the Serial Monitor port speed for debugging information and define the temperature levels for operation. Next, we'll define the I/O pin for the MOSFET and define the pins for the TFT display. We'll set the update time to 1 second and assign the values for the remaining variables.

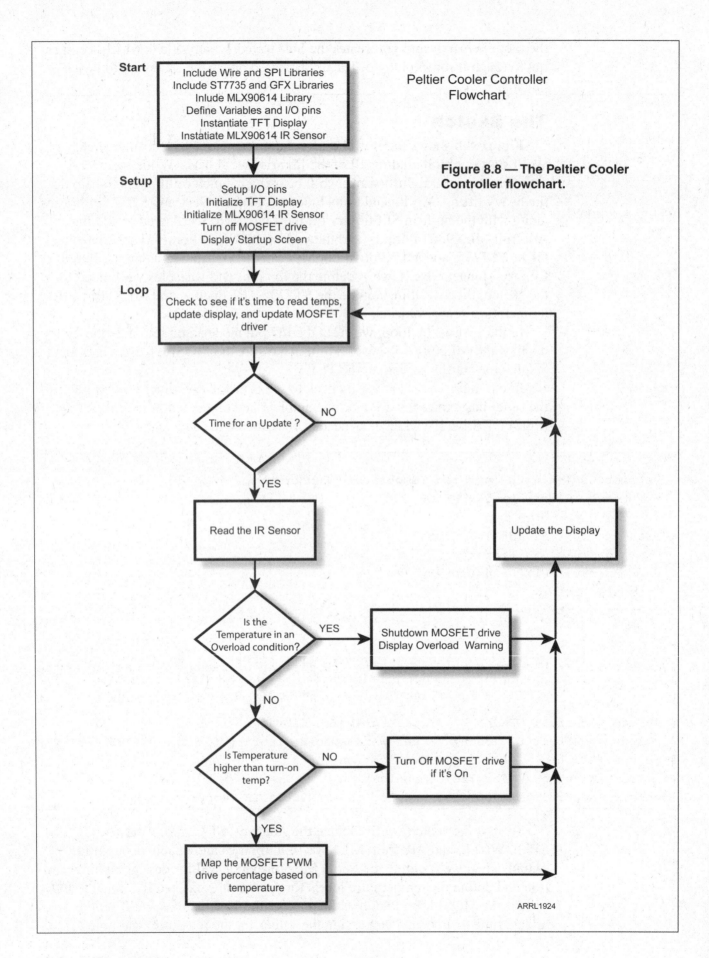

Figure 8.8 — The Peltier Cooler Controller flowchart.

Finally, we'll instantiate (create) the ST7735 TFT and MLX90614 IR temperature sensor objects.

```
#define comm_speed  9600  // Set the Serial Monitor Baud Rate

#define on_temp 90   // Define the Peltier MOSFET turn-on temperature
#define full_temp 150 // Define the full PWM duty cycle temp
#define overload_temp 180 //Define the max temp to shut it all down

#define MOSFET 3  // Define the MOSFET/Peltier Cooler drive pin - use a PWM pin

#define TFT_CS    10  // Assign the TFT CS to pin 10
#define TFT_RST   9   // Assign the TFT RST to pin 9
#define TFT_DC    8   // Assign the TFT DC to pin 8

#define update_delay 1000 // Set the Display update delay to one second

#define tft_delay 10  // set the TFT command delay to 10ms

boolean first_pass = true;  // Variable used to determine if first pass through
the main loop
boolean overload = false; // Variable used to show overload status
int long update_time = 0; // Set the Display update time counter to zero
int MOSFET_pwm = 0; // Set the MOSFET Drive PWM duty cycle
String MOSFET_status="Normal";
float ambient_temp, object_temp;

Adafruit_ST7735 tft = Adafruit_ST7735(TFT_CS,  TFT_DC, TFT_RST);
  // Initialize the TFT display
Adafruit_MLX90614 mlx = Adafruit_MLX90614();
  // Initialize the temperature sensor
```

In the setup() function, we'll start the serial port so that it can be used for debugging information, then we'll set the pin mode for the MOSFET PWM drive I/O pin and turn the MOSFET drive off. Next, we'll initialize the ST7735 TFT, start the MLX90614 sensor, and display a startup screen on the TFT display to let us know everything's working.

```cpp
void setup()
{
    Serial.begin(comm_speed);   // Start the Serial port

    // initialize the digital pins
    pinMode(MOSFET, OUTPUT);

    // Turn off the MOSFET
    digitalWrite(MOSFET, LOW);

    tft.initR(INITR_18BLACKTAB);    // initialize a 1.8" TFT with ST7735S chip,
                                    // black tab
    delay(tft_delay);

    mlx.begin();    // Start the temperature sensor

    tft.fillScreen(ST7735_BLUE); // Clear the display - fill with Blue background
    delay(tft_delay);
    tft.setRotation(1); // Set the screen rotation
    delay(tft_delay);
    tft.setTextWrap(false); // Turn off Text Wrap
    delay(tft_delay);
    tft.setTextSize(3); // Set the Font Size
    delay(tft_delay);
    tft.setTextColor(ST7735_GREEN); //Set the Text Color
    delay(tft_delay);
    tft.setCursor(40, 10);   //Set the Cursor and display the startup screen
    delay(tft_delay);
    tft.print("KW5GP");
    delay(tft_delay);
    tft.setTextSize(2);
    delay(tft_delay);
    tft.setCursor(45, 60);
    delay(tft_delay);
    tft.print("Peltier");
    delay(tft_delay);
    tft.setCursor(50, 80);
    delay(tft_delay);
    tft.print("Cooler");
    delay(tft_delay);
    tft.setCursor(25, 100);
    delay(tft_delay);
    tft.print("Controller" );

    delay(5000);   //Wait 5 seconds then clear the startup message
    tft.fillScreen(ST7735_BLUE); // Clear the display
    delay(tft_delay);

    update_display(); // update the display
    first_pass = false; // turn off the update display first pass flag
}
```

In the main `loop()`, we first check to see if the update timer has expired. If it hasn't, we stay in this loop until it's time for an update.

```
void loop()
{
  // Check to see if it's time to update the data
if (abs(millis()) > abs(update_time + update_delay))
  // check to see if update time has expired
```

Once the update timer has expired, we'll first read the ambient and object temperatures from the temperature sensor, and check for an overtemperature condition. Next, we check to see if the object temperature is above the cooling start temperature. If it is, we map the object temperature to a PWM percentage setting, turn on the MOSFET to the Peltier device, and display "Active" on the TFT display.

```
{
    read_temps(); // Read the temperature sensor

    if (object_temp < overload_temp)  // Check to be sure we're not in an
                                      // overheat condition
    {
      MOSFET_status = "Off";
      MOSFET_pwm = 0;
      if (object_temp >= on_temp) // Start cooling things down if temp is higher
                                  // than specified starting temperature
      {
        MOSFET_pwm = map(object_temp, on_temp, full_temp, 1, 100);
          // Map the percentage of PWM needed
        if (MOSFET_pwm <0)
        {
          MOSFET_pwm = 0;
        }
        if (MOSFET_pwm >100)
        {
          MOSFET_pwm = 100;
        }
        MOSFET_status = "Active";
      }
```

Finally, if there is an overtemperature condition, we shut the MOSFET drive off, removing power to the Peltier device, and display an overload warning on the TFT display, otherwise, we set the MOSFET PWM pin to the correct drive percentage and update the TFT display as shown in Figure 8.1 at the beginning of this chapter.

```
  } else
    {
      MOSFET_pwm = 0;  // Shutdown the cooling on an overload
      overload = true;
      MOSFET_status = "OVERLOAD"; // Indicate an overload temperature on the
                                  // display if max temp has been exceeded
    }
    analogWrite(MOSFET, MOSFET_pwm);  // Set the cooling rate
    update_display(); // update the display
#ifdef debug
    Serial.print("Ambient = ");
    Serial.print(ambient_temp);
    Serial.print("*F   Object = ");
    Serial.print(object_temp);
    Serial.print("*F");
    Serial.print("   MOSFET Drive : ");
    Serial.print(MOSFET_pwm);
    Serial.println(" %");
#endif
    update_time = millis(); // Reset the display update timer
  }
}
```

This sketch has three functions. The update_display() function, which is used to update the temperature and the Peltier module MOSFET drive status.

```
// update_display() function - Updates the TFT display
void update_display()
{
  if (first_pass) // Only do this the first time the function is called
  {
    // Clear the screen and display normal operation
    tft.fillScreen(ST7735_BLUE); // Clear the display
    delay(tft_delay);
    tft.setTextSize(1); // Set the text size to 1
    delay(tft_delay);
    tft.setTextColor(ST7735_GREEN); // Set the text color to green
    delay(tft_delay);
    tft.setTextSize(1); // Set the text size to 2
    delay(tft_delay);
    tft.setCursor(20, 5);
    delay(tft_delay);
    tft.print("Peltier Cooler Status");  // Display screen title
    tft.setCursor(5, 40);
    delay(tft_delay);
    tft.print("Ambient Temp :");  // Display ambient temperature template text
    tft.setCursor(5, 60);
```

```
    delay(tft_delay);
    tft.print("Object Temp :");  // Display the object temperature template text
    delay(tft_delay);
    tft.setCursor(5, 80);
    delay(tft_delay);
    tft.print("MOSFET PWM  :");  // Display the MOSFET PWM Drive value
    delay(tft_delay);
    tft.setCursor(5, 110);
    delay(tft_delay);
    tft.print("MOSFET Mode :");  // Display the controller status
    delay(tft_delay);
  }

  clear_temps();  // Clear the temperature display area

  tft.setCursor(100, 40);
  delay(tft_delay);
  tft.print(ambient_temp, 1); // Display the ambient temperature

  tft.setCursor(100, 60);
  delay(tft_delay);
  tft.print(object_temp, 1); // Display the ambient temperature

  tft.setCursor(100, 80);
  delay(tft_delay);
  tft.print(MOSFET_pwm);
  tft.print(" %");

  tft.setCursor(100, 110);
    delay(tft_delay);
  tft.print(MOSFET_status);
}
```

The second function is the `clear_temps()` function, which is used to erase only the temperature and MOSFET PWM value display areas to help speed up screen updates and reduce display flicker.

```
// clear_temps() function - Clears the temperature area of the display
void clear_temps()
{
  tft.fillRect(90, 40, 70, 50, ST7735_BLUE); // Clear the temperature values and
                                             // the MOSFET % status
  delay(tft_delay);

  tft.fillRect(90, 110, 70, 10, ST7735_BLUE); // Clear the controller status line
  delay(tft_delay);

}
```

And finally, the `read_temps()` function is used to read the ambient and object temperatures from the MLX90614 temperature sensor.

```
//read_temps() function - Reads the IR temperature sensor
void read_temps()
{
  //Read the temperature sensor
  ambient_temp = mlx.readAmbientTempF();
  object_temp = mlx.readObjectTempF();
}
```

Enhancement Ideas

There is a lot of room left on the TFT display to show more information such as the Peltier module voltage and current. The voltage could be read on one of the Arduino analog pins using a voltage divider, and the current can be read using a 30 A Hall Effect current sensor attached to another analog input pin. You could also use a rotary encoder to change the minimum and maximum cooling settings. If your project is generating some serious heat, you could also add a fan and temperature-based speed control as well. And, because the TFT is a color graphic display, you can add colors to the temperatures, and add lines and boxes to enhance the display output.

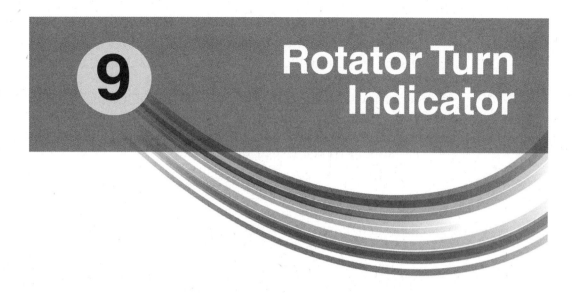

Rotator Turn Indicator

Rotator controllers seem to be a natural target for my Arduino projects. In the case of the CDE/HyGain rotators, their basic electronics haven't changed since they were first designed in the 1950s, and there's a lot of opportunity to improve on things with that original design. In fact, the CDE/HyGain rotator controllers are pretty much interchangeable among the various rotator versions, with the primary exception being the inclusion of, or lack of, a braking system.

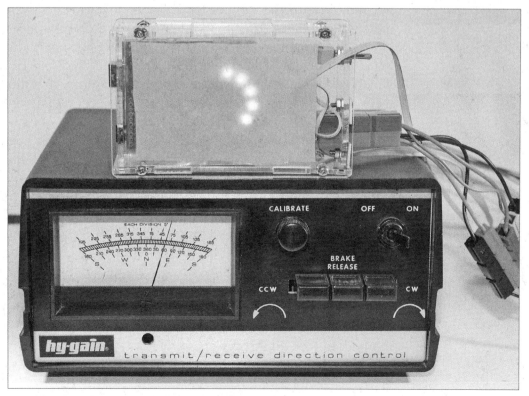

Figure 9.1 — The sequential Rotator Turn Indicator.

It seems that I always have a rotator control box on my workbench to tinker and play with, looking for new things to do with it and the Arduino. In my first book, I interfaced the control box to a PC so that it could be controlled by *Ham Radio Deluxe* software, and I even built a CDE/HyGain control box from scratch with PC control. In that book, I also interfaced the older Yaesu G-5400/5500 azimuth-elevation rotators to a PC running *SatPC32* or *Ham Radio Deluxe* software. In my second book, I did the same thing for the Yaesu G-450/800/1000 series of rotators.

This book is no exception, and there are several rotator controller projects here as well. In a later chapter, we'll be adding direct position control using a keypad to enter the desired antenna bearing to add some serious functionality to the rotator controller. We'll also be giving a complete makeover to the CDE/HyGain AR-40 in other chapters.

But the Arduino is also supposed to be fun, and I really wanted to include some less complex projects in this book. It's all too easy to get wrapped up and create some monster complex Arduino projects, but not everyone is at that level. A lot of hams are looking for projects that they can build easily and, more importantly, understand. One of the big takeaways I got from talking to my fellow Arduino enthusiasts that were just starting out, is that they wanted some smaller and less complex ham radio Arduino projects to use as an educational stepping-stone to the more advanced projects. With that thought in mind, I came up with several simple and fun Arduino projects for the CDE and Yaesu rotator controllers to add some Arduino-powered "flash" to your ham shack.

Those of us who are children of the 1960s may remember the fun TV advertisements featuring the sequential blinking taillights on the Mercury Cougar. While not the first car to employ the sequential taillights, the Cougar is the most memorable to me because of the TV advertising with a live cougar — the commercial linked the sequential taillights to the cat's moving tail.

And yet again, as if on cue, while thinking up some of the projects for this book, the circular addressable RGB NeoPixel/WS281x LED rings became widely available at an inexpensive price. One of my first thoughts when I saw these was that I just had to create several projects putting these fun-looking LEDs to use. The LED rings are available in a variety of sizes, and there's even one that's almost identical in size to the dial face on the Yaesu G-450/800/1000-series rotator controller. Harking back to the old sequential taillight ads, I thought it would be fun to jazz up the old CDE/HyGain and Yaesu rotator controllers with an LED ring to indicate when the rotator was turning using sequential lights. I could also show the direction of rotation in the process.

There are actually three versions of this project, for the CDE/HyGain, Yaesu G-450A, and Yaesu G-800SA/1000SA rotators. There are only minor differences hardware-wise and sketch-wise among the three, so we'll go ahead and discuss all three as we go through the rest of this chapter. The most difficult thing about interfacing the Arduino to antenna rotators is that as a general rule, these rotators were never designed with computers and 5 V logic-level signals in mind. In fact, all three of these rotator types have a different voltage level, and only the Yaesu G-800SA/1000SA rotator controllers use a dc voltage to drive the rotator motor.

On each of the rotator controllers, we will need to find a location where we

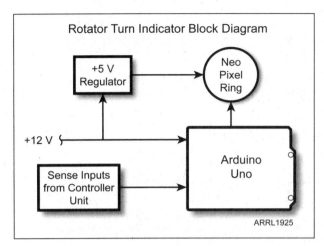

Figure 9.2 — The Rotator Turn Indicator block diagram.

can access a signal that indicates that the rotator is turning, and in which direction. In the case of the CDE/HyGain rotator, we'll also need to be able to determine when the brake release is active.

Once we have determined those locations, we'll need to interface to them in such a way that the Arduino can read the inputs safely, because these controllers generally operate at voltages much higher than the Arduino's 5 V limit. We won't be able to use a simple voltage divider circuit as the ground potential may be floating, which would cause major issues if we directly tied the controller chassis ground and the Arduino unit grounds together. We'll also have to deal with the fact that the CDE/HyGain and Yaesu G-450 rotators power the rotator motors using ac voltage.

Figure 9.2 shows the block diagram for the project. The cool thing about a block diagram is that it is so general, that for this project, it can encompass all three types of rotator controllers. One thing you will note on the block diagram is that we will be using a separate 5 V regulator to power the LED NeoPixel Ring. The LED rings come in a variety of sizes, from a small 1.5-inch-diameter version with 12 LEDs, up to a 4.4-inch-diameter version with 32 LEDs and

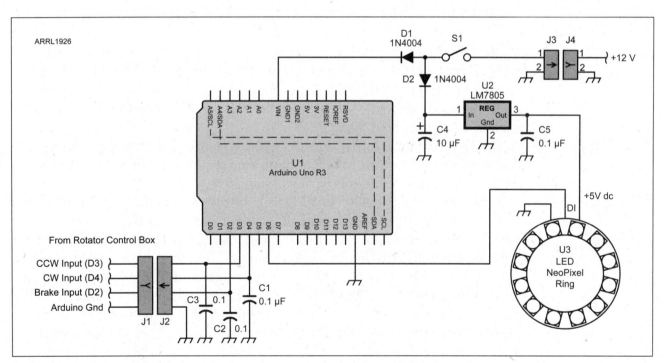

Figure 9.3 — The CDE/HyGain Rotator Turn Indicator schematic.
C1-C3, C5 — 0.1 µF, 50 V capacitor
C4 — 10 µF, 25 V electrolytic capacitor
D1, D2 — 1N4004 diode
J1 — 4-pin DuPont-style female header socket
J2 — 4-pin DuPont-style male header
J3 — 2-pin DuPont-style male header
J4 — 2-pin DuPont-style female header socket
S1 — SPST switch
U1 — Arduino Uno
U2 — LM7805 5-V voltage regulator
U3 — RGB NeoPixel/WS281x-compatible LED ring
Enclosure — Solarbotics Mega SAFE

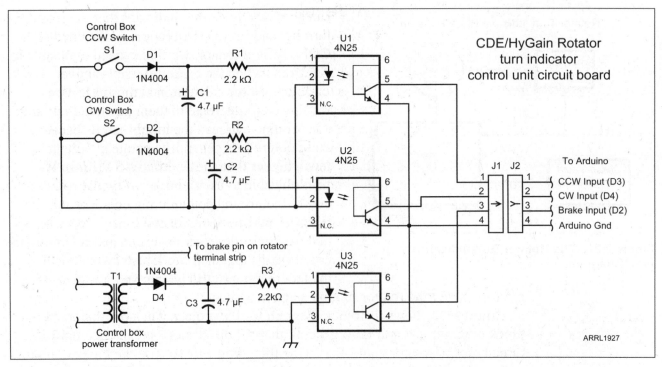

Figure 9.4 — The CDE/HyGain interface circuit schematic. S1 and S2 are the switches built into the control box.
C1-C3 — 4.7 µF, 50 V electrolytic capacitor
D1-D3 — 1N4004 diode
J1 — 4-pin DuPont-style male header
J2 — 2-pin DuPont-style female header socket
R1-R3 — 2.2 kΩ, 1 W resistor
U1-U3 — 4N25 optocoupler

beyond. It all pretty much comes down to what size LED ring you prefer for your project. The 1.5-inch version fits nicely inside a Solarbotics Mega SAFE enclosure with the Uno and the rest of the electronics. The 4.4-inch diameter version is nearly the same size as the outer ring on the Yaesu G-series rotators and it could be mounted there.

However, the more LEDs you add to the equation, the more of a current draw you need to plan for. While multiplexing the LEDs does reduce the necessary current, it is still possible to draw up to 50 mA per LED. While this would be okay to power the smaller 12 LED ring, this works out to about 1.6 A needed to power the 32 LED ring, far more than the maximum 1 A the Arduino onboard voltage regulator can provide. Also, you really never want to draw a lot of power using the Arduino's onboard regulator. The more current you draw, the hotter that regulator gets. Eventually it can eventually fail, rendering your entire Arduino board useless. I'll always opt to use an external voltage regulator once the onboard current draw exceeds 300 to 400 mA, due to the fact that the Nano only can handle 500 mA, and I like to be able to use the Uno and Nano interchangeably in my projects.

Figure 9.3 shows the schematic for the main electronics of the CDE/HyGain Rotator Turn Indicator. As you can see, this is a very simple project to build. Interfacing to the controller unit is done using the circuit shown in **Figure 9.4**.

The CDE/HyGain Version

As I mentioned earlier in this chapter, the CDE/HyGain has remained relatively unchanged since its original design back in the 1950s. It actually has two internal power supplies, a small 12 V dc "instrumentation" supply to power the meter circuitry, and a 28 V ac circuit to power the rotator motor itself. Because the 12 V dc power supply ground is separate from the chassis ground that also serves as the common connection for the ac motor windings, the 12 V supply cannot be used to power the Arduino.

Because we don't want to have to worry about the difference in ground potentials in the various circuits, we'll use optocouplers to isolate the rotator controller side of things from the Arduino side. An optocoupler is a neat little component. It's actually an LED mounted inside a chip and linked with a phototransistor. This provides 5000 V of isolation between the two circuits, and yet gives us an easy way to pass digital signals to the Arduino from the rotator controller. And, because an optocoupler is LED-based, all we need to do is find a way to light the LED with the signal we want to transfer.

Figure 9.5 — The CDE/HyGain control unit board installation.

The best place to get the clockwise and counterclockwise rotation indicator signal is directly from the switches on the control unit itself. The voltage on these switches is about 28 V ac, so we'll first need to rectify it and then use the appropriate current limiting resistor for the LED portion of the optocoupler. An interesting thing about the CDE/HyGain rotator controller design is that the 28 V ac circuit is not powered on until the brake release switch is pressed, which is why you often hear and feel that loud clunking noise as the power-on voltage surge hits the 28 V ac transformer. Because the readings on the CW and CCW switches are not valid until the 28 V ac power is turned on, we also need to sense the brake release voltage. All of the sensing circuitry is mounted on a small piece of copper-clad perfboard and mounted on the underside of the rotator controller as shown in **Figure 9.5** and **Figure 9.6**. If you look closely at the switches on the upper left of Figure 9.5, you can see the connecting points for the optocoupler inputs. The Powerpole connector to the right of the controller is the connector to the main electronics enclosure that houses the Arduino and LED ring.

Figure 9.6 — The CDE/HyGain interface board.

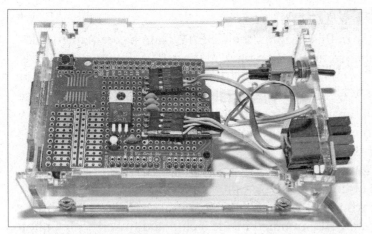

Figure 9.7 — The CDE/HyGain Arduino enclosure with the top cover and LED ring removed.

Figure 9.7 shows the CDE/HyGain Rotator Turn Indicator Arduino and LED board mounted in a Solarbotics Mega SAFE enclosure. The LED ring will be mounted in the enclosure cover behind a sheet of silver tissue paper that I used to create a mask to hide the LED ring, so that only the light from the LEDs shows through as shown in Figure 9.1 at the beginning of this chapter.

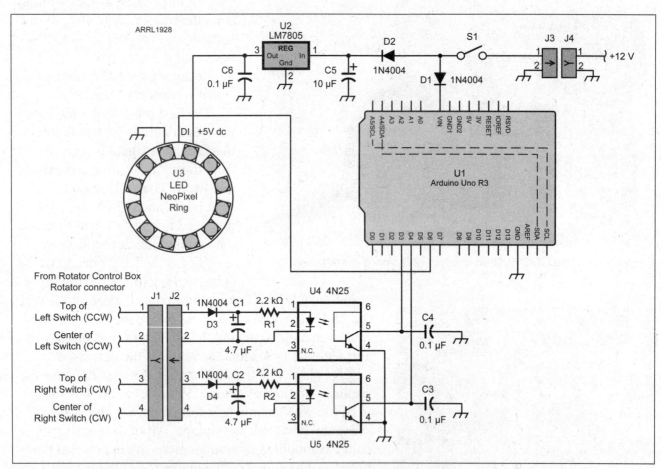

Figure 9.8 — The Yaesu G-450A Rotator Turn Indicator schematic.
C1, C2 — 4.7 µF, 50 V electrolytic capacitor
C3, C4, C6 — 0.1 µF, 50 V capacitor
C5 — 10 µF, 25 V electrolytic capacitor
D1-D4 — 1N4004 diode
J1 — 4-pin DuPont-style female header socket
J2 — 4-pin DuPont-style male header
J3 — 2-pin DuPont-style male header
J4 — 2-pin DuPont-style female header socket

R1, R2 — 2.2 kΩ, 1 W resistor
S1 — SPST switch
U1 — Arduino Uno
U2 — LM7805 5-V voltage regulator
U3 — RGB NeoPixel/WS281x-compatible LED ring
U4, U5 — 4N25 optocoupler
Enclosure — Solarbotics Mega SAFE

The Yaesu G-450A Version

Figure 9.8 shows the schematic for the Yaesu G-450A version of the Rotator Turn Indicator. The Yaesu version of this project is actually simpler than the CDE/HyGain version, and all of the needed components can be built inside a Solarbotics Mega SAFE enclosure as shown in **Figure 9.9**. In fact the G-450A and the G-800SA/1000SA versions are nearly identical in construction. The differences are:
- the diodes going to the optocouplers have different orientation
- a different current-limiting resistor on one of the inputs
- different control unit signal connections

The rotator motor on the Yaesu G-450A uses an ac voltage and switches between two motor windings to control the direction. Because of this, there is no easy way to access the rotation direction signals, so we'll pull these directly from the front panel switches as we did with the CDE/HyGain version. **Figure 9.10** shows where to attach the optocoupler inputs on the switches on

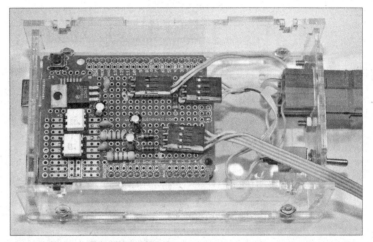

Figure 9.9 — Yaesu G-450A Arduino enclosure with the top cover and LED ring removed.

Figure 9.10 — Yaesu G-450A switch wiring connections.

the G-450A control unit. Because the G-450A front-panel switches are directly wired to the ac power transformer and the rotator motor windings, these wires carry 28 V ac. We'll need to rectify these inputs using a diode and capacitor on each input before connecting them to the current-limiting resistor and the optocouplers.

The Yaesu G-800SA/1000SA Version

Figure 9.11 shows the schematic for the Yaesu G-800SA/1000SA version. The G-800SA/1000SA rotator is the only one of the three rotator types that runs completely from dc power. However, we still can't use the voltage divider method for the signal inputs because the rotator motor dc polarity is reversed to rotate in the opposite direction. So, we'll have to use the same input diode rectification method that we used with the G-450A rotator, except this time, we can tap directly off of the connections going to the rotator motor for the rotation sig-

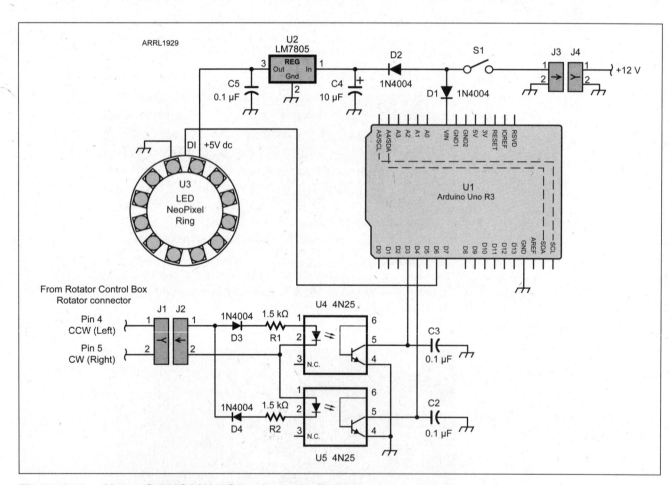

Figure 9.11 — Yaesu G-800SA/1000SA schematic diagram.
C1 — not used
C2, C3, C5 — 0.1 μF, 50 V electrolytic capacitor
C4 — 10 μF, 250 V electrolytic capacitor
D1-D4 — 1N4004 diode
J1 — 2-pin DuPont-style female header socket
J2 — 2-pin DuPont-style male header
J3 — 2-pin DuPont-style male header
J4 — 2-pin DuPont-style female header socket
R1, R2 — 1.5 kΩ, 1 W resistor
S1 — SPST switch
U1 — Arduino Uno
U2 — LM7805 5-V voltage regulator
U3 — RGB NeoPixel/WS281x-compatible LED ring
U4, U5 — 4N25 optocoupler
Enclosure — Solarbotics Mega SAFE

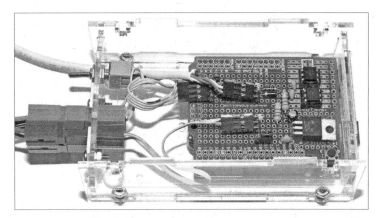

Figure 9.12 — The Yaesu G-800 Arduino enclosure with the top cover and LED ring removed.

nals. Because the dc voltage differs slightly between clockwise and counterclockwise rotation, we'll use a 1.5 kΩ resistor on the counterclockwise input instead of the 2.2 kΩ one we use for the clockwise input. We'll also have the diodes working in opposite directions, so that the optocouplers always see the correct signal polarity.

Figure 9.12 shows the G-800SA/1000SA version of the Arduino enclosure. The rotation input signals are spliced to the wires going to the rotator motor connectors on the inside of the back cover of the control unit. With my unit, I have replaced the troublesome Molex connector that comes with the controller with a much more reliable twist-lock connector, and spliced that connector into the wires going to the original connector (**Figure 9.13**). This allowed me to connect my Rotator Turn Indicator unit to the G-800SA using the original Molex connector.

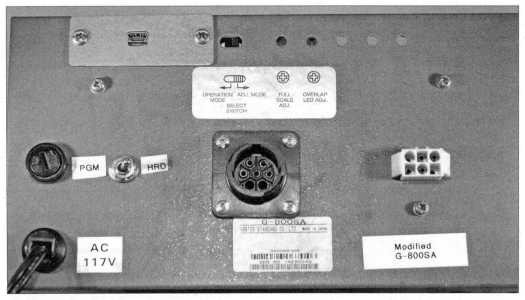

Figure 9.13 — Rear view of the modified G-800SA control unit showing both the Molex and twist-lock connectors.

The Flowchart

The flowchart for the Rotator Turn Indicator is shown in **Figure 9.14**. One of the nice things about flowcharts is that they show the sketch at a much more general level and help keep you from getting wrapped up in details. This way, whenever you do start to get bogged down in your sketch, you can fall back to

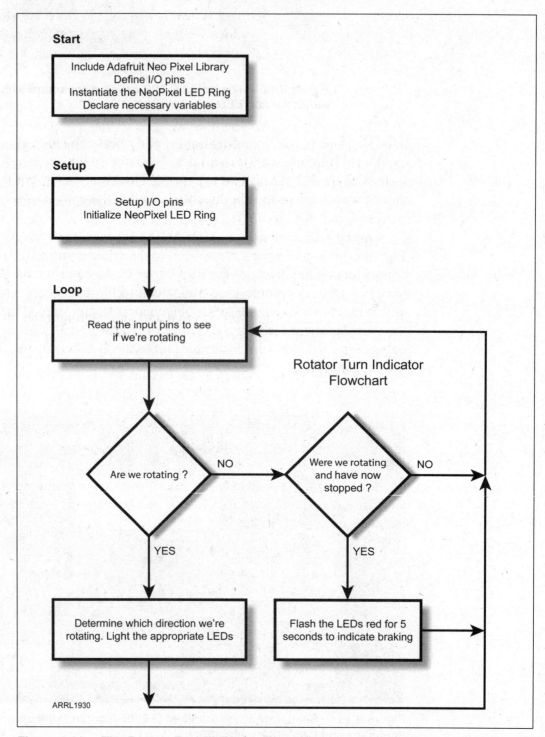

Figure 9.14 — The Rotator Turn Indicator flowchart.

the flowchart and see what function the block of code was actually planned to do, and possibly help keep you on track with what you're actually trying to accomplish. As with the block diagram, the same flowchart for the Rotator Turn Indicator project can be used for all three versions.

As the flowchart shows, this will be a relatively simple and straightforward sketch for all three rotator types. The sketch will only need one library, the Adafruit NeoPixel library. We'll start out by including the NeoPixel library and defining the I/O pins. Next we'll instantiate (create) the NeoPixel object and declare the remaining variables we'll need in the sketch. Note, there are two choices for creating the NeoPixel object. We'll cover those when we go through the sketch in detail.

In the `setup()` function, all we need to do is set the pin modes for the I/O pins and initialize the NeoPixel LED ring.

In the main `loop()` function, we'll read the input pins waiting for a switch press. Once we receive a switch press, the direction of the rotation is determined and the LED ring will start lighting LEDs sequentially to match the direction of the rotation. When the switch is released, all of the LEDs will blink red for 5 seconds to indicate a braking condition. The braking sequence will occur whether the control unit has a physical brake or not, to indicate that the rotation has stopped.

The Sketch

Because the sketches are so similar among the three rotator types, we'll go through the CDE/HyGain version in detail, and highlight the differences between that version and the other versions without going through those sketches in as much detail. There's not a whole lot of difference among the various sketches other than the process where we determine which switch is pressed to let us know which direction to rotate on the LED ring. Sketches, libraries, Fritzing diagrams and other useful files for the projects in this book can be found at **www.sunriseinnovators.com** and **www.kw5gp.com**.

We start out the Rotator Turn Indicator by uncommenting the debugging information preprocessor directive if we want to include the debugging code in the sketch. Then we'll include the Adafruit NeoPixel library and define the number of LEDs on the RGB LED ring and the I/O pins we'll use for the input pins.

```
/*
  Antenna Rotator Turn Indicator
  CDE/HyGain version

  Glen Popiel - KW5GP

  Uses the AdaFruit NeoPixel Library

  Released under the GPLv3 license
*/

//#define debug // Uncomment this to enable the display of debug information on the Serial Monitor

#include <Adafruit_NeoPixel.h>  // Include the Adafruit NeoPixel Library

#define led_pin 6 // Define the Neopixel I/O pin

#define num_pixels 12 // NeoPixel Strip/Ring size

#define brake_pin 2 //define the brake input pin
#define ccw_pin 3 //define the counterclockwise rotation input pin
#define cw_pin 4 // define the clockwise rotation input pin
#define debounce 20 // Set a 20ms debounce period
```

Figure 9.15 — An RGBW LED pixel.

Next, we'll instantiate the RGB LED NeoPixel object. Note that there are two options here. The RGB LED rings are available in two different types, a standard red-green-blue (RGB) and the newer red-green-blue-white (RGBW) that include a bright white LED inside each RGB LED pixel. You can tell the difference between the versions by looking closely at an individual pixel. The RGBW LEDs have a small dark chip inside as well as half of the LED masked off as shown in **Figure 9.15**. You'll need to uncomment the proper line in the sketch for the type of LED ring you are using.

```
// Setup the NeoPixel library
// Uncomment the following for standard RGB Neopixel/WS281x LED Strips
//Adafruit_NeoPixel pixels(num_pixels, led_pin, NEO_GRB + NEO_KHZ800);

// Uncomment the following for RGBW Neopixel/WS281x LED Strips (RGB w/White LED)
// If you look at the unlit LEDs, half of each LED is covered with yellow
Adafruit_NeoPixel pixels(num_pixels, led_pin, NEO_GRBW + NEO_KHZ800);
```

Finally, we'll define the time delay between pixels when turning them on sequentially and initializing the variables needed in the sketch.

```
#define delay_time 500 // Time (in milliseconds) to pause between pixels

int current_pixel, last_pixel; // Pixel variables
bool cw_in, ccw_in, brake_in, moving; // Rotation variables
```

In the `setup()` function, we'll start the serial port if debugging is enabled, set the pin modes for the input pins, initialize the LED ring, and turn off the rotation indication flag.

```
void setup()
{
#ifdef debug   // Start the serial port only if debugging is enabled
  Serial.begin(9600);
#endif
  pinMode(cw_pin, INPUT_PULLUP);    // Set the clockwise pin mode
  pinMode(ccw_pin, INPUT_PULLUP);   // Set the counterclockwise pin mode
  pinMode(brake_pin, INPUT_PULLUP); // Set the brake pin mode

  pixels.begin(); // Initialize the NeoPixel display
  pixels.clear(); // Set all pixel colors to 'off'
  pixels.show();  // Send the updated pixel colors to the hardware
  moving = false; // Clear the moving flag
}
```

For the CDE/HyGain rotators, a LOW condition indicates the switch is pressed. Because the actual rotation motor power is turned on when the brake release switch is pressed, the input to the brake release optocoupler is off until the switch is pressed, which will then cause the brake input to the Arduino to go LOW. We'll use that condition as the starting point in the sketch. Because these are old switches, we'll need to debounce them. Unfortunately, because we're working with ac voltage, we can't put capacitors across these switches to do it in hardware, so we'll implement a brief delay in software to handle it for us.

```
void loop()
{
  brake_in = digitalRead(brake_pin);  // Check to see if the brake release
                                      // switch is pressed
  delay(debounce);
  if (brake_in == LOW)   // If the brake release is pressed
  {
#ifdef debug
    Serial.println("Brake enabled");
#endif
```

After the debounce period, we'll read all three switch inputs to determine which direction the rotator is turning. We'll set the moving flag to indicate that we are rotating, and then call either the `turn_cw()` or `turn_ccw()` function once we determine the direction of rotation. When rotation stops and the switches are released, all of the LEDs in the RGB LED ring are blinked red for five seconds to indicate a braking period.

```
    if ((brake_in == HIGH && cw_in == HIGH && ccw_in == HIGH) || ((brake_in ==
LOW) && (cw_in == HIGH && ccw_in == HIGH)))
    {
       // A switch transition delay caused invalid condition
       moving = false; // Turn off the rotating flag
    } else {

       if ((cw_in == HIGH || ccw_in == HIGH))   // One or both direction switches
                                                // is pressed
       {
#ifdef debug
         Serial.print( "Brake = ");
         Serial.print(brake_in);
         Serial.print("   CCW = ");
         Serial.print(ccw_in);
         Serial.print("   CW = ");
         Serial.println(cw_in);
         Serial.println("CW or CCW Switch pressed");
#endif

         moving = true;   // Set the rotating flag
         if (cw_in == HIGH)
         {
           // we're turning clockwise

#ifdef debug
         Serial.println("Turning Clockwise");
#endif
         turn_cw();   // Call the turn_cw() function
         }

         if (ccw_in == HIGH)
         {
           // We're turning counterclockwise

#ifdef debug
         Serial.println("Turning CounterClockwise");
#endif

         turn_ccw();   // Call the turn_ccw() function
         }
```

```
    } else
    {
      // Have we been moving? If so, flash the brake lights
      if (moving)
      {
        moving = false; // turn off the rotating flag
      }
      // No more rotation - clear the LED's
      pixels.clear(); // Set all pixel colors to 'off'
      pixels.show();  // Send the updated pixel colors to the hardware
    }
  }
}
```

There are three functions in the sketch. The turn_cw() function is used to process the clockwise (right) rotation. The turn_cw() function will remain in a loop until the CW switch is released and then call the brake() function. While in the loop, every 500 ms (as defined in the beginning of the sketch), an LED pixel is added to the lit LEDs in the direction of the rotation. When all of the LEDs are lit, the LEDs are cleared and the process repeats until the CW switch is released.

```
void turn_cw()    // The turn_cw() function
{
  int current_pixel = 0;
  delay(debounce);
  // Repeat the loop while the cw (right) switch is pressed
  while (digitalRead(cw_pin) == HIGH && digitalRead(brake_pin) == LOW)
  {
    pixels.setPixelColor(current_pixel, pixels.Color(0, 150, 0));
      // For each pixel set LED color to green
    pixels.show();   // Send the updated pixel colors to the hardware.

    if (current_pixel >= num_pixels)  // Check to see if we've exceeding the
                                      // number of pixels in the ring
    {
      // clear all pixels
      pixels.clear(); // Set all pixel colors to 'off'
      pixels.show();   // Send the updated pixel colors to the hardware
      current_pixel = current_pixel - num_pixels; // Reset the pixel counter
    } else
    {
      current_pixel = current_pixel + 1;  // increment the pixel to light
    }
    delay(delay_time); // Pause before next pass through loop
  }
  brake(); // Call the brake() function when the cw switch is released
}
```

The turn_ccw() function performs the same function as the turn_cw() function, except it lights the LEDs in the opposite direction.

```
void turn_ccw()    // The turn_ccw() function
{
  int current_pixel = num_pixels - 1; // Start at the top and light the pixels
                                      // counterclockwise
  delay(debounce);
  // Repeat the loop while the ccw (left) switch is pressed
  while (digitalRead(ccw_pin) == HIGH  && digitalRead(brake_pin) == LOW)
  {
    pixels.setPixelColor(current_pixel, pixels.Color(0, 150, 0));
    // For each pixel set LED color to green
    pixels.show();   // Send the updated pixel colors to the hardware.

    if (current_pixel < 0)  // Check to see if we've exceeding the number of
                            // pixels in the ring
    {
      // clear all pixels
      pixels.clear(); // Set all pixel colors to 'off'
      pixels.show();    // Send the updated pixel colors to the hardware

      current_pixel = num_pixels - 1; // decrement the pixel to light
    } else
    {
      current_pixel = current_pixel - 1;   // turn on the next LED
    }
    delay(delay_time); // Pause before next pass through loop
  }
  brake();  // Call the brake() function when the ccw switch is released
}
```

Finally, when either the CW or CCW switch is released, the LEDs are all turned off, and the LED ring will then blink all the LEDs in red for 5 seconds to indicate a braking cycle.

```
void brake()    // The brake() function
{
#ifdef debug
  Serial.println("Braking");
#endif

  pixels.fill(0, 0, num_pixels); // Turn off all the LEDs
  pixels.show();   // Send the updated pixel colors to the hardware.
  delay(delay_time);
  for (int i = 0; i < 5; i = i + 1)  // Flash all LEDs for 5 seconds for braking
                                     // indicator
  {
    pixels.fill(pixels.Color(150, 0, 0), 0, num_pixels);  // Light all LEDs with Red
    pixels.show();    // Send the updated pixel colors to the hardware.
    delay(delay_time * 2);
```

```
    pixels.fill(0, 0, num_pixels); // Turn off all the LEDs
    pixels.show();    // Send the updated pixel colors to the hardware.
    delay(delay_time * 2);
  }
}
```

The Yaesu G-450A Sketch

The Yaesu G-450A sketch is very similar to the CDE/HyGain sketch with just a few exceptions. The Yaesu G-450A and G-800SA/1000SA rotators do not have a brake switch, just CW and CCW switches, so we don't need a brake input pin. Aside from that, the initialization and setup are virtually the same as with the CDE/HyGain. However, things get a little different in the `loop()` function. With the Yaesu G-450A, the front panel switches are interconnected and don't give a simple LOW signal when a switch is pressed. With the G-450A, when both inputs are LOW, the rotator is idle. When rotating CCW (left), only the CCW input goes HIGH. However, when rotating clockwise (right), both the CW and CCW inputs go HIGH. This will require changes in the direction determination logic in the sketch.

```
/*
   The Yaesu G-450 inputs are different from the G-800/1000. The G-450 has an
AC-powered rotator motor
   The only good place to get the sensing information is across the front panel
switches
   These switches are connected in such as way as to give an unexpected set of
signal outputs to the Arduino Board

    When idle both inputs are high
    When rotating Left (CCW) only the CCW input goes high
    When rotating right (CW) but inputs go high
 */

  if ((cw_in == HIGH))   // This goes high regardless of direction so we only
                         // need to check it
  {
#ifdef debug
    Serial.print("   CCW = ");
    Serial.print(ccw_in);
    Serial.print("   CW = ");
    Serial.println(cw_in);
    Serial.println("CW or CCW Switch pressed");
#endif
```

```
      moving = true;  // Set the rotation flag
      if (ccw_in == HIGH && cw_in == HIGH)  // If both inputs are high we're
                                            // rotating clockwise (right)
      {
         // we're turning clockwise

#ifdef debug
         Serial.println("Turning Clockwise");
#endif
         turn_cw();  // Call the turn_cw function
      }

      if ( ccw_in == LOW && cw_in == HIGH)  // If ccw is low and cw is high we're
                                            // rotating counterclockwise (left)
      {

         // We're turning counterclockwise

#ifdef debug
         Serial.println("Turning Counterclockwise");
#endif

         turn_ccw(); // Call the turn_ccw function
      }
   } else
   {
      if (moving) // Have we been moving and aren't anymore?
                  // If so, flash the brake lights
      {
         moving = false; // Turn off the rotating flag
      }
      //nothing happening- clear the LED's
      pixels.clear(); // Set all pixel colors to 'off'
      pixels.show();    // Send the updated pixel colors to the hardware
   }
}
```

Likewise, the `turn_cw()` and `turn_ccw()` functions will need to use different logic conditions to continue to sequentially light LEDs while rotating.

```
void turn_cw()   // The turn_cw function
{
  int current_pixel = 0;   // Select the starting LED
  delay(debounce);
  while (digitalRead(ccw_pin) == HIGH && digitalRead(cw_pin == HIGH))
    // Repeat the loop as long as we're rotating clockwise


void turn_ccw()   // The turn_ccw function
{
  int current_pixel = num_pixels - 1;   // Select the starting LED
  delay(debounce);
  while (digitalRead(ccw_pin) == LOW && digitalRead(cw_pin) == HIGH)
  {
```

The Yaesu G-800SA/1000SA Sketch

The sketch for the Yaesu G800SA/1000SA is nearly identical to the G-450A. The only difference is that the CW and CCW inputs act more like you would think they should. When the rotator is idle, both inputs are HIGH. When rotating clockwise, the CW (right) input is LOW, and when rotating counterclockwise, the CCW (left) input goes low.

```
  if ((cw_in == LOW || ccw_in == LOW))   // If one of them is low, we're rotating
  {
#ifdef debug
    Serial.print("   CCW = ");
    Serial.print(ccw_in);
    Serial.print("   CW = ");
    Serial.println(cw_in);
    Serial.println("CW or CCW Switch pressed");
#endif

    moving = true;   // Set the rotation flag

    if (cw_in == LOW) // If the cw input is low, we're turning clockwise (right)
    {
      // we're turning clockwise

#ifdef debug
      Serial.println("Turning Clockwise");
#endif

      turn_cw();   // Call the turn_cw() function
    }

    if (ccw_in == LOW) // If the ccw input is low, we're turning
                      // counterclockwise (left)
```

Rotator Turn Indicator

```
    {
      // We're turning counterclockwise

#ifdef debug
      Serial.println("Turning CounterClockwise");
#endif

      turn_ccw();  // Call the turn_ccw() function
    }
  } else
  {
    if (moving) // Have we been moving and aren't anymore?
                // If so, flash the brake lights
    {
      moving = false; // Turn off the rotation flag
    }
    //nothing happening- clear the LED's
    pixels.clear(); // Set all pixel colors to 'off'
    pixels.show();   // Send the updated pixel colors to the hardware
  }
}
```

Finally, the same logic applies with the `turn_cw()` and `turn_ccw()` functions:

```
void turn_cw()   // The turn_cw function
{
  int current_pixel = 0;  // Set the starting LED point
  delay(debounce);
  while (digitalRead(cw_pin) == LOW)  // Repeat the loop as long as the cw
                                      // switch is pressed

void turn_ccw()   // The turn_ccw function
{
  int current_pixel = num_pixels - 1; // Set the starting LED point
  delay(debounce);
  while (digitalRead(ccw_pin) == LOW)  // Repeat the loop as long as the cw
                                       // switch is pressed
```

Enhancement Ideas

This project is pretty complete as it is, but it would be very easy to combine this project with the next project, the Rotator Position Indicator, and have the Arduino and sketch do the functions of both projects.

10 Rotator Position Indicator

Similar in concept to the Rotator Turn Indicator in the previous chapter, the Rotator Position Indicator also uses an RGB NeoPixel LED ring. The big difference is that this time, we'll use the LED ring to indicate the current rotator position. Amazingly, the build for this project is extremely simple and lives up to what I'm always saying about the Arduino, "With just a handful of parts, wires, and a sketch, you can create some cool projects."

For this project, we'll need to access the voltage value of the rotator position potentiometer that's housed in the rotator motor assembly. Fortunately, the wiring for this potentiometer is connected to the rotator control unit, where we can access it to determine the rotator's position. As with the project in the previous chapter, this device can be used with both the CDE/HyGain and Yaesu G450A/800SA/1000SA rotators with only minor differences between the two rotator types. We'll be using what we learn in this project to build the CDE/HyGain Keypad Entry Rotator Controller in a later chapter.

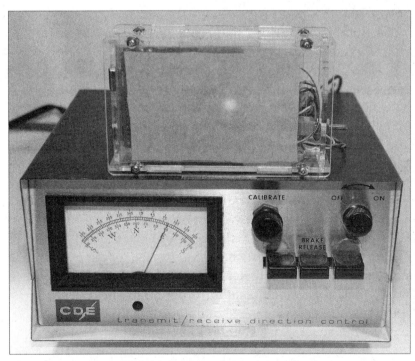

Figure 10.1 — The Rotator Position Indicator. See the end of this chapter for ideas on adding a compass image to show direction.

Overview

When it comes to computerizing a rotator controller, once you understand the two basic functions of making the controller rotate and reading the rotator position, you can easily add a ton of Arduino functionality to your rotator controller. From those two basic functions, it's actually only a small step to interface your rotator to a PC and an antenna control program such as *HRD Rotator* that's built into *Ham Radio Deluxe* software. We'll revisit this with the AR-40 rotator projects in the next two chapters.

The block diagram for this chapter's project, shown in **Figure 10.2**, is about as simple as an Arduino project can get. All we need to do is safely connect the rotator position input from the rotator control unit to an analog input pin on the Arduino. The position voltage on the CDE/HyGain rotators ranges from zero to approximately 12 V dc, which means we'll need a voltage divider circuit to reduce the positioning voltage to a safe level for the Arduino. For the Yaesu G-450A/800SA/1000SA rotator controllers, this voltage already ranges between zero and 5 V dc, so no voltage divider is necessary.

As you can see from the CDE/HyGain schematic in **Figure 10.3** and the Yaesu G450A/800SA/1000SA schematic in **Figure 10.4**, there's very little construction needed for this project. Because we're using a 12 LED NeoPixel ring, each pixel maps to 30 degrees of positioning. You could easily use a larger LED ring with more LED pixels, but even if you went with a 60 LED ring, each pixel would only represent 6 degrees. This means that you don't need any real accuracy from the 10-bit A/D in the Arduino, as each pixel in the 12 LED ring would amount to roughly 85 A/D counts for the 12 LED ring, and 17 for the 60 LED ring.

Sketch-wise, the only difference between a 12 pixel LED ring and a 60 pixel ring is changing the number of LEDs at the beginning of the sketch. I chose to use the 12 LED version because it fits nicely inside the top cover of the Solarbotics enclosure. As with the Rotator Turn Indicator, I mounted the LED ring behind a sheet of silver tissue paper to help mask out the LED electronics and let only the LED light shine through. Note that with this project, unlike with the Rotator Turn Indicator, we do not need a separate 5 V regulator to power the LED ring due to the fact that only one LED pixel will be lit at any given time.

An interesting side benefit from using this project with the CDE/HyGain and Yaesu rotators is the position indicator smoothing caused by the 10 µF capacitor on the rotator position potentiometer. Even with the resistance of the 100 kΩ resistor, you will still see a small benefit from a smoothing effect caused by the capacitor slightly delaying the signal transition as the rotator turns. This helps to make the meter move slightly more smoothly.

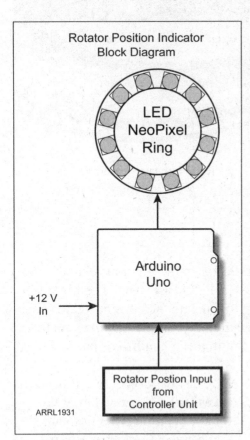

Figure 10.2 — The Rotator Position Indicator block diagram.

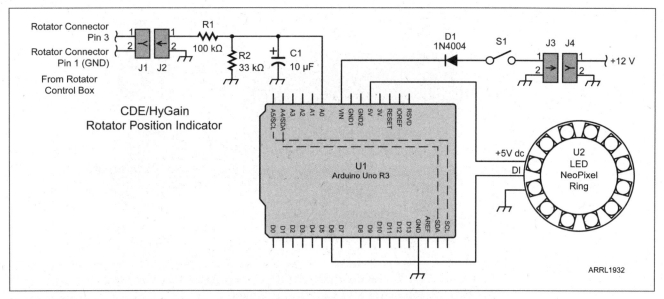

Figure 10.3 — The CDE/HyGain Rotator Position Indicator schematic.
C1 — 10 µF, 25 V electrolytic capacitor
D1 — 1N4004 diode
J1, J4 — 2-pin DuPont-style female header socket
J2, J3 — 2-pin DuPont-style male header
R1 — 100 kΩ, ⅛ W resistor
R2 — 33 kΩ, ⅛ W resistor
S1 — SPST switch
U1 — Arduino Uno
U2 — RGB NeoPixel/WS281x-compatible LED ring
Enclosure — Solarbotics Mega SAFE

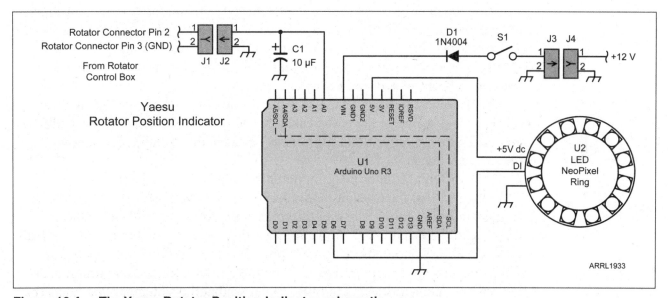

Figure 10.4 — The Yaesu Rotator Position Indicator schematic.
C1 — 10 µF, 25 V electrolytic capacitor
D1 — 1N4004 diode
J1, J4 — 2-pin DuPont-style female header socket
J2, J3 — 2-pin DuPont-style male header
S1 — SPST switch
U1 — Arduino Uno
U2 — RGB NeoPixel/WS281x-compatible LED ring
Enclosure — Solarbotics Mega SAFE

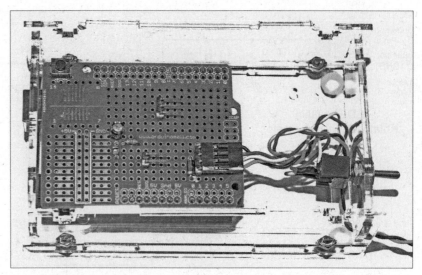

Figure 10.5 — The finished CDE/HyGain Rotator Position Indicator with the top cover and LED ring removed.

Flowchart

As mentioned earlier, the build for this project uses very few components and wires as shown in the finished CDE/HyGain version shown in **Figure 10.5**. We can easily access the rotator position signal from the rotator motor connector on the control unit for both the CDE/HyGain and the Yaesu controllers. In the case of the CDE/HyGain unit, you can attach to the screw terminals in the back of the control unit. For the Yaesu controller, the easiest thing to do would be to splice into the rotator position and ground wires in the cable going between the control box and the rotator motor.

The flowchart for the Rotator Position Indicator is shown in **Figure 10.6**. This is probably one of the easiest sketches you'll see for a ham radio Arduino project. There are only slight differences between the sketches for the two rotator types because of

1) the different rotator position potentiometer versions, and

2) the fact that the Yaesu rotators can rotate an extra 90° beyond 360° into what is known as an *overlap condition* between 360° and 450° of rotation.

As with the Rotator Turn Indicator, we'll start out by including the Adafruit NeoPixel library and defining the I/O pins. Because no two rotator position potentiometers are exactly alike, we'll define a minimum and maximum position potentiometer voltage calibration point so that the statement that maps the analog input voltage to the LED pixel can do so with reasonable accuracy. Next, we'll instantiate (create) the NeoPixel ring object and declare the necessary variables. In the `setup()` function, all we need to do is initialize the LED ring.

In the `loop()` function, we'll read the analog input pin for the rotator position potentiometer and map that voltage to a pixel on the LED ring. If the pixel has changed, we'll turn off the old pixel and light the new pixel. This is a very short and sweet flowchart and sketch.

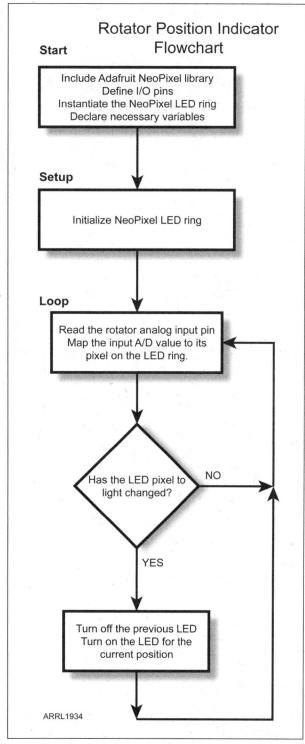

Figure 10.6 — The CDE/HyGain Rotator Position Indicator flowchart.

The Sketch

Because the differences are slight, we'll just go through the CDE/HyGain sketch in detail and highlight the differences with the Yaesu version. Sketches, libraries, Fritzing diagrams and other useful files for the projects in this book can be found at **www.sunriseinnovators.com** and **www.kw5gp.com**.

We'll start out with including the Adafruit NeoPixel library and defining the I/O pin for the NeoPixel LED ring and the rotator position potentiometer analog input pin. Because we're dealing with an analog voltage from the rotator position potentiometer, we'll define zero and maximum position voltage calibration settings. Next, we'll define the number of pixels in the LED ring. As mentioned before, I chose to use a 12 pixel LED ring because it fits perfectly inside the top cover of the Solarbotics enclosure.

```
/*
  Antenna Rotator Position Indicator
  CDE/HyGain version

  Glen Popiel - KW5GP

  Uses the Adafruit NeoPixel library

  Released under the GPLv3 license
*/

//#define debug // Uncomment this to enable the display of debug information on
              // the Serial Monitor

#include <Adafruit_NeoPixel.h>  // Include the Adafruit NeoPixel library

#define led_pin 6 // Define the Neopixel I/O pin

#define rotator_pot A0 // Define the rotator position input

#define cal_zero 0  // define the zero calibration point

#define cal_max 660 // define the max calibration point

#define num_pixels 12 // NeoPixel Ring size
```

Figure 10.7 — An RGBW LED pixel.

Next, we'll instantiate (create) the NeoPixel LED object. Note that there are two possible options for this process. The RGB LED rings are available in two different types — a standard red-green-blue (RGB), and the newer red-green-blue-white (RGBW) that include a bright white LED inside each RGB LED pixel. You can tell the difference between the versions by looking closely at an individual pixel. The RGBW LEDs have a small dark chip inside, as well as half of the LED masked off, as shown in **Figure 10.7**. You'll need to uncomment the proper line in the sketch for the type of LED ring you are using. Finally, we'll declare the variables needed in the sketch.

```
// Setup the NeoPixel library
// Uncomment the following for standard RGB Neopixel/WS281x LED Strips
//Adafruit_NeoPixel pixels(num_pixels, led_pin, NEO_GRB + NEO_KHZ800);

// Uncomment the following for RGBW Neopixel/WS281x LED Strips (RGB w/White LED)
// If you look at the unlit LEDs, half of each LED is covered with yellow
Adafruit_NeoPixel pixels(num_pixels, led_pin, NEO_GRBW + NEO_KHZ800);

#define delay_time 100 // Time (in milliseconds) to pause between LED updates

int rotator_input = 0, current_pixel = 0, last_pixel = 0; // sketch variables
```

The `setup()` function for the sketch is short and sweet. We'll start the serial port if debugging is enabled and initialize the NeoPixel LED ring.

```
void setup()
{
#ifdef debug
  Serial.begin(9600); // Start the serial port if debugging is enabled
#endif

  pixels.begin(); // Initialize the NeoPixel display
  pixels.clear(); // Set all pixel colors to 'off'
  pixels.show();  // Send the updated pixel colors to the hardware
}
```

The `loop()` function is also short and simple. We'll read the analog pin for the rotator position potentiometer and map that A/D value to the LED pixel we need to light.

```
void loop()
{

  rotator_input = analogRead(rotator_pot);  // Read the Rotator Position
                                            // Indicator

#ifdef debug
  Serial.print("Rotator Input = ");
  Serial.print(rotator_input);
#endif

  current_pixel = map(rotator_input, cal_zero, cal_max, 0, num_pixels - 1);
  // Map the current rotator position to the appropriate pixel

#ifdef debug
  Serial.print("   Mapped to = ");
  Serial.println(current_pixel);
#endif
```

Finally, we'll turn off the previous pixel and light the current pixel representing the rotator direction.

```
pixels.setPixelColor(last_pixel, pixels.Color(0, 0, 0));
  // Turn off the previous pixel

pixels.setPixelColor(current_pixel, pixels.Color(0, 150, 0));
  // Turn on the corresponding LED pixel
  pixels.show();

if (last_pixel != current_pixel)  // Only update the LED ring if the LED to
                                  // light has changed
{
  last_pixel = current_pixel;
}
delay(delay_time);
}
```

The Yaesu version of the sketch is nearly identical, with just a few minor differences. The Yaesu rotators can rotate 450° instead of the CDE/HyGain's 360° limit. The area between 360 and 450° is known as "overlap." We'll need to take that overlap condition into account when we map the LED pixel to light. In the initialization area, where we define the number of pixels in the LED ring, we'll also define how many pixels will be part of the overlap area.

```
#define num_pixels 12 // NeoPixel Ring size

#define overlap 4 // For Yaesu we need to define number of overlap rotation pixels
              // for 450 degrees
```

In the `loop()` function, the only extra thing we'll need to do is handle the overlap condition.

```
current_pixel = map(rotator_input, cal_zero, cal_max, 0, (num_pixels + overlap)
- 1);  // Map the current rotator position to the appropriate pixel

  if (current_pixel > num_pixels - 1) // Check to see if the rotator has gone
                                      // beyond 360 degrees (overlap)
  {
    // We're in overlap - subtract a full rotation

#ifdef debug
  Serial.print("   Overlap Condition = ");
  Serial.print(current_pixel);
#endif

    current_pixel = (current_pixel - num_pixels); // Subtract a full rotation to
                                // get the relative value of the pixel to light

  }
```

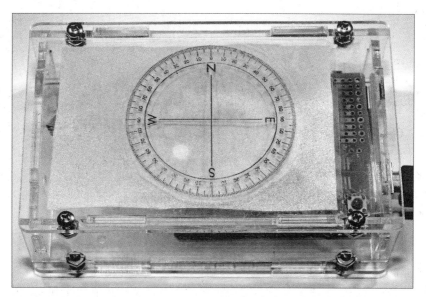

Figure 10.8 — Close-up of the compass dial and LED ring.

As a final touch, I downloaded an image of a compass dial and printed it on printable decal paper that you can get online or from a craft store such as Hobby Lobby. These decals are identical to the ones you apply to model airplanes, ships, and the like. Depending on the choice of decal paper, you can print them using either an inkjet or laser printer. (If you use the inkjet version, be sure to spray the finished decal with the recommended sealer before wetting it, to prevent the ink from smearing.) Using just water, I applied the decal to the top of the Solarbotics enclosure, as shown in **Figure 10.8**.

Enhancement Ideas

There are several things you can do to enhance the Rotator Position Indicator project. As the sketch is currently written, the zero and maximum rotation values are hard-coded in the sketch. You could add two pushbutton switches to read and dynamically save the calibration settings, and write them to EEPROM for a more semi-permanent calibration process. You can also combine this project with the Rotator Turn Indicator in the previous project and have it perform the functions of both projects. You could even have the LEDs change color when rotation is detected.

Also, you could implement this project on a Yaesu G-5400/5500 azimuth-elevation rotator. You could use the NeoPixel LED ring for the azimuth position, and a vertical NeoPixel LED stick to indicate the elevation.

Build Your Own AR-40 Rotator Controller

The HyGain AR-40 is another antenna rotator that's been around for what seems like forever. The AR-40 is a popular light-duty rotator that is closely related to the CDE/HyGain Ham-M/II/III/IV series of rotators, and is currently manufactured and sold by MFJ Enterprises. I'm unable to find out what year the AR-40 was designed, but it does have some interesting features that we have to take into account for our project design. For example, the AR-40 is completely powered off until the **START** button on the controller unit is pressed. The rest of the circuit design has that same basic look and feel of the other CDE/HyGain rotators, but a 741 operational amplifier (op amp) is part of the rotator position sensing circuit. Because the 741 op amp wasn't created until 1968, the AR-40 design would appear to be later than that of the Ham-series rotators. Still, it feels like a much older and simpler design than the Ham series, so its design seems to be a bit out of time and place.

Because the AR-40 is a popular rotator and so different from the Ham series CDE/HyGain product line, I felt it would be a good candidate for an Arduino-powered project. Like the CDE/HyGain projects in my first Arduino book, we'll build a 100% Arduino-powered control unit in this chapter, and will also modify an existing controller that you'll see in the next chapter. We'll also interface this controller to a PC for control by *Ham Radio Deluxe* (*HRD*) software. I would like to thank Martin Jue, K5FLU, and the folks at MFJ Enterprises for providing me with an AR-40 unit to tear apart and put back together for the AR-40 projects presented here.

Figure 11.1 — The AR-40 Rotator Controller in its case.

Needless to say, this project is somewhat more complex than the previous projects in this book. It looks harder than it really is, simply because there are so many moving pieces and devices that we have to account for. The plan is to completely replace the AR-40 controller with a control unit of our own construction and use an Arduino to power it. We'll add a 1.8-inch color graphic TFT display, and use a high-precision 16-bit programmable-gain analog-to-digital (A/D) converter to give us much better resolution on the positioning information coming from the rotator position potentiometer.

In many ways, it's easier to build this AR-40 controller from scratch than to modify a standard AR-40 control unit. That's mainly because we're able to redesign the entire internal operation of the control unit and plan for Arduino control from the ground up.

Probably the trickiest part will be acquiring the power transformer. I contacted MFJ and they were able to send me one, so that would be the best place to start. The original AR-40 transformer has 16 V and 24 V secondaries (the lower voltage winding doesn't need to be 16 V for this project — see discussion below). I was also able to locate several dual-winding, 24 V ac center-tapped transformers on Digi-Key's website (**www.digikey.com**). You may end up having to use two separate transformers as is done in the CDE/HyGain Ham-series control units.

One winding (or transformer) will need to be 24 V ac at 1.6 A or better to power the motor, and the other should be about 12 V ac at 1 A or so to power the electronics. Note that you will need separate transformers or a single transformer with separate windings to provide the 24 V ac needed to power the rotator motor. The 24 V ac motor common and the dc power ground cannot be tied together.

This second transformer (or second winding, in the case of a dual-winding transformer) will be regulated down to +5 V dc using a 7805 regulator, so the voltage is not critical — 12 V ac is a popular choice, and the 7805 can handle up to 35 V input. However, the Arduino Nano board typically uses an AMS1117 voltage regulator. The AMS1117 has a maximum rating of 15 V dc, so check the supply voltage at the output of the bridge rectifier (U5) if you are using the Vin pin for power to the Nano as shown in the schematic. If it's higher than 15 V, power the Nano using the 5 V from the 7805 output.

When we're finished with this project, you'll have all the building blocks you'll need to interface just about any rotator controller. We do have a few special design considerations to take into account. While we want to add a TFT display, PC control, and a manual clockwise/counterclockwise (CW/CCW) rotation switch, we also want to retain all of the features of the original AR-40, with the position selection dial and the START button on the top of the box, so this project will have a lot of moving pieces.

The Block Diagram

Figure 11.2 shows the block diagram of the scratch-built AR-40 rotator controller. Rather than calling the unit scratch-built, from here on out, we'll refer to it as the AR-40D, with the D standing for "Digital version."

As you can see from the block diagram, we have a lot more happening in

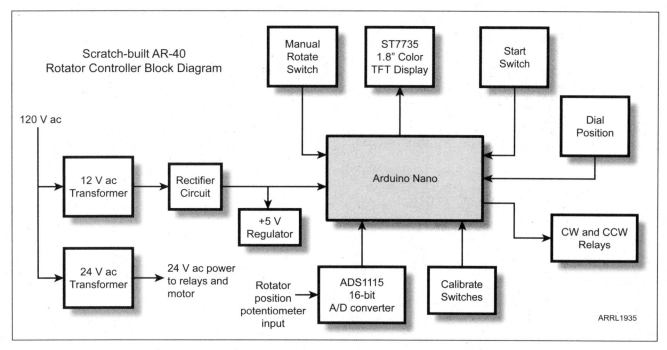

Figure 11.2 — The AR-40D Rotator Controller block diagram.

this project than the previous projects. We'll have multiple switch inputs, multiple analog inputs, an SPI color TFT display, a 16-bit programmable gain I²C analog-to-digital converter (A/D), and relay outputs to control the rotator motors. We'll also communicate with *Ham Radio Deluxe* software on a PC using the Nano's USB/serial port. So in other words, we're going to use just about every I/O method you can use with an Arduino.

As further proof of how powerful the little Arduino really is, we're going to fit the entire AR-40D sketch in an Arduino Nano, and even have some room left over. We'll build the entire project into a metal enclosure with most of the components mounted on a piece of copper-clad perfboard. We'll even custom print the dial label using clear decal paper in an effort to keep the look and feel as close to the original AR-40 as we can.

The Schematic Diagram

Figure 11.3 shows the schematic for the AR-40D. This project is a completely self-contained replacement unit for the AR-40 controller, and is mounted in a metal enclosure as shown in **Figure 11.4**. As you can see from the photo, there's a lot of room left on the circuit board, and it could probably be built on a board half the size of the one I used. But, when you start out with an unknown prototype, it's always better to have room left over than end up trying to shoehorn the last couple of parts into place.

One special note regarding the power transformer used with this project: The original AR-40 transformer assembly from MFJ uses a thermal switch to protect the transformer from overheating. I have not experienced any heating issues with the AR-40 rotator controller transformer at all, and you probably could leave this component out of the circuit without creating a safety concern.

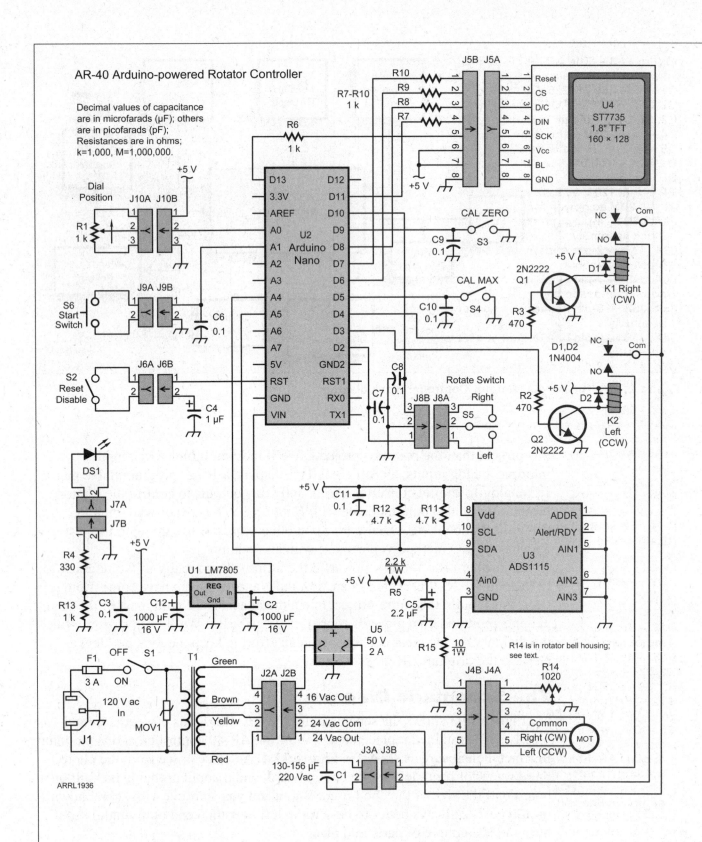

Figure 11.3 — The AR-40D Rotator Controller schematic diagram.
C1 — 130-156 µF, 220 V ac motor capacitor
C2, C12 — 1000 µF, 35 V electrolytic capacitor
C3, C6-C11 — 0.1 µF ceramic capacitor
C4 — 1 µF, 16 V electrolytic capacitor
C5 — 2.2 µF, 16 V electrolytic capacitor
D1, D2 — 1N4004 diode
DS1 — LED
F1 — 3 A, 120 V ac fuse
J1 — 120 V ac power socket
J2A, J2B — 4-pin Anderson Powerpole connector block assembly
J3A, J3B — 2-pin Anderson Powerpole connector block assembly
J4A, J4B — 7-pin twist-lock male and female connector assembly
J5A, J5B — 8-pin header and socket connector assembly
J6A, J6B — 2-pin header and socket connector assembly
J7A, J7B — 2-pin header and socket connector assembly
J8A, J8B — 3-pin header and socket connector assembly
J9A, J9B — 2-pin header and socket connector assembly
J10A, J10B — 3-pin header and socket connector assembly
K1, K2 — SPST relay, 3 A contacts, 5 V dc coil
MOV1 — 150 V metal oxide varistor (MOV)
Q1, Q2 — 2N2222A NPN transistor
R1 — 1 kΩ potentiometer
R2, R3 — 470 Ω, ⅛ W resistor
R4 — 330 Ω, ⅛ W resistor
R5 — 2.2 kΩ, 1 W resistor
R6-R10 — 1 kΩ, ⅛ W resistor
R11, R12 — 4.7 kΩ, ⅛ W resistor
R13 — 1 kΩ, ¼ W resistor
R14 — Rotator position potentiometer (part of the motor assembly; see text)
R15 — 10 Ω, 1 W resistor
S1, S2 — SPST toggle switch
S3, S4 — SPST pushbutton switch
S5 — SPDT center-off momentary toggle switch
S6 — SPST momentary pushbutton switch
T1 — Dual-winding transformer with 120 V ac primary, 16 V and 24 V secondaries (specs for the original AR-40 transformer). Or look for a dual-winding transformer or separate transformers with secondaries of 12 V at 1 A and 24 V at 1.6 A or higher. See text
U1 — LM7805 5 V regulator
U2 — Arduino Nano
U3 — ADS1115 16-bit analog-to-digital converter (A/D)
U4 — ST7735-compatible 1.8-inch color TFT display
U5 — 50 V, 2 A bridge rectifier module (2W10 or equiv.)

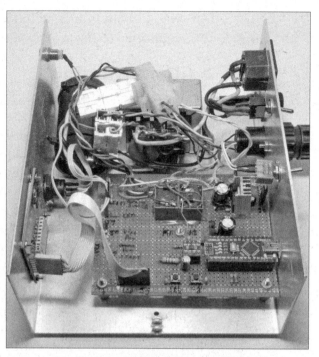

Figure 11.4 — The AR-40D Rotator Controller assembly and enclosure.

This switch is not shown in the schematic, but you could add this functionality to a standard power transformer by using a thermal cutoff switch that is used in microwave ovens, clothes dryers, gas heaters, and other common equipment. You can find them in various temperature cutoff values from most parts suppliers. I found some 167 °F and higher "thermal breaker cutoff fuses" that are actually small resettable thermal circuit breakers on eBay for $1.50 each. If you go this route, you do need to realize that there are two types of thermal cutoff switches. One is like a fuse, and if it trips, the cutoff switch is dead and needs to be replaced. The other type is a self-resetting thermostat type that will reset when the temperature cools to 10° or so below its cutoff value. I would recommend getting the self-resetting version if you use a thermal cutoff switch for this project.

Also note that the rotator position sensor potentiometer has been assigned the designator R14 but it is actually part of the rotator bell housing motor unit, so you don't need to worry about it. It is included in the schematic so you can see how everything is connected to the project assembly.

Construction

The entire project is constructed inside an 8 inch wide, 6 inch deep, 4 inch high metal enclosure. The power transformer and ac motor capacitor are mounted to the bottom of the enclosure, along with a 4.5 inch long by 3.5 inch wide piece of copper-clad perfboard mounted on standoffs to hold the project electronics. The TFT display, power indicator LED, and the manual **ROTATE** switch are mounted to the front side of the enclosure as shown in **Figure 11.5**.

Figure 11.6 shows the rear of the AR-40D enclosure. The leftmost switch in the photo is the **RESET DISABLE** switch (S2 on the schematic) that we use to prevent *Ham Radio Deluxe* from resetting the Arduino when it starts up and first attempts to connect to the Arduino rotator functions. You can see the small hole for the Nano's USB port cut out in the rear panel of the enclosure. I used a metal nibbling tool, and filed the opening to the correct size and shape. I have had great success using the twist-lock type connectors for connecting a small pigtail cable with a an Anderson Powerpole connector block at the other end to connect to the rotator cable, so we'll use one of those for the AR-40D as well. Figure 11.6 also shows how the 1.8 inch color TFT display is mounted to the front of the enclosure using 2 mm hardware. To help add to the look of the AR-40D, you can also see that a 5 mm LED bezel mount was used for the front panel power indicator LED.

Any high current connections, such as the rotator motor power and the power transformer connections to the electronics board, were done using either Powerpole connector blocks or Molex connectors. The different sizes and types of connector blocks used in this project were chosen so that there would be no way to connect something incorrectly while building and testing the prototype unit.

The control board for the AR-40D is shown in **Figure 11.7**. You can see how the Nano is mounted in a pair of SIP headers to create a socket at the edge of the board, allowing for the Nano's USB port to fit the hole in the rear of the enclosure for connecting the USB cable to the workstation. The ADS1115 16-bit pro-

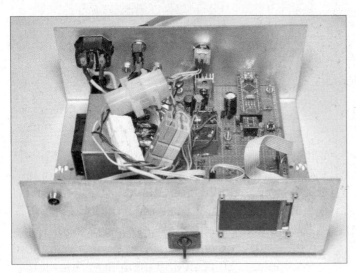

Figure 11.5 — The AR-40D front panel.

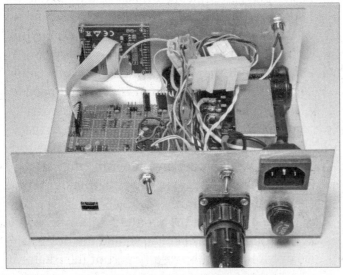

Figure 11.6 — The rear of the AR-40D enclosure.

Figure 11.7 — The AR-40D control board.

grammable gain A/D (U3) can be seen to the left of the Nano. Note that your ADS1115 module may not look the same as the one used in this project. I prefer to use this version because it has two rows of pins, making it easy to fashion a small socket for it using two 4-pin header sockets. The schematic diagram for the AR-40D project takes into account all versions of the ADS1115, so you should have no problem wiring it in the circuit, regardless of which version of the module you use.

One special note here is that R5, the voltage dropping resistor for the rotator potentiometer, needs to be 1 W to handle the heat dissipation caused by the voltage divider. We could have chosen a higher value for this resistor, but a higher resistance value would open the door for noise on the positioning signal caused by induction from the 24 V ac motor drive on the rotator cable or possibly from RF coming in on the long rotator cable (which acts like an inductor and antenna). In this design, the 2.2 kΩ value for R5 is more of a voltage dropping resistor than a divider, which is why this part of the circuit generates enough heat to warrant a 1 W resistor.

In the prototyping and testing phase of this project, multiple values of dropping resistors were used, with varying degrees of success. We'll get into the details later, but basically a nonlinearity in the rotation position potentiometer (R14 in the bell housing) caused the position readings to be slightly inaccurate as the rotator turned. The primary reason for this is that the AR-40 rotator position potentiometer is not wired as a potentiometer, but instead as a variable resistor. This causes changes in the current through the circuit and results in nonlinearity in the voltage reading on the position resistor as the rotator turns. As it turns out, the 2.2 kΩ resistor I chose to use for R5 had the least nonlinearity, and even then, corrections had to be made in the sketch, as you'll see later on.

Build Your Own AR-40 Rotator Controller 11-7

The 10 Ω resistor (R15) is in this circuit to raise the minimum voltage point of the rotation position potentiometer just slightly above 0 V. This keeps the voltage going to the A/D input positive, regardless of any negative noise spikes or induced voltage on the long rotator cable that could damage the A/D converter.

The LM7805 +5 V voltage regulator (U1) can be seen in the upper right of the photo, with a small TO-220 heatsink mounted to it. While the LM7805 is a rugged and reliable chip, it's always best to use a heatsink to help dissipate any heat that is generated. The relays in the top center of the control board (K1 and K2) are used to control the rotator motor. As you can see in the photo, the clamping diodes (D1, D2) to prevent reverse inductive voltage spikes from the relay coils are soldered directly to the relay coil pins. These diodes will only conduct when a reverse voltage spike occurs as the relay coils are de-energized, protecting the 2N2222A transistors (Q1, Q2) we use to drive the relays. In this photo, you can also see how male header pins are used to connect the ribbon cables to the rest of the enclosure electronics in order for the control board to be easily removed for testing and modification.

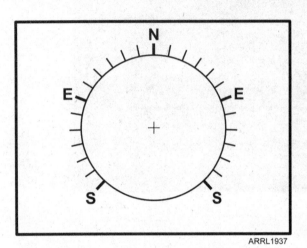

Figure 11.8 — The AR-40D position dial label.

As a finishing touch, the **START** button (S6) and **DIAL POSITION** potentiometer (R1) were mounted in the top cover of the enclosure. Using inkjet-printable clear decal paper from Hobby Lobby, I printed a photo of the actual dial label on the original AR-40 control unit (**Figure 11.8**). The decal was then applied to the top cover of the enclosure. The finished AR-40D controller can be seen in Figure 11.1 at the beginning of this chapter.

The Flowchart

Figure 11.9 shows the flowchart for the AR-40D project. This is a great example of why you should use flowcharts to plan your sketch. The full sketch for the AR-40D project is about as big as you can fit in an Arduino Nano, and has a lot of possible functions and processes that we need to handle. Without a flowchart or similar design document, it would be very easy to forget a command or feature that you wanted to include in your project. A high level flowchart of this type will help keep you on track, and can also work to simplify your sketch creation process.

As you'll see when we get into the sketch itself, the sketch main `loop()` is about as short as possible, with all of the work offloaded to functions that are called from the `loop()`. This allows the `loop()` to cycle through the operations quickly, and allows your project to respond quickly and smoothly to commands and operations.

By moving a lot of the workload into functions, you also make things easier in several other ways. A function essentially compartmentalizes a task or operation, allowing you to break your sketches into smaller, more easily managed

Figure 11.9 — The AR-40D flowchart.

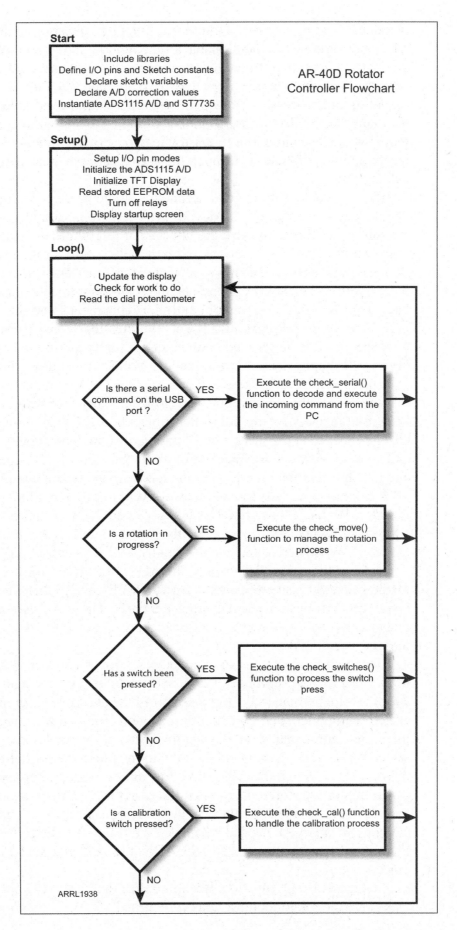

pieces in keeping with the "Divide and Conquer" philosophy mentioned earlier. Also, any variables declared inside a function are known as local variables, and do not exist outside of the function. When you exit the function, these variables cease to exist, and the resources they used are then freed up for use by other functions and processes. You can still interact with the variables you created in the main sketch, also known as global variables, from within a function, with global variables continuing to exist when you exit the functions. You can also create multi-use functions through the use of parameters as part of the function call process.

Rather than go into the whole function usage process, we're going to keep it simple here when it comes to dealing with functions, and recommend that you consult one of the many Arduino tutorials or references on how to use functions. Suffice it to say, functions are extremely handy for the more complex sketches such as this. One major advantage of functions is that you can reuse them in other sketches. For example, the rotator commands for this project emulate the Yaesu GS-232A rotator controller. This means that the workstation running *Ham Radio Deluxe* thinks it's communicating with a Yaesu GS-232A rotator controller, and has no clue that it's actually talking to an Arduino. This keeps *HRD* happy and all is good on the workstation side of things.

The function that processes the Yaesu GS-232A commands is the same one that I wrote for the rotator controller projects in my first Arduino book. It has been used in every rotator controller project because then, with only minor changes needed. This saves a huge amount of time when creating the sketch, and gives me a whole toolbox full of these "black box" functions that I can mix and match for various projects, without having to rewrite them from scratch.

Using the flowchart as a guide, we'll start out by including the needed libraries. We use a lot of libraries in this sketch, but that makes the entire project a lot easier by not having to handle the minute details that the library takes care of for us. We'll need to include the Adafruit GFX, Adafruit ST7735, and SPI libraries for the color TFT display, the ADS1115 library for the 16-bit programmable gain A/D converter, and the EEPROM library so that we can save the calibration settings on a semi-permanent basis. There are also several libraries that are used by the libraries mentioned above that we'll talk about when we get into the sketch in detail.

Because the sketch has so many pieces, a lot of constants and variables are used. I like to use preprocessor directives to define any constant variables. This way, they are permanently stored as part of the sketch in flash memory when the sketch is compiled, and they don't use any precious SRAM memory. Also, in the initialization area, we'll declare the necessary variables and constants. This sketch will utilize constants in addition to the preprocessor defined values. The reason for this is that the A/D correction values needed to properly position the rotator motor are stored in an array. It's easier to save them in an array, rather than define them and store them in the array on startup. We'll explain why these calibration values are needed when we go through the sketch itself in detail. Finally, we'll instantiate (create) the ADS1115 A/D and the ST7735 display objects.

In the `setup()` function, we'll initialize the digital I/O pin modes and make sure the rotator control relays are turned off. We'll then initialize the I^{2}C

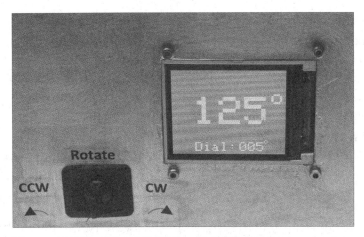

Figure 11.10 — The AR-40D color TFT display.

bus and the ADS1115 A/D converter, set the A/D gain, sample rate and the analog input mode. Next we'll initialize the ST7735 color TFT display and start the serial port. After that, we'll read the EEPROM for the stored calibration data, and display a brief startup screen on the TFT display.

For this project, you can think of the main `loop()` function as more of a process/function manager. The `loop()` will continually call a series of functions to see if any of them have any work to do or have work in progress, and do some housekeeping such as reading the **DIAL POSITION** potentiometer and repeat the loop indefinitely.

The first thing the `loop()` will do is update the color TFT display to show any changes in the information that it displays, such as a change in the position of the **DIAL POSITION** potentiometer. An example of the TFT display is shown in **Figure 11.10**. Next, we'll check the serial port to see if there are any commands from *Ham Radio Deluxe*. If a valid command is received, the command is decoded and executed by the `check_serial()` function.

In the next part of the loop, we'll check to see if we're already executing a rotation function using the `check_move()` function. This function will perform the rotation operations and update the positioning information that is sent back to *Ham Radio Deluxe* on the workstation as shown in **Figure 11.11**.

Next, we'll check the switches on the AR-40D enclosure itself, such as the **START** button and the left/right manual **ROTATE** switch (S5). If either switch is pressed, the `check_switches()` function will perform the requested operation based on which switch is activated. Finally, we'll check to see if either of the calibration switches (S3, S4) have been pressed, and update the calibration

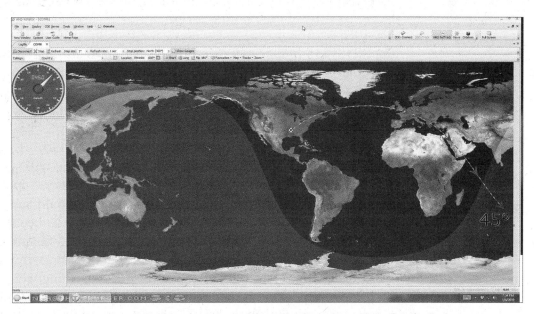

Figure 11.11 — The AR-40D rotator position displayed on *Ham Radio Deluxe*.

data stored in the EEPROM accordingly. As you can see, the functions are the real heart of this sketch, and we'll go into them in detail as we go through the sketch itself.

The Sketch

This is probably one of the more involved sketches you'll do with the Arduino because of all the features and functionality we've given to the AR-40D project. Instead of the usual debug mode, there are three levels of the debug mode, to allow you to select detailed debug information for the area you're troubleshooting.

```
/*
  AR-40D Rotator Controller
  written by Glen Popiel - KW5GP
*/

// #define debug   // Enables Debug Mode - Sends Debug output to Serial Monitor
// #define debug1  // Sends adc debug data to Serial Monitor
// #define debug2  // Sends switch debug data to Serial Monitor

#include <Adafruit_GFX.h>      // Core graphics library
#include <Adafruit_ST7735.h>   // Hardware-specific library
#include <SPI.h>
#include <EEPROM.h>   // Include EEPROM Library
#include <ADS1115.h>

const int comm_speed = 19200;  // Set the Serial Monitor Baud Rate
```

In addition to the multiple levels of debug mode, we'll start by including the Adafruit_GFX and Adafruit_ST7735 libraries, as well as the built-in SPI library for the ST7735 TFT display. Next we'll include the EEPROM library so we can access the EEPROM memory inside the Arduino Nano to save and recall the AR-40D's calibration settings. Next, we'll include the ADS1115.h A/D library. The ADS1115 library will also install several other libraries that it will need to function — the I2Cdev and Wire libraries for I²C bus communication. Because the ADS1115 library includes them for us automatically, we don't need to worry about them other than to be sure the I2Cdev library is installed, as the Wire library is built into the IDE.

The Adafruit libraries can be installed using the IDE's Library Manager, and the SPI and Wire libraries are built in to the IDE. You will need to manually install the ADS1115 and I2Cdev libraries. If you prefer, you could use Adafruit's ADS1X15 library for the A/D, but you would have to change the parts of the sketch that interact with the ADS1115 A/D because the Adafruit library does not allow you access all of the ADS1115 features, such as sample rate. To start out with, use the ADS1115 library that I used with the sketch and make any changes you want after you've got everything working.

Finally, we define the serial communications speed as 19200 baud. This will allow us to communicate with a workstation running *Ham Radio Deluxe*. In the

HRD Rotator program, be sure to select the rotator model as a Yaesu GS-232A because those are the commands this sketch will decode and process.

Next, we'll use preprocessor statements to define the various I/O pins. You'll note that this project requires so many digital I/O pins that we've had to use the A1 pin that is normally an analog input pin as a digital I/O pin used for the **START** switch.

```
#define rotate_right 10  // Assign Rotate Switch right input to Pin 10
#define rotate_left 2   // Assign Rotate Switch right input to Pin 2
#define left 3     // Assign Left (Counter Clockwise) Relay to Pin 3
#define right 4    // Assign Right (Clockwise) Relay to Pin 4
#define cal_zero 9    // Assign Zero Calibrate switch to Pin 9
#define cal_360 5     // Assign 360 Calibrate Switch to Pin 5
#define start_switch A1 // Assign the Start switch to pin A1
#define dial_pot A0 // Assign the Dial pot to pin A0
#define TFT_CS    6  // Assign the TFT CS to pin 6
#define TFT_RST   7  // Assign the TFT RST to pin 7
#define TFT_DC    8  // Assign the TFT DC to pin 8

#define tft_delay 10   // set the TFT command delay to 10ms
#define debounce 10  // set the switch debounce delay to 10ms
```

The EEPROM used in the Arduino Nano is accessed as a sequential-read type of memory. We'll be using the EEPROM to store our zero and maximum rotator position potentiometer calibration values, so we don't have to hardcode or redo the calibration every time the Arduino is reset or powered off. To use the EEPROM in this manner, we'll need to define each byte of storage used, so that we can properly extract the information we store. To ensure that we have valid data in the EEPROM, we set the first byte to a known value, such as the 54 used in this sketch. This value is used to confirm that the data in the EEPROM actually contains valid calibration data and is not just random data. In case we haven't saved any calibration data, we'll also define some calibration default values based on the results of the prototype testing.

```
#define EEPROM_ID_BYTE 1   // EEPROM ID to validate EEPROM data location
#define EEPROM_ID 54  // EEPROM ID Value
#define EEPROM_AZ_CAL_0 2    // Azimuth Zero Calibration EEPROM location
#define EEPROM_AZ_CAL_MAX 4  // Azimuth Max Calibration Data EEPROM location

#define AZ_CAL_0_DEFAULT 30 // Set the default CAL zero point to 30
#define AZ_CAL_MAX_DEFAULT 25000 // Set the default CAL MAX point to 25000
```

Next, we'll define the Azimuth Tolerance, which is used to determine the accuracy of the rotation process. The default is 1 degree, but you may want to change that to a higher value if you are experiencing excessive noise on your rotator position potentiometer signal. We'll also declare some of the variables needed by the sketch.

```
#define AZ_Tolerance 1   // Set the Azimuth Accuracy Tolerance

int long current_AZ;   // Variable for current Azimuth ADC Value
int AZ_Degrees;        // Variable for Current Azimuth in Degrees
boolean moving = false;  // Variable to let us know if the rotor is moving
int long previous_AZ = -1;  // Variable to track the previous AZ reading
String Azimuth = "   ";
boolean calibrate = false;  // Variable used to determine if calibrate switch
                            // has been pressed
boolean first_pass = true;  // Variable used to determine if first pass through
                            // the main loop
```

As it turns out, even though the rotator position potentiometer (R14, in the rotator motor housing) is a linear 1020 Ω potentiometer that smoothly changes value as the rotator motor turns, the value it returns is not as linear as we would like, as shown in **Figure 11.12**. This nonlinearity occurs because the rotator position potentiometer is not wired like a normal potentiometer would be, with a voltage on one side of the potentiometer, a ground on the other side, and with the wiper used to provide the position information. Instead, the AR-40 rotator position sensor is wired more like a rheostat, and it uses the position sensor like a variable resistor. This means that the current flow through the circuit, however slight, will have an effect when trying to read the position voltage. In Figure 11.12, you can see that the graph of the actual A/D values has several "bumps" in the data as compared to the ideal value. If left uncorrected, this can cause our rotator positioning to be off by several degrees. To make our project as accurate as possible, we'll need to define a series of internal correction points we can use to smooth out the bumps in the data curve. The correction values are stored in three arrays, and the 360° rotation angle of the rotator is divided up into twelve "correction" zones of 30° each.

To make things even more interesting, these correction values are different based on the direction the rotator is turning, due to the inductive noise on the rotator cable from

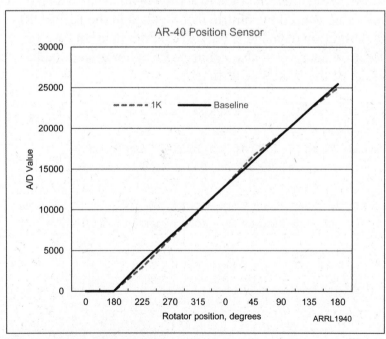

Figure 11.12 — Graph of the actual rotator position potentiometer A/D values.

the 24 V ac used to turn the rotator. So, we'll have three sets of correction values. One for when the rotator is idle, one for when it's rotating clockwise, and yet another for when it's rotating counterclockwise. And we're still not done with smoothing out the data. We'll also need to adjust the minimum and maximum A/D calibration values based on which direction we're rotating. All this may sound complicated and possibly even overkill — especially when you realize that no antenna has a beamwidth of 1°. But, if we're going to use computer control, why not take a little extra effort make things as accurate as possible. The good news is all of this work has already been done for you, so you shouldn't have to change the zone correction values.

```
// A/D Corrections for Position Sensor when not moving
const int   AZ_Correction[12] = { -550, -1500, -2100, -1900, -2350, -2400, -2800,
-2500, -2000 , -1850, -1000, 27};

// A/D Corrections for Position Sensor when moving right (CW)
const int   R_correct[12] = { -700, -1600, -2300, -2100, -2500, -2600, -3000,
-2700, -2200 , -2000, -1300, -50};

// A/D Corrections for Position Sensor when moving left (CCW)
const int   L_correct[12] = { -550, -1350, -1900, -1700, -2200, -2200, -2600,
-2300, -1800 , -1730, -700, -900};

const int r_zero_adj = 0; // A/D zero point correction when moving right
const int l_zero_adj = -210; // A/D zero point correction when moving left
const int r_max_adj = 0; // A/D max point correction when moving right
const int l_max_adj = 0; // A/D max point correction when moving right

const int AZ_Max_Correction = 30; // Correction to reduce MAX cal point to
                                  // prevent hitting stop
```

Continuing in the initialization portion of the sketch, we'll next declare all the variables needed in the sketch, and finally, we'll instantiate (create) the ADS1115 A/D and the ST7735 color TFT display objects.

```
int set_AZ;   // Azimuth set value
int AZ_0;     // Azimuth Zero Value from EEPROM
int AZ_MAX;   // Azimuth Max Value from EEPROMs
int zone_size; // Approximate number of A/D counts per zone
int zone = 0;  // Current rotation zone

int long update_time = 0; // Set the Display update time counter to zero
int update_delay = 1000;  // Set the Display update delay to one second
boolean turn_signal = false;  // Turn off the turn signl indicator
String previous_direction = "S";  // Set the previous direction to "S" - stop

String rotate_direction;  // Variable for the rotation direction

byte inByte = 0;   // incoming serial byte
byte serial_buffer[50];   // incoming serial byte buffer
int serial_buffer_index = 0;   // The index pointer variable for the Serial buffer
String Serial_Send_Data; // Data to send to Serial Port
String Requested_AZ; // RS232 Requested Azimuth - M and short W command
int AZ_To; // Requested AZ Move
int AZ_Distance; // Distance to move AZ

boolean manual_move = false;  // Variable to indicate this is a manually
                              // inititiated move
boolean right_status; // Variable to indicate the move right (CW) switch status
boolean left_status; // Variable to indicate the move left (CCW) switch status
boolean start_status; // Variable to indicate the Start switch status
boolean start_move; // Variable to indicate that a move has been started
boolean start_display = false;  // Variable to indicate the status of the
                                // display for a manual move

int previous_dial_pos = -100; // Variable for the previous dial pot position
int previous_dial_degrees = -100; // Variable for the previous dial pot position
                                  // in degrees
int dial_degrees; // Variable for the current dial position in degrees
int dial_pos; // Variable for the current dial position
String dial_display = "    ";  // Variable for the dial display string
String dial_rotate = "    ";   // Variable for the dial rotating string

ADS1115 adc;   // Define the A/D as adc

Adafruit_ST7735 tft = Adafruit_ST7735(TFT_CS,  TFT_DC, TFT_RST);
   // Initialize the TFT display
```

In the `setup()` function, we'll initialize all of the digital I/O pin modes and turn off the rotator control relays.

```
void setup()
{
  // initialize the digital pins
  pinMode(right, OUTPUT);
  pinMode(left, OUTPUT);
  pinMode(cal_zero, INPUT_PULLUP);
  pinMode(cal_360, INPUT_PULLUP);
  pinMode(rotate_right, INPUT_PULLUP);
  pinMode(rotate_left, INPUT_PULLUP);
  pinMode(start_switch, INPUT_PULLUP);

  // Turn off the rotator control relays
  digitalWrite(right, LOW);
  digitalWrite(left, LOW);
```

Next, we'll start the I²C bus and the ADS1115 16-bit programmable gain A/D. Then we'll set the A/D operating mode to continuous sampling at a sample rate of 128 samples per second with a range of 0 to 2.048 V. Then, we'll set the input type to a single-mode signal on analog input 0.

```
Wire.begin(); // Join the I2C Bus

adc.initialize(); // Initialize the ADS1115 16 bit A/D module
Wire.beginTransmission(0x48); // Begin direct communication with ADC
Wire.write(0x1);   // Connect to the ADC and send two bytes - Set the config
                   // register to all 1's
Wire.write(0x7F); // MSB
Wire.write(0xFF); // LSB
Wire.endTransmission(); // End the direct ADC comms

adc.setMode(ADS1115_MODE_CONTINUOUS); // free running conversion
adc.setGain(ADS1115_PGA_2P048);  // set adc gain to 2.048v range,
                                 // .0625 mv/step
adc.setRate(ADS1115_RATE_128);   // set adc sample rate to 128 samples
                                 // per second
adc.setMultiplexer(ADS1115_MUX_P0_NG);  // AN0+ Vs ground - Single mode
                                        // input on A0
```

Then we'll initialize the ST7735 color TFT display and start the serial port to enable communication with *Ham Radio Deluxe* running on the workstation. Then we'll display a startup screen on the TFT display.

```
tft.initR(INITR_18BLACKTAB);    // initialize a 1.8" TFT with ST7735S
                                // chip, black tab
delay(tft_delay);

Serial.begin(comm_speed);   // Start the Serial port

tft.fillScreen(ST7735_BLUE); // Clear the display
delay(tft_delay);
tft.setRotation(1); // Set the screen rotation
delay(tft_delay);
tft.setTextWrap(false);
delay(tft_delay);
tft.setTextSize(3);
delay(tft_delay);
tft.setTextColor(ST7735_GREEN);
delay(tft_delay);
tft.setCursor(40, 10);
delay(tft_delay);
tft.print("KW5GP");
delay(tft_delay);
tft.setTextSize(2);
delay(tft_delay);
tft.setCursor(55, 60);
delay(tft_delay);
tft.print("AR-40D");
delay(tft_delay);
tft.setCursor(45, 80);
delay(tft_delay);
tft.print("Rotator" );
delay(tft_delay);
tft.setCursor(25, 100);
delay(tft_delay);
tft.print("Controller");
```

Finally, we'll read the EEPROM calibration data by calling the read_eeprom_cal_data() function and then set up the correction zone areas. Next, we'll clear the variable that indicates that a rotation has been commanded, and read the **DIAL POSITION** potentiometer information so that it can be displayed on the TFT display. Then, we'll clear the TFT and display the basic template used for the display, so that we'll only need to update the information on the display that changes between updates.

```
  read_eeprom_cal_data();   // Read the EEPROM calibration data

#ifdef debug   // Display the calibration data when in debug mode
  Serial.print("ROM Cal - Zero: ");
  Serial.print(AZ_0);
  Serial.print(" Max: ");
  Serial.print(AZ_MAX);
#endif

  zone_size = AZ_MAX / 12;  // Divide the rotation position into 12 zones
                            // for calilbration correction
  AZ_MAX = AZ_MAX - AZ_Max_Correction;   // Subract a bit from the MAX position
                                         // to keep from hitting the stop

#ifdef debug
  Serial.print(" Adj Max: ");
  Serial.print(AZ_MAX);
  Serial.print(" Zone Size: ");
  Serial.println(zone_size);
#endif

  delay(5000);   //Wait 5 seconds then clear the startup message
  tft.fillScreen(ST7735_BLUE); // Clear the display
  delay(tft_delay);

  set_AZ = -1;   // Preset the Azimuth Move Variable

  read_dial_pot(); // Read the dial pot

  update_display(); // update the display
  first_pass = false; // turn off the update display first pass flag

}  // End Setup Loop
```

Build Your Own AR-40 Rotator Controller

Because we have moved the majority of the sketch operations into functions, the main `loop()` function is quite short, only having to do a quick check of the controller's switches and processes, then calling the function that is required to perform the desired operations.

```
void loop() // Start the Main Loop
{
  update_display(); // update the display
  check_serial(); // Check for a Serial command on the USB port
  check_move(); // Check to see if we are moving and process the move
  check_switches(); // Check the switches for manual commands

  // If not moving, read the dial pot and check if a calibration switch is
  //pressed
  if (!moving)
  {
    read_dial_pot();
    check_cal();  // Check to see if either calibration switch is pressed
  }

#ifdef debug1
  delay(250);
#endif

} // End Main Loop
```

The Functions

As we've been saying throughout this project, the functions are where all of the major work is done. We'll go through each function, step by step, so you can see how all of the various pieces are put together. The functions allow this sketch to do quite a lot, yet remain relatively easy to work with, mainly because everything is broken out into smaller building blocks.

The `check_cal()` function will check to see if one of the two calibration switches on the electronics board has been pressed. It will determine which switch has been pressed, and then call the appropriate function to save the new calibration values to EEPROM and to the running sketch calibration variables.

```cpp
// check_cal() function - Checks to see if a calibration switch is pressed and
// set calibration accordingly
void check_cal()
{
  if (digitalRead(cal_zero) == 0) // Cal Zero button pressed
  {
    set_0_az_cal(); // Set the Cal zero point
  }
  if (digitalRead(cal_360) == 0) // Cal Max button pressed
  {
    set_max_az_cal(); // Set the Cal max point
  }
}
```

The `all_stop()` function is used to turn off all of the rotation control relays, incorporate a 3-second brake delay, and then turn off the brake indication shown on the TFT display.

```cpp
// all_stop() function - turns off relays, delays 3 seconds then turns off Brake
// Time indicator
void all_stop()
{
  tft.setTextSize(2);
  delay(tft_delay);
  clear_top_line();
  delay(tft_delay);
  tft.setCursor(40, 3); // Display message on LCD
  delay(tft_delay);
  tft.print("Braking");
  delay(tft_delay);
  digitalWrite(right, LOW);  // Turn off CW Relay
  delay(tft_delay);
  digitalWrite(left, LOW);   // Turn off CCW Relay
  delay(tft_delay);
  moving = false;
  rotate_direction = "S";   // Set direction to S (Stop)
  delay(3000);
  clear_top_line();
  delay(tft_delay);
}
```

The `read_eeprom_cal_data()` function will read the Nano's EEPROM and save the calibration data (if any) to the system calibration variables. If calibration data has not been saved yet, the system calibration variables are set to default values and written to the EEPROM.

```
// read_eeprom_cal_data() function - Reads the EEPROM Calibration data
void read_eeprom_cal_data()
{
  if (EEPROM.read(EEPROM_ID_BYTE) == EEPROM_ID)    // Verify the EEPROM has
                                                   // valid data
  {

#ifdef debug // Display the Calibration data in debug mode
    Serial.print("Read EEPROM Cal Data Valid - AZ_CAL_0: ");
    Serial.print((EEPROM.read(EEPROM_AZ_CAL_0) * 256) + EEPROM.read(EEPROM_AZ_CAL_0 + 1), DEC);
    Serial.print("  AZ_CAL_MAX: ");
    Serial.println((EEPROM.read(EEPROM_AZ_CAL_MAX) * 256) + EEPROM.read(EEPROM_AZ_CAL_MAX + 1), DEC);
#endif

    AZ_0 = (EEPROM.read(EEPROM_AZ_CAL_0) * 256) + EEPROM.read(EEPROM_AZ_CAL_0 + 1); // Set the Zero degree Calibration Point
    AZ_MAX = (EEPROM.read(EEPROM_AZ_CAL_MAX) * 256) + EEPROM.read(EEPROM_AZ_CAL_MAX + 1); // Set the 360 degree Calibration Point

  } else
  { // EEPROM has no Calibration data - initialize eeprom to default values

#ifdef debug
    // Send status message in debug mode
    Serial.println("Read EEPROM Cal Data Invalid - set to defaults");
#endif

    AZ_0 = AZ_CAL_0_DEFAULT;   // Set the Calibration data to default values
    AZ_MAX = AZ_CAL_MAX_DEFAULT;
    write_eeprom_cal_data();   // Write the data to the EEPROM
  }
}
```

The `write_eeprom_cal_data()` function will write the current system calibration variables to the Nano's EEPROM.

```
// write_eeprom_cal_data() function - Writes the Calibration data to the EEPROM
void write_eeprom_cal_data() // Write the Calibration data to the EEPROM
{
#ifdef debug
  Serial.println("Writing EEPROM Cal Data");  // Display status in debug mode
#endif

  EEPROM.write(EEPROM_ID_BYTE, EEPROM_ID); // Write the EEPROM ID to the EEPROM
  EEPROM.write(EEPROM_AZ_CAL_0, highByte(AZ_0)); // Write Zero Calibration Data
                                                 // High Order Byte
  EEPROM.write(EEPROM_AZ_CAL_0 + 1, lowByte(AZ_0)); // Write Zero Calibration
                                                    // Data Low Order Byte
  EEPROM.write(EEPROM_AZ_CAL_MAX, highByte(AZ_MAX)); // Write 360 Calibration
                                                     // Data High Order Byte
  EEPROM.write(EEPROM_AZ_CAL_MAX + 1, lowByte(AZ_MAX)); // Write 360 Calibration
                                                        //  Data Low Order Byte
}
```

The `check_serial()` function is one of the main functions in the sketch. This function will check the serial port for data and commands from *Ham Radio Deluxe* running on an attached workstation. The function starts out by checking to see if there is a character waiting to be read on the serial port, and building the command one character at a time until a carriage return is received to indicate the end of the command sequence.

```
// check_serial() function - Checks for data on the Serial/USB port
void check_serial()
{
  if (Serial.available() > 0) // Get the Serial Data if available
  {
    inByte = Serial.read();   // Get the Serial Data

    // You may need to uncomment the following line if your PC software
    // will not communicate properly with the controller

    //   Serial.print(char(inByte));  // Echo back to the PC

    if (inByte == 10)  // ignore Line Feeds
    {
      return;
    }
    if (inByte != 13) // Add to buffer if not CR
    {
      serial_buffer[serial_buffer_index] = inByte;

#ifdef debug // Print the Character received if in Debug mode
      Serial.print("RX: ");
      Serial.print(" *** ");
      Serial.print(serial_buffer[serial_buffer_index]);
      Serial.println(" *** ");
#endif

      serial_buffer_index++;   // Increment the Serial Buffer pointer

    } else
    { // It's a Carriage Return, execute command
```

Next, the function will convert the command to all uppercase characters and decode the received command. The commands are the standard Yaesu GS-232A commands, and you can manually simulate these commands using the IDE's Serial Monitor to send manual commands to test the various operations of your sketch.

```
      if ((serial_buffer[0] > 96) && (serial_buffer[0] < 123))
        // If first character of command is lowercase, convert to uppercase
      {
        serial_buffer[0] = serial_buffer[0] - 32;
      }

      switch (serial_buffer[0]) {  // Decode first character of command
```

The first command we check for is the "A" command, which will stop all rotation. We do this by using a `Switch...Case` command based on the decimal ASCII value of the first character of the received command string. If this is the decoded command, the `az_rotate_stop()` function is called and all antenna rotation is stopped.

```
        case 65:    // A Command - Stop the Azimuth Rotation

#ifdef debug
        Serial.println("A RX");
#endif

        az_rotate_stop();
        break;
```

The next command we check for is the "C" command, which will call the `send_current_az()` function to encode the current azimuth value in degrees in Yaesu GS-232A format and send the response back to the workstation.

```
        case 67:      // C - return current azimuth

#ifdef debug
        Serial.println("C RX ");
#endif

        send_current_az();   // Return Azimuth if C Command
        break;
```

Next, we'll check for the "F" command, which will call the `set_max_az_cal()` function to set the maximum azimuth calibration point. You can use this command from the IDE's Serial Monitor to set the maximum azimuth calibration point, in the same way as if you had pressed the **CAL MAX** switch on the electronics board.

```
        case 70:   // F - Set the Max Calibration

#ifdef debug
        Serial.println("F RX");
        Serial.println(serial_buffer_index);
#endif

        set_max_az_cal();   // F - Set the Max Azimuth Calibration
        break;
```

The next command we'll decode is the "L" command. This command will call the `rotate_az_ccw()` function to rotate counterclockwise. Note that this rotation will continue until commanded to stop.

```
        case 76:    // L - Rotate Azimuth CCW

#ifdef debug
          Serial.println("L RX");
#endif

          rotate_az_ccw();   // Call the Rotate Azimuth CCW Function
          break;
```

The next command on the list is the "M" command. This command will call the `rotate_to()` function, which will instruct the rotator to rotate to the three character bearing contained in the remainder of the received command string.

```
        case 77:    // M - Rotate to Set Point

#ifdef debug
          Serial.println("M RX");
#endif

          rotate_to();   // Call the Rotate to Set Point Command
          break;
```

The next command we check for is the "O" command. When received, this command will call the `set_0_az_cal()` function and set the zero calibration point, just as if you has pressed the **CAL ZERO** button on the electronics board.

```
        case 79:    // O - Set Zero Calibration

#ifdef debug
          Serial.println("O RX");
          Serial.println(serial_buffer_index);
#endif

          set_0_az_cal();   // O - Set the Azimuth Zero Calibration
          break;
```

Next, we'll check for the "R" command, which will call the `rotate_az_cw()` function to rotate clockwise until commanded to stop.

```
    case 82:   // R - Rotate Azimuth CW
#ifdef debug
        Serial.println("R RX");
#endif

        rotate_az_cw();    // Call the Rotate Azimuth CW Function
        break;
```

The final command we decode is the "S" command, which will call the `az_rotate_stop()` function to stop all rotation.

```
    case 83:   // S - Stop All Rotation
#ifdef debug
        Serial.println("S RX");
#endif

        az_rotate_stop();   // Call the Stop Azimuth Rotation Function
        break;

    }
```

If there is no command decoded in the received character string, the command is ignored, the serial receive buffer is cleared, the buffer index pointer is reset, and the sketch is now ready to receive the next command.

```
    serial_buffer_index = 0;   // Clear the Serial Buffer and Reset the Buffer
                               // Index Pointer
    serial_buffer[0] = 0;
  }
 }
}
```

The next major function is the `check_move()` function. This function will check for an active rotation command and stop when the rotator has reached the specified destination. If there is no rotation in progress, the function will do nothing and return back to the main `loop()`. If there is a rotation command in process, the function will first read the A/D converter and map the received A/D count value to degrees.

```
// check_move() function - Checks to see if we've been commanded to move
void check_move()
{
  if (set_AZ != -1)
  {
    // We're moving - check and stop as needed
    update_delay = 100; // Change the display update delay to 100ms
    read_adc(); // Read the ADC

    // Map the Azimuth to degrees

#ifdef debug
    Serial.print(" *** check_move *** AZ_To: ");
    Serial.print(AZ_To);
    Serial.print("  Zone: ");
    Serial.print(zone);
    Serial.print("  Zone Adj: ");
    if (rotate_direction == "R")
    {
      Serial.print(R_correct[zone]);
    } else
    {
      Serial.print(L_correct[zone]);
    }
    Serial.print(" Current_AZ: ");
    Serial.print(current_AZ);
    Serial.print(" AZ_Deg: ");
    Serial.print(AZ_Degrees);
#endif
```

Next, we'll check to see how far we have remaining to move, and call the `fix_180()` function to correct the bearing if we've crossed the 0°/360° point.

```
    if (set_AZ != -1) // If we're moving
    {
      AZ_Distance = set_AZ - AZ_Degrees;   // Check how far we have to move

#ifdef debug
      Serial.print("  AZ_Distance: ");
      Serial.print(AZ_Distance);
#endif
```

```
    fix_180();   // Adjust for North Centering

#ifdef debug
    Serial.print("   Adj AZ_Distance: ");
    Serial.print(AZ_Distance);
    Serial.print("   Direction: ");
    Serial.println(rotate_direction);
#endif
```

Next, we'll check to see if we've reached the commanded destination. If we have, then we call the `az_rotate_stop()` function to stop the rotation, turn off the move operation, and reset the display update time to one second. Otherwise, we'll continue the rotation process.

```
      if (abs(AZ_Distance) <= AZ_Tolerance)   // No move needed if we're within
                                              // the tolerance range
      {
        az_rotate_stop();   // Stop the Azimuth Rotation
        set_AZ = -1;   // Turn off the Azimuth Move Command
      } else
      { // Move Azimuth - figure out which way
        if (AZ_Distance > 0)    //We need to move CW
        {
          rotate_az_cw();   // Rotate CW if positive
        } else {
          rotate_az_ccw();  // Rotate CCW if negative
        }
      }
    }
  } else
  {
    if (!manual_move && !start_move)   // Reset the display update delay to one
                                       // second if we're not moving
    {
      update_delay = 1000;
    }
  }
}
```

Build Your Own AR-40 Rotator Controller

The `send_current_az()` function is used to send the current antenna position in degrees back to the workstation via the serial port. This function will read the A/D converter, map the A/D count value to degrees, and then send the formatted data out the serial port to the workstation.

```
// send_current_az() function - Sends the Current Azimuth Function
void send_current_az()
{
  read_adc();   // Read the ADC

#ifdef debug
  Serial.println();
  Serial.print("Deg: ");
  Serial.print(AZ_Degrees);
  Serial.print("  Return Value: ");
#endif

  // Send it back via serial/USB port
  Serial_Send_Data = "";
  if (AZ_Degrees < 100)   // pad with 0's if needed
  {
    Serial_Send_Data = "0";
  }
  if (AZ_Degrees < 10)
  {
    Serial_Send_Data = "00";
  }
  Serial_Send_Data = "+0" + Serial_Send_Data + String(AZ_Degrees);
     // Send the Azimuth in Degrees
  Serial.println(Serial_Send_Data);   // Return value via serial/USB port
}
```

The `set_max_az_cal()` function is used to set the current position as the maximum rotation calibration point. This function will call the `read_adc()` function to read the A/D converter and map the current position to degrees. This function will then call the `write_eeprom_cal_data()` function to write the calibration data to the Nano's EEPROM.

```
void set_max_az_cal()  // Set the Max Azimuth Calibration Function
{
#ifdef debug
  Serial.println("Cal Max AZ Function");
#endif

  calibrate = true;
  read_adc();   // Read the ADC

  // save current az value to EEPROM - Zero Calibration
```

```
#ifdef debug
  Serial.println(current_AZ);
#endif

  AZ_MAX = current_AZ;   // Set the Azimuth Maximum Calibration to Current
                         // Azimuth Reading
  write_eeprom_cal_data();  // Write the Calibration Data to EEPROM

#ifdef debug
  Serial.println("Max Azimuth Cal Complete");
#endif

  calibrate = false;
}
```

The `rotate_az_ccw()` function is used to start rotation counterclockwise. It will energize the counterclockwise rotator relay, and display a left arrow on the TFT display to indicate the director the rotator is turning. It will also turn on the moving flag so that the `check_move()` function can manage the remainder of the rotation process.

```
// rotate_az_ccw() function - Rotate Azimuth CCW
void rotate_az_ccw()
{
  digitalWrite(left, HIGH);  // Set the Rotate Left Pin High
  digitalWrite(right, LOW);  // Make sure the Rotate Right Pin is Low
  rotate_direction = "L";    // set the direction flag to "L" (Left - CCW)

  // Turn on the turn signal on the display
  if (!turn_signal || (previous_direction != rotate_direction))
  {
    tft.setTextSize(2);
    delay(tft_delay);
    clear_top_line();
    delay(tft_delay);
    tft.setCursor(3, 10); // display the left arrow on the LCD
    delay(tft_delay);
    tft.print("<==");
    delay(tft_delay);
    turn_signal = true;
    previous_direction = rotate_direction;
  }
  moving = true;   // Set the moving flag
}
```

The `rotate_az_cw()` function does the same thing as the `rotate_az_ccw()` function we just looked at, only this time, the rotation is clockwise and a right arrow is displayed on the TFT display while rotation is occurring.

```
// rotate_az_cw() function - Rotate Azimuth CW
void rotate_az_cw()
{
  digitalWrite(right, HIGH);    // Set the Rotate Right Pin High
  digitalWrite(left, LOW);      // Make sure the Rotate Left Pin Low
  rotate_direction = "R";   // set the direction flag to "R" (Right - CW)
  if (!turn_signal || (previous_direction != rotate_direction))

    // Turn on the turn signal on the display
  {
    tft.setTextSize(2);
    delay(tft_delay);
    clear_top_line();
    delay(tft_delay);
    tft.setCursor(120, 10); // display the left arrow on the LCD
    delay(tft_delay);
    tft.print("==>");
    delay(tft_delay);
    turn_signal = true;
    previous_direction = rotate_direction;
  }
  moving = true;   // Set the moving flag
}
```

The `az_rotate_stop()` function is used to stop rotation. This function will de-energize the rotator relays and display a three-second braking message on the TFT display. If the move was commanded by the **START** button on the top of the AR-40 control unit, it will clear the TFT display rotation indicator, and then the function will turn off the rotation process flags.

```
// az_rotate_stop() function - Stop Azimuth Rotation
void az_rotate_stop()
{
  digitalWrite(right, LOW);    // Turn off the Rotate Right Pin
  digitalWrite(left, LOW);     // Turn off the Rotate Left Pin

  // update the display
  tft.setTextSize(2);
  delay(tft_delay);
  clear_top_line();
  delay(tft_delay);
  tft.setCursor(40, 3); // Display Braking message on LCD
  delay(tft_delay);
  tft.print("Braking");
```

```
    delay(tft_delay);
    set_AZ = -1;

    if (start_move) // If this is a move commanded by the start button
    {
      // Clear the moving display
      tft.fillRect(15, 91, 140, 8, ST7735_BLUE);
      delay(tft_delay);
    }
    turn_signal = false;
    start_move = false;
    start_display = false;
    delay(3000);
    clear_top_line(); // Clear the top line of the display
    moving = false; // Clear the moving flag
    rotate_direction = "S";  // Set direction to S (Stop)
    previous_direction = rotate_direction;
}
```

The `rotate_to()` function is used to rotate to a set point. This is part of the "M" command we decoded earlier. This function will decode the remainder of the "M" command to extract the desired bearing. It will then call the `read_adc()` function to get the current position and figure out how far we need to rotate. If the rotation will cross the 0°/360° point, we'll adjust the rotation distance and direction by calling the `fix_180()` function. The function will then start the rotation in the appropriate direction and set the rotation flag, so that the `check_move()` function can manage the remainder of the rotation process.

```
// rotate_to() function - Rotate to Set Point
void rotate_to()
{
#ifdef debug
  Serial.println("M Command -  rotate_to Function");
  Serial.print("  Chars RX: ");
  Serial.print(serial_buffer_index);
#endif

  // Decode Command - Format Mxxx - xxx = Degrees to Move to
  if (serial_buffer_index == 4)   // Verify the Command is the proper length
  {
#ifdef debug
    Serial.print("  Value [1] to [3]: ");
#endif

    Requested_AZ = (String(char(serial_buffer[1])) + String(char(serial_buffer[2])) + String(char(serial_buffer[3]))) ;  // Decode the Azimuth Value
    AZ_To = (Requested_AZ.toInt()); // AZ Degrees to Move to as integer
```

```
    // Flush the buffer to allow the next command
    serial_buffer_index = 0;

#ifdef debug
    Serial.println(Requested_AZ);
    Serial.print("AZ_To: ");
    Serial.print(AZ_To);
#endif

    read_adc();   // Read the ADC

#ifdef debug
    Serial.print("  Zone: ");
    Serial.print(zone);
    Serial.print("  Current Deg: ");
    Serial.print(AZ_Degrees);
#endif

    AZ_Distance = AZ_To - AZ_Degrees;   // Figure out how far we have to move

#ifdef debug
    Serial.print("  AZ_Dist: ");
    Serial.print(AZ_Distance);
#endif

    fix_180();   // Correct for North Centering

    set_AZ = AZ_To; // Set the Azimuth to move to
    moving = true;   // Set the moving flag

#ifdef debug
    Serial.print("  Adj AZ_Dist: ");
    Serial.print(AZ_Distance);
#endif

    if (abs(AZ_Distance) <= AZ_Tolerance)   // No move needed if we're within the
                                            // Tolerance Range
    {
      az_rotate_stop();   // Stop the Azimuth Rotation
      set_AZ = -1;   // Turn off the Move Command
    } else
    { // Move Azimuth - figure out which way
      if (AZ_Distance > 0)    //We need to move CW
        {

#ifdef debug
        Serial.println("  Turn right ");
#endif
```

```
        rotate_az_cw();   // If the distance is positive, move CW
      } else
      {

#ifdef debug
      Serial.println(" Turn left ");
#endif
      rotate_az_ccw();   // Otherwise, move counterclockwise
      }
    }
  }
}
```

The `set_0_az_cal()` function is used to save the current position as the zero calibration point for both the system calibration variables and the Nano's EEPROM. This function will call the `read_adc()` function to get the current position and map it to a degree value. The function will then call the `write_eeprom_cal_data()` function to write the calibration values to EEPROM.

```
// set_0_az_cal() function - Sets Azimuth Zero Calibration
void set_0_az_cal()
{

#ifdef debug
  Serial.println("Cal Zero Function");
#endif

  calibrate = true;
  read_adc();   // Read the ADC
  // save current Azimuth value to EEPROM - Zero Calibration

#ifdef debug
  Serial.println(current_AZ);
#endif

  AZ_0 = current_AZ;   // Set the Azimuth Zero Calibration to current position
  write_eeprom_cal_data();   // Write the Calibration Data to EEPROM

#ifdef debug
  Serial.println("Zero AZ Cal Complete");
#endif

  calibrate = false;
}
```

The `fix_az_string()` function is used to format the azimuth value to a three character string as called for in the Yaesu GS-232A protocol. The character string is padded with leading zeroes as necessary to get a three character value.

```
// fix_az_string() function - Converts azimuth to string and pad to 3 characters
void fix_az_string()
{
  Azimuth = "00" + String(AZ_Degrees);
  Azimuth = Azimuth.substring(Azimuth.length() - 3);
}
```

The `update_display()` function is used to update the information displayed on the color TFT display. The first thing this function will do is check to see if this is the first time we've run this function. If so, we'll clear the TFT display and display a template of the display information that doesn't change. This will help keep display flickering to a minimum and speed up regular screen updates.

```
// update_display() function - Updates the TFT display
void update_display()
{
  if (first_pass) // Only do this the first time the function is called
  {
    // Clear the screen and display normal operation
    tft.fillScreen(ST7735_BLUE); // Clear the display
    delay(tft_delay);
    tft.setTextSize(1); // Set the text size to 1
    delay(tft_delay);
    tft.setTextColor(ST7735_GREEN); // Set the text color to green
    delay(tft_delay);
    tft.setTextSize(2); // Set the text size to 2
    delay(tft_delay);
    tft.setCursor(32, 110);
    delay(tft_delay);
    tft.print("Dial: ");  // Display the dial position template text
    delay(tft_delay);

    previous_AZ = -1;  // Reset the previous AZ before starting
    first_pass = false; // turn off the first pass flag
  }
```

Next, we'll check to see if the update time delay has expired, indicating that it is time to update the TFT display data. First, we'll read the A/D converter and map the count value to degrees. If the bearing has changed because the last update, we'll display the new information, rather than needlessly erase and rewrite the same information. The reason we do it this way is because you can't just overwrite data with a TFT display. You have to first fill the area you want to erase with the background color and then write the new data. Otherwise, the characters will keep writing on top of each other, and the displayed data quickly turns into an unreadable mess. We'll also check to see if it's a manually commanded **DIAL POSITION** potentiometer move and display that status on the TFT display.

```
// Check to see if it's time to update the display data
if (abs(millis()) > abs(update_time + update_delay))
    // check to see if update time has expired
{
  read_adc(); // Read the ADC
  if (AZ_Degrees != previous_AZ)   // If the ADC reading has change update
                                   // the display
  {
    tft.setTextSize(6); // Set the text size to 6
    delay(tft_delay);
    tft.fillRect(30, 30, 130, 55, ST7735_BLUE); // Clear the position
                                                // display area
    delay(tft_delay);
    tft.setTextColor(ST7735_GREEN); // Set the text color to green
    delay(tft_delay);
    tft.setCursor(30, 40);
    delay(tft_delay);
    tft.print(Azimuth); // Display the azimuth in degrees
    delay(tft_delay);
    tft.setCursor(140, 30);
    delay(tft_delay);
    tft.setTextSize(3); // Set the text size to 3
    delay(tft_delay);
    tft.print("o"); // display the degree symbol
    delay(tft_delay);
    previous_AZ = AZ_Degrees; // set the previous azimuth reading to the
                              // current reading

    // If it's the start of a manually commanded move update the display
    // to reflect this
    if (start_move && !start_display)
    {
      // It's a manual dial move
      tft.setTextSize(1);
      delay(tft_delay);
      tft.setCursor(15, 91);
      delay(tft_delay);
      tft.print("Moving to Dial Position");
      delay(tft_delay);
      start_display = true;
    }

  }
```

Next, if the **DIAL POSITION** potentiometer has changed position, we'll update that part of the TFT display with the new **DIAL POSITION** potentiometer position and reset the display update timer.

```
    if (previous_dial_degrees != dial_degrees)   // if the dial pot position
                                                 // has changed
    {
       tft.fillRect(95, 110, 40, 20, ST7735_BLUE); // Clear the dial pot
                                                   // display area
       delay(tft_delay);
       tft.setCursor(95, 110);
       delay(tft_delay);
       tft.setTextSize(2); // Set the text size to 2
       delay(tft_delay);
       tft.print(dial_display);  // Display the current dial pot position
       delay(tft_delay);
       tft.setTextSize(1); // Set the text size to 1
       delay(tft_delay);
       tft.setCursor(130, 105);
       delay(tft_delay);
       tft.print("o"); // display the degree symbol
       delay(tft_delay);
       previous_dial_degrees = dial_degrees; // set the previous dial reading
                                             // to the current reading
    }
    update_time = millis(); // Reset the display update timer
  }

#ifdef debug1
   Serial.print("    ");
   Serial.print(" Deg: ");
   Serial.println(AZ_Degrees); // Send position to Serial Monitor in debug mode
#endif

}
```

The `read_adc()` function is used to read the ADS1115 16-bit programmable gain A/D converter. The function starts out by setting the gain to the 2.048 V range and a sample rate of 128 samples per second. We'll then read channel 0, which is where the rotator position potentiometer is connected. If we're performing a calibration operation, we'll exit the function at this point without further processing.

```
// read_adc() function - Reads the A/D Converter
void read_adc()
{
  // Read ADC and display position
  adc.setGain(ADS1115_PGA_2P048);   // set adc gain to 2.048v range, .0625 mv/step
```

```
  adc.setRate(ADS1115_RATE_128);   // set adc sample rate to 128 samples
                                   // per second

  // Read adc
  current_AZ = adc.getDiff0();   // Read ADC channel 0

#ifdef debug1
  Serial.print("   Read ADC: "); // display ADC read status in debug mode
  Serial.print(current_AZ);      // Display ADC value in debug mode
#endif

  if (calibrate)  // Exit if we're calibrating
  {
    return;
  }
```

Next, we'll check to see if the A/D reading is below the **CAL ZERO** or above the **CAL MAX** points and reset them to the correct boundary value. Then, we'll call the `zone_correct()` function to apply the rotation zone correction values to the A/D reading and then call the `map_az()` function to map the corrected A/D value to degrees. Finally, the `fix_az_string()` function is called to convert the azimuth value to a three-character value that we'll use in the rest of the sketch.

```
  if (current_AZ < 0) // Force the reading to zero if negative
  {
    current_AZ = 0;
  }

  if (current_AZ > AZ_MAX)  // Force the reading to max if we're above max
  {
    current_AZ = AZ_MAX;
  }

  zone_correct(); // Adjust A/D based on rotation zone

  map_az(); // Map the azimuth to degrees

  fix_az_string();  // Convert azimuth to string then pad to 3 characters
                    // and save in Azimuth

#ifdef debug1
  Serial.print(" Moving: ");
  Serial.print(moving);
  Serial.print(" set_AZ: ");
  Serial.print(set_AZ);
#endif

}
```

The `clear_top_line()` function is used to clear the top area of the TFT display, which is where the rotation direction arrow is displayed.

```
// clear_top_line() function - Clears the top line of the display
void clear_top_line()
{
  tft.fillRect(0, 0, 160, 30, ST7735_BLUE);
  delay(tft_delay);
}
```

The `zone_correct()` function is used to apply correction values to the A/D reading based on the current rotation zone. The first thing this function will do is determine which of the 12 rotation zones we're currently in. Because we'll be using this zone value as a pointer to the array containing the correction values and Arduino array indexes start at 0, our zones are numbered 0 through 11.

```
// zone_correct() function - Adjusts A/D value based on rotation zone
void zone_correct()
{
  int start_zone; // Variable to hold the start of the zone
  int end_zone;   // Variable to hold the end of the zone
  int correction; // Variable to hold the correction value
  int start_pos;  // Start of current zone
  int end_pos;    // End of current zone
  int current_pos;  // current position in zone
  int calculated_AZ;  // Used to adjust current AZ at zero point

  zone = current_AZ / zone_size;  // Figure out what rotation zone we're in

  if (zone < 0) // Force the zone to zero if below zero
  {
    zone = 0;
  }
  if (zone > 11)  // Force the zone to 11 if we're above 11
  {
#ifdef debug1
    Serial.println(" Zone forced to 11");
#endif
    zone = 11;
  }
```

The correction values are based on the slope-intercept equation to give us the proper correction values based on the actual position within the rotation zone. We'll need to calculate the starting and ending A/D values of the current rotation zone to fit into the equation. Because of the way we use an array to get the starting and ending correction values, we need to manually provide the start position parameter for zone 0, and the ending position parameter for zone 11. Also, because we use a different set of correction values based on whether we're rotating or not, and which direction we're rotating if we are rotating, we'll need to get the correction values from the proper correction array.

```
  start_pos = (zone * zone_size); // Calulate the start of the zone
  end_pos = start_pos + zone_size;  // Calculate the end of the zone

#ifdef debug1
  Serial.print(" Zone: ");
  Serial.print(zone);
#endif

  // Handle Zone zero differently
  if (zone == 0)
  {
    start_zone = 0;
    if (!moving)  // If we're not moving, use static correction values
    {
      start_pos = AZ_0;
      end_zone = AZ_Correction[0];
    } else
    {
      // we're moving - use moving correction factors
      if (rotate_direction == "R")
      {
        // We're turning right
        start_pos = AZ_0;
        end_zone = R_correct[0];
        current_AZ = current_AZ - r_zero_adj ;
      } else
      {
        // We're turning left
        start_pos = AZ_0;
        end_zone = L_correct[0];
        current_AZ = current_AZ - l_zero_adj ;
      }
    }
  }

  // Handle zone 11 differently
  if (zone == 11)
  {
    if (moving)
    {
      // we're moving - use moving correction factors
      if (rotate_direction == "R")
      {
        start_zone = R_correct[zone - 1];
        end_zone = R_correct[zone];
```

Build Your Own AR-40 Rotator Controller

```
      current_AZ = current_AZ  - r_max_adj;
    } else
    {
      // We're turning left - use left corrections
      end_pos = AZ_MAX;
      start_zone = L_correct[11];
      end_zone = l_max_adj;
      current_AZ = current_AZ  - l_max_adj;
    }
  } else
  {
    start_zone = AZ_Correction[zone - 1];
    end_zone = AZ_Correction[zone];
    calculated_AZ = current_AZ - AZ_0;
  }
}
```

Now that we've handled the special conditions for zones 0 and 11, we'll use the values in the correct array to get the zone starting and ending A/D values for the other zones.

```
calculated_AZ = current_AZ;

  // Handle zones 1 thru 10
  if (zone > 0 && zone < 11)
    if (moving)
    {
      // Add the moving correction value when moving
      if (rotate_direction == "R")
      {
        start_zone = AZ_Correction[zone - 1];
        end_zone = R_correct[zone];
      } else
      {
        // We're turning left - use left corrections
        start_zone = AZ_Correction[zone - 1];
        end_zone =  L_correct[zone];
      }
    } else
    {
      start_zone = AZ_Correction[zone - 1];
      end_zone = AZ_Correction[zone];
      calculated_AZ = current_AZ - AZ_0;
    }

#ifdef debug1
  Serial.print("  Zone data: ");
  Serial.print(start_zone);
```

```
    Serial.print(" ");
    Serial.print(end_zone);
    Serial.print("  Zone range: ");
    Serial.print(start_pos);
    Serial.print(" ");
    Serial.print(end_pos);
#endif
```

Next, we'll use the correction values we just determined to map the current position in the zone to its calculated correction value, and then apply that calculated correction value to the current azimuth A/D reading. I know this all sounds super complicated, but it's actually a relatively simple solution to handling the nonlinearity issues with the rotator position potentiometer. You can treat this as one of those black-box things I've been talking about regarding libraries and function. It works, it does what we need it to do, and how it does its magic isn't as much of a concern to us as getting the magic to happen in the first place. It's kind of like making sausage. We really don't know how it's made, nor do we really want to know. All we care about is that it comes out right. (I had a full head of hair before I started working on this function, and now there's not a whole lot left. But at least it gets the job done and that's all that matters.)

```
  // Map the correction factor based on position in zone

  correction = map(calculated_AZ, start_pos, end_pos, start_zone, end_zone);
    // Map the correction

  current_AZ = current_AZ + correction;

  if (current_AZ < 0) // Force the current azimuth value to zero if negative
  {
    current_AZ = 0;
  }

#ifdef debug1
    Serial.print(" Adj: ");
    Serial.print(correction);
    Serial.print(" Final AZ: ");
    Serial.print(current_AZ);
#endif

}
```

The `map_az()` function is used to map the azimuth A/D value to degrees and adjusts the result to account for the north-centering positioning on the AR-40 rotator.

```
// map_az() function - maps azimuth value to degrees
void map_az()
{
  AZ_Degrees = map(current_AZ, AZ_0, AZ_MAX, 0, 360); // Map the current_AZ
                                                     // to calibrated degrees

  if (AZ_Degrees < 0) // Limit Zero point to 0
  {
    AZ_Degrees = 0;
  }

  if (AZ_Degrees > 360) // Limit max point to 360
  {
    AZ_Degrees = 360;
  }

  if (AZ_Degrees < 180)  // Adjust for North-centered controller
  {
    AZ_Degrees = 180 + AZ_Degrees;
  } else
  {
    AZ_Degrees = AZ_Degrees - 180;
  }

}
```

The `fix_180()` function is used to correct for when we rotate past the 0°/360° point to take into account the North centering rotation of the AR-40.

```
// fix_180() function - Correct for North Centering
void fix_180()
{
  if (AZ_Degrees >= 180 && AZ_Degrees <= 360  && zone != 11)
    // Make sure we don't rotate left beyond 180 and we're not in zone 11
  {
    if ((AZ_Degrees + AZ_Distance < 180) && (AZ_Distance < 0))
    {
      AZ_Distance = 360 + AZ_Distance;
    }
  } else {
    // We're in zone 4 thru 12 - Make sure we don't rotate right beyond 180

#ifdef debug
    Serial.print(" In Zone 4 thru 12");
```

```
#endif

    if ((AZ_Degrees + AZ_Distance > 180) && (AZ_Distance > 0))
    {
      AZ_Distance = -360 + AZ_Distance;
    }
  }
}
```

The `check_switches()` function is another of the major control functions we call from the main `loop()`. This function will check the status of the left/right manual **ROTATE** switch, along with the **START** button on the rotator enclosure. If the manual **ROTATE** switch is pushed to the left to turn counter-clockwise, the `rotate_az_ccw()` function is called to process the rotation until the switch is released. If the manual **ROTATE** switch is pushed to the right to turn clockwise, the `rotate_az_cw()` function is called to process the rotation until the switch is released.

```
// check_switches() function - checks the left, right, and start move switches
void check_switches()
{
  // Checks the Rotate and Start switches
  right_status = digitalRead(rotate_right); // Read the Right move switch
  left_status = digitalRead(rotate_left); // Read the Left move switch
  start_status = digitalRead(start_switch); // Read the Start move switch

  // If any of the move switches are pressed
  if (left_status == LOW || right_status == LOW || start_status == LOW)
  {
    // manual left, right or start active
    if (!manual_move) // If we're not already moving
    {
      update_delay = 100; // Set the display update delay to 100ms
      if (left_status == LOW) // If the Left switch is pressed
      {
        // We're moving left
        rotate_az_ccw();   // Rotate Left
        manual_move = true; // Set the manual move flag

#ifdef debug2
        Serial.println("Rotate Left pressed");
#endif

      }

      if (right_status == LOW)// If the Right switch is pressed
      {
```

```
        // We're moving right
        rotate_az_cw();   // Rotate Right
        manual_move = true; // Set the manual move flag

#ifdef debug2
        Serial.println("Rotate Right pressed");
#endif

      }
```

If the **START** button is pressed, we build a simulated serial command to rotate to the position indicated by the **DIAL POSITION** potentiometer. Doing it this way allows us to use the previous functions we used to handle the rotation, without having to write a bunch of extra code. This function will also take into account the special condition of there being two 180° points on the rotator position dial, full counterclockwise and full clockwise. When either 180° point is selected, the function will calculate which is the closer of the two, and select the appropriate rotation direction to complete the command. The function will then create the "M" move command with the appropriate destination heading and call the `rotate_to()` function to process the rotation as if it had been commanded by the serial port.

```
      if (start_status == LOW && !start_move) // If the Start switch is pressed
                                              // and we're not already moving
      {
        // Start switch is pressed - we need to move to dial position
        // Fake a serial command and stuff the buffer with the Move command data
        dial_rotate = dial_display; // Set the position to rotate to
        // Determine which way to turn if we're going to 180 degrees -
        // move to closest 180 position
        if (dial_rotate == "180")
        {
          if (dial_pos > 500)
          {
            // Force right turn no matter what
            dial_rotate = "179";
          } else
          {
            // Force Left turn no matter what
            dial_rotate = "181";
          }
        }
        // Create the rotate_to function command string
        serial_buffer[0] = 'M';
        serial_buffer[1] = dial_rotate[0];
        serial_buffer[2] = dial_rotate[1];
        serial_buffer[3] = dial_rotate[2];
        serial_buffer_index = 4;
        start_move = true;  // Set the manual start command flag
```

```
#ifdef debug2
        Serial.print("Start switch pressed - moving to dial position - Degrees: ");
        Serial.print(dial_display);
        Serial.print("  Buffer: ");
        Serial.print(serial_buffer[0]);
        Serial.print(serial_buffer[1]);
        Serial.print(serial_buffer[2]);
        Serial.print(serial_buffer[3]);
        Serial.print("  Length: ");
        Serial.println(serial_buffer_index);
#endif

        rotate_to();    // Call the rotate_to function
      }
      delay(debounce);   // Delay for switch debounce
    }
```

As part of the manual **ROTATE** left/right switch being pressed, when it is released, we'll call the `az_rotate_stop()` function to stop all rotation and end the rotation process.

```
    } else
    {
      // We're not moving (anymore)
      if (manual_move)
      {
        az_rotate_stop(); // call the rotate stop function
        manual_move = false;   // Turn off the manual move flag

#ifdef debug2
        Serial.println("Switch Released");
#endif

      }
    }
}
```

The `read_dial_pot()` function will read the dial position potentiometer A/D value. If the value has changed from the previous read, the new value will be mapped to degrees, formatted to a three-character value, and the TFT display updated with the new value.

```
// read_dial_pot() function - Reads the dial pot position
void read_dial_pot()
{
  // Reads the pot and updates display is it has changed
  dial_pos = analogRead(dial_pot);
  if (dial_pos > (previous_dial_pos + 1) || dial_pos < (previous_dial_pos - 1))
// If the dial pot has changed
  {
    // Dial pot has moved
    dial_degrees = map(dial_pos, 0, 1023, 0, 360) + 180;
       // Map the dial position to degrees
    if (dial_degrees > 359) // Adjust for North Centering
    {
      dial_degrees = dial_degrees - 360;
    }
    previous_dial_pos = dial_pos; // Update the previous dial position

    fix_dial_string();  // Pads the dial string to 3 characters

#ifdef debug2
    Serial.print("Dial Pot : ");
    Serial.print(dial_pos);
    Serial.print("   Dial Degrees : ");
    Serial.println(dial_degrees);
#endif

    update_display(); // Update the display
  }
}
```

The last function in the sketch, the `fix_dial_string()` function, is used to convert the dial potentiometer value to a three-character string that is padded with leading zeros to display on the TFT display.

```
// fix_dial_string() function - Converts azimuth to string and pads to
// 3 characters
void fix_dial_string()
{
  dial_display = "00" + String(dial_degrees);
  dial_display = dial_display.substring(dial_display.length() - 3);
}
```

And that's the end of the sketch. As you can see, by breaking down all of the major pieces into functions, we're able to simplify the sketch, and get everything working one building block at a time. This is why so much emphasis is placed on breaking things into smaller pieces using the flowchart in the initial design, and into groups of functions when we get down to the writing of the sketch. I'm not sure this project could have even been made to work properly if it had not been designed in this way. As it is now, once you learn how all the various pieces and functions work together, making changes and modifications shouldn't be that difficult at all.

Sketches, libraries, Fritzing diagrams, and other useful files for the projects in this book can be found at **www.sunriseinnovators.com** and **www.kw5gp.com**.

Enhancement Ideas

While you may think this project is so complete that there's no room for improvement, there are several things that could be done to add even more functionality. This is a common theme for me. Just as I finish up a project, there's some new Arduino module I find out about that would have made the project even cooler.

This was the case with the 4×4 matrix keypads. By the time I got one in my hands, this project had been completed and there was no time to go back and add in the direct-entry keypad. So, I'll leave that to you. Get one of the 4×4 matrix keypads and give the AR-40D a keypad entry rotator position capability, just as we'll do with the CDE/HyGain rotator controller in a later chapter.

Also, while the color TFT display is functional and adequate, it could benefit from a makeover and some personal touches to the displayed information. You could have some fun with the display, and rather than use directional arrows to indicate rotation, you could simulate the LED ring concept from the earlier rotator projects and display that on the TFT while it is rotating.

12 Modified AR-40 Rotator Controller

As we discussed in Chapter 11, the HyGain AR-40 is a light-duty rotator manufactured and sold by MFJ Enterprises. For this project, we'll take an existing AR-40 control unit and modify it for control by *Ham Radio Deluxe* software. The Arduino controller for this project will emulate the Yaesu GS-232A rotator controller. *HRD Rotator* knows how to communicate with the Yaesu GS-232A, so it interfaces quite nicely with the Arduino controller.

Again, I'd like to thank Martin Jue, K5FLU, and the gang at MFJ for providing me with an AR-40 rotator controller to use for the AR-40 projects in this book. The AR-40 is a bit different than most rotator controllers in several ways. The major difference is that the AR-40 rotator control unit is powered off until you set the rotator position dial and press the **START** button. As soon as you press the **START** button, the rotation direction control relay either energizes or de-energizes depending on the desired direction of rotation. This means that there is no idle position for the rotator motor when powered on, and no easy way to get positioning information to send to *Ham Radio Deluxe* without modifying the control unit. So, let's break out the soldering iron and start modifying.

As it turns out, the modification for the AR-40 controller isn't really as bad as you might think. We'll want to retain the original AR-40 functionality while giving it Arduino-powered capability with the flip of a switch. For this project, we'll

Figure 12.1 — The AR-40 rotator and controller.

Figure 12.2 — The AR-40 Arduino control unit.

need to cut a few circuit board traces and add a few wires on the AR-40 control board. We'll mount a switch in the back of the AR-40 control box to allow us to switch between the standard and Arduino-powered modes. We need to switch things this way to take into account the AR-40's rotation direction control logic, as well as switching the position sensor between the AR-40 controller and the Arduino control unit — the two don't work well together when both are active. We'll also install a small relay control board inside the AR-40 control box. We'll interface everything to an Arduino Nano mounted inside a Solarbotics Mega SAFE Enclosure that will also contain the 1.8-inch color TFT display and the 16-bit ADS1115 programmable-gain analog-to-digital converter. The finished Arduino-powered controller is shown in **Figure 12.2**.

The Modified AR-40 Block Diagram

Many of the pieces of this project are similar to the AR-40D project in the previous chapter. We're only adding the computer control by *Ham Radio Deluxe* functionality to this project, so it will actually be much less complex than the AR-40D version. Before we start the modification process, let's take a look at what the modification will encompass.

Figure 12.3 shows the block diagram for the AR-40 modification. For this

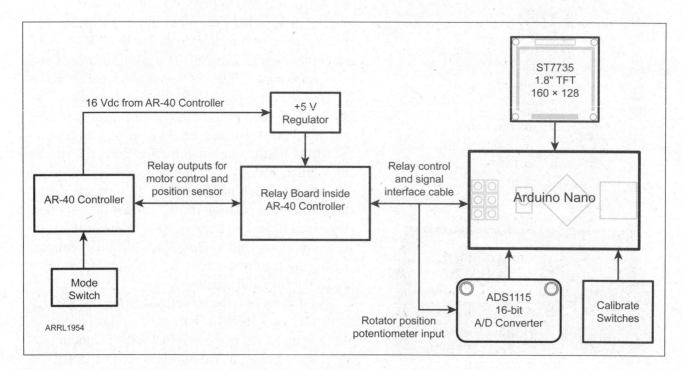

Figure 12.3 — The Modified AR-40 Rotator Controller block diagram.

project, we can use the 16 V dc on the AR-40 control board to power the new relay board that will be mounted inside the original AR-40 control unit. This relay board will connect to the Arduino Nano controller and display unit using an 8-conductor cable. The Arduino enclosure will also house the ADS1115 16-bit programmable A/D and the position calibration switches. A 4PDT toggle switch will be added to the AR-40 controller enclosure to allow switching between the standard AR-40 and Arduino-powered modes.

Making the AR-40 Control Unit Modifications

Because the AR-40 is not powered on until you press the START button, we'll need a way to have the power on all the time without the rotator turning until we command it to. This means we'll have to make a number of circuit board trace cuts and wiring changes to turn the AR-40 power on without rotating until commanded to, and to switch the rotator positioning sensor output between the AR-40 control board and the Arduino board. We'll start with the modifications needed to the AR-40 control unit and then add the new relay control board.

Figure 12.4 shows the schematic for the original, unmodified AR-40 rotator controller. **Figure 12.5** shows the modified controller schematic, including dashed lines showing modifications to the stock controller and its circuit board (added wires and circuit board trace cuts we'll need to make). Circled numbers correspond to trace cuts and added wires described in a step-by-step process below. **Figure 12.6** shows the schematic for the new relay board that we'll mount inside the controller enclosure, and **Figure 12.7** shows the schematic for the new Arduino board and TFT display circuitry.

Step by Step Controller Modifications and Relay Board

IMPORTANT! Before you start making these modifications, please make sure you have the AR-40 control unit unplugged from ac power. Several of the modification steps involve rewiring parts of the ac power to the control unit. Please exercise caution when making these changes, and verify your work carefully. Also, the Manual/Arduino mode switch will have ac power on one section of the switch. Please make sure the wires connecting to this switch are properly insulated and don't create a shock hazard.

Figure 12.8 shows the bottom of the AR-40 control board. You will notice a series of arrows and numbers indicating the various steps and locations we will need to make modifications. First, we'll need to cut a series of circuit board traces. Carefully remove the control board from the enclosure and turn it upside down.

Step 1. On the upper left side of the board, where Arrow 1 is pointing, you'll need to cut the circuit board trace between diode CR7 and resistor R8.

Step 2. On the mid-upper right side of the board, cut the trace between diode CR4 and the Position Dial potentiometer R1C as shown by Arrow 2. The A, B, and C pins of Position Dial potentiometer R1 are labeled on the top side of the AR-40 control board.

Step 3. On the right-center part of the board, cut the trace between relay K1 and capacitor C2 as shown by Arrow 3.

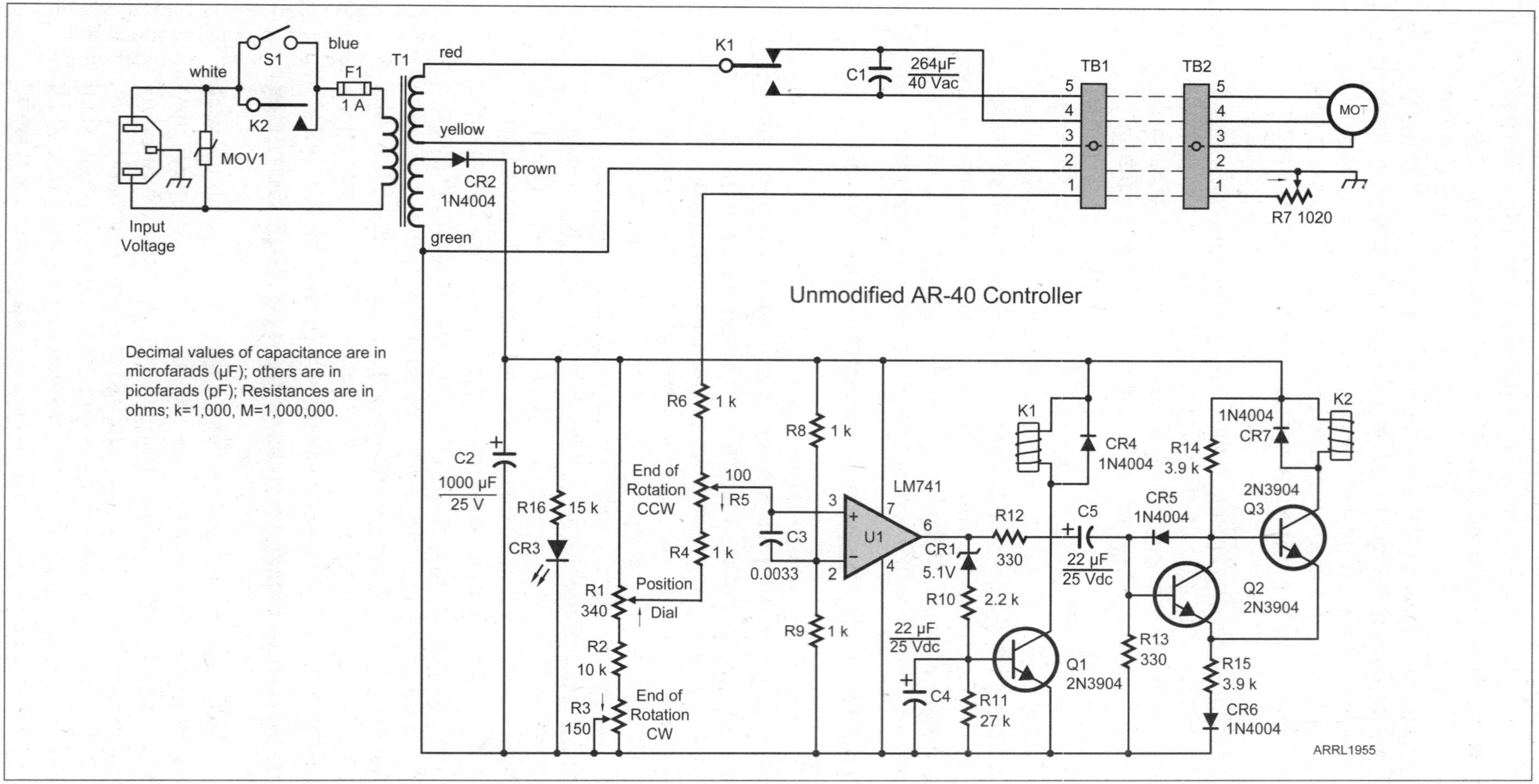

Figure 12.4 — The original (unmodified) AR-40 rotator controller schematic.

12-4   Chapter 12

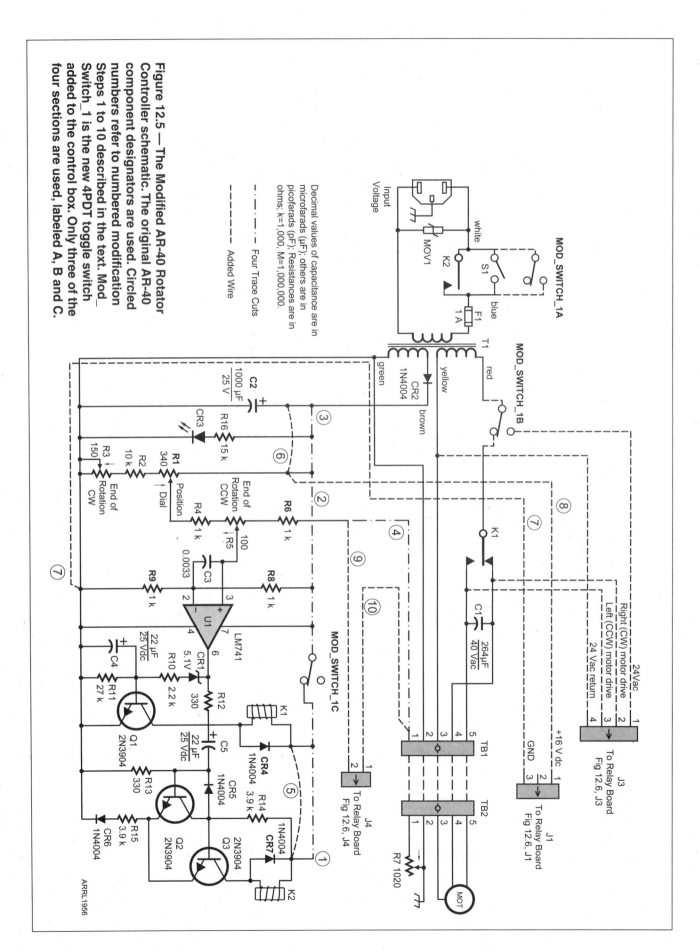

Figure 12.5 — The Modified AR-40 Rotator Controller schematic. The original AR-40 component designators are used. Circled numbers refer to numbered modification Steps 1 to 10 described in the text. Mod_Switch_1 is the new 4PDT toggle switch added to the control box. Only three of the four sections are used, labeled A, B and C.

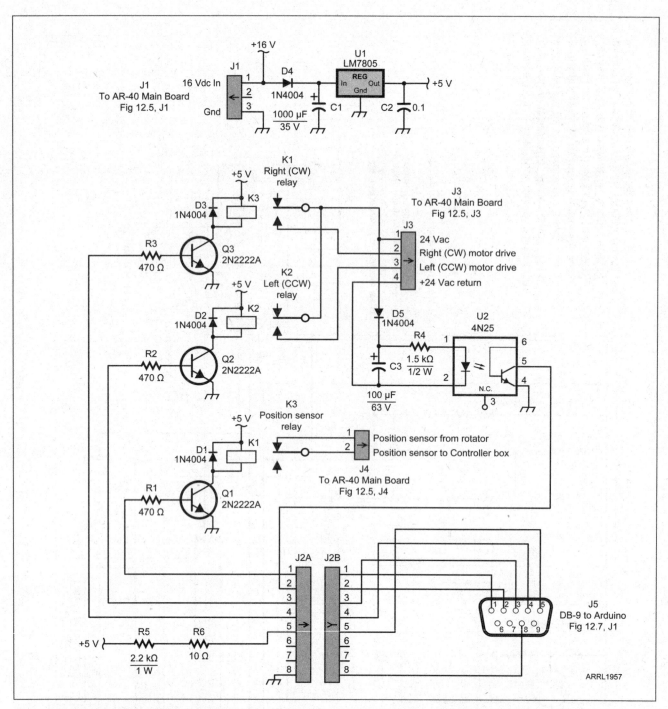

Figure 12.6 — Modified AR-40 Rotator Controller relay board schematic. Note that K1, the Position Sensor relay, could be replaced by using the unused section of the new 4PDT switch (Mod_Switch_1) in Figure 12.5
C1 — 1000 µF, 35 V electrolytic capacitor
C2 — 0.1 µF ceramic capacitor
C3 — 100 µF, 63 V electrolytic capacitor
D1-D5 — 1N4004 diode
J1 — 3-pin header connector
J2A, J2B — 8-pin header and socket connector assembly
J3 — 4-pin Anderson Powerpole connector block assembly
J4 — 2-pin header and socket connector assembly
J5 — DB-9 female connector

K1-K3 — 5 V SPDT relay
Q1-Q3 — 2N2222A NPN transistor
R1-R3 — 470 Ω, ⅛ W resistor
R4 — 1.5 kΩ, ½ W resistor
R5 — 2.2 kΩ, 1 W resistor
R6 — 10 Ω, 1 W resistor
U1 — LM7805 +5 V regulator
U2 — 4N25 optocoupler

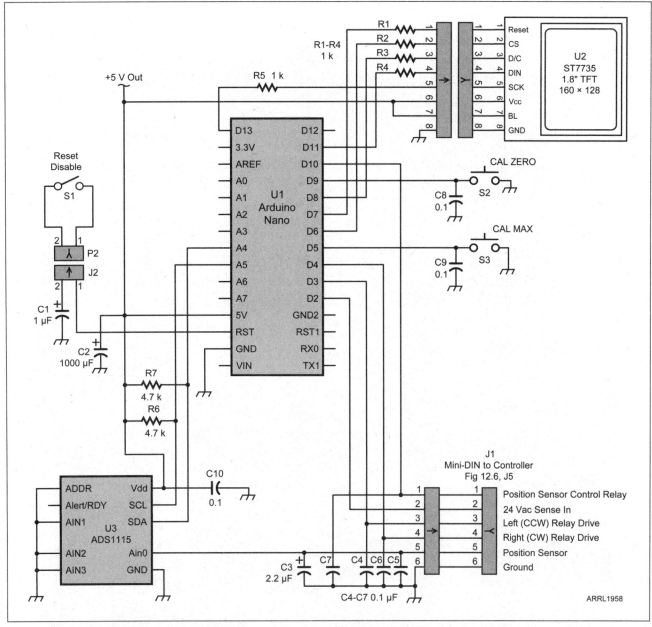

Figure 12.7 — The Modified AR-40 Rotator Controller Arduino board schematic.
C1 — 1 µF, 16 V electrolytic capacitor
C2 — 1000 µF, 16 V electrolytic capacitor
C3 — 2.2 µF, 35 V electrolytic capacitor (see text for part placement)
C4-C10 — 0.1 µF ceramic capacitor
J1 — 6-pin header connector
J2 — 2-pin header connector
P1 — 6-pin header connector
P2 — 2-pin socket connector
R1-R5 — 1 kΩ, 1/8 W resistor
R6, R7 — 4.7 kΩ, 1/8 W resistor
S1 — SPST toggle switch
S2, S3 — SPST momentary contact, normally open pushbutton switch
U1 — Arduino Nano
U2 — ST7735-compatible 1.8-inch color TFT display
U3 — ADS1115 analog-to-digital converter

Figure 12.8 — The AR-40 control board modifications. The arrows and labels correspond to step-by-step instructions described in the text. Note that there is a wire soldered to the board and labeled with Arrow 10 in the photo. This wire was used during project development and testing but is not used in the final version.

Step 4. Cut the trace between terminal T1 and resistor R6 as shown by Arrow 4 on the top right of the board.

These are all of the circuit board trace cuts you'll need to make. Next, we'll need to add two wires to the bottom of the board.

Step 5. Add a wire between diode CR7 and CR4 as shown by Arrow 5A at the top left of the board, and Arrow 5B at the mid-upper right of the board.

Step 6. Add a wire between Dial potentiometer R1C and capacitor C2 as shown by Arrow 6A at the top center of the board and Arrow 6B in the center of the board.

This completes wires we need to add to the AR-40 control unit main board. Next, we'll need to add the connections to the relay board.

Step 7. Solder a wire from relay board connector J1 pin 3 to resistor R9 as shown by Arrow 7 in the top center of the board. This is the ground wire to the relay board.

Step 8. Solder a wire from relay board connector J1 pin 1 to R1C as shown by Arrow 6B in the center of Figure 12.8. This is the +16 V dc from the AR-40 control board to the relay board.

Step 9. Solder a wire from relay board connector J4 pin 2 to resistor R6 as shown by Arrow 8 as shown in the top right of Figure 12.8. This is used to switch the position sensor signal from the AR-40 main board to the Arduino Control unit.

Step 10. Solder a wire from relay board connector J4 pin 1 to terminal lug 1 as shown by Arrow 9 in the top right of Figure 12.8. This is also used to switch the position sensor signal from the AR-40 main board to the Arduino control unit.

Note that there is a wire soldered to the board and labeled with Arrow 10 in the photo. This wire was used during project development and testing but is not used in the final version.

Next, we'll need to add the Manual/Arduino Mode switch to the AR-40 enclosure. This switch is used to disable parts of the AR-40 circuit board when we're running under Arduino control. Carefully drill a hole in the rear of the AR-40 enclosure. The location directly above the ac power cable as shown in **Figure 12.9** works best for the placement of this switch.

The Manual/Arduino Mode switch is a 4PDT switch labeled Mod_Switch_1 on the schematic. We're only using three poles of the switch, but it's a whole lot easier to find a 4-pole switch than a 3-pole one. Now that the switch is mounted, we can continue the modifications.

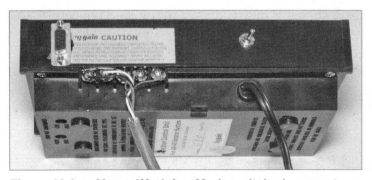

Figure 12.9 — Manual/Arduino Mode switch placement on the rear of the AR-40 control unit. The DB-9 connector (J5) for the cable from the relay board to the Arduino board is on the left.

Step 11. Solder a wire from the normally closed contact of Mod_Switch_1C to diode CR4 as shown by Arrow 5C in Figure 12.8. (This is the same trace that the wire in Step 5, Arrow 5B is soldered to.)

Step 12. Solder a wire from the wiper of Mod_Switch_1C to capacitor

Modified AR-40 Rotator Controller   12-9

C2 as shown by Arrow 6B in Figure 12.8.

Step 13. Remove the white wire coming from the power transformer that connects to spade lug #7 on the top side of the controller board. Build a small adapter cable consisting of a male spade lug to plug into the female spade lug we just removed, a female spade lug to plug into spade lug #7 on the control board, and a wire to connect to the wiper of Mod_Switch_1A. Essentially, we're adding Mod_Switch_1A in parallel to the **START** switch to keep the AR-40 ac power on when we're in Arduino Mode.

Step 14. Remove the blue wire coming from the power transformer that connects to spade lug #6 on the AR-40 board. Build a small adapter cable consisting of a male spade lug to plug into the female spade lug we just removed, a female spade lug to plug into spade lug #6 on the control board, and a wire to connect to the normally open contact of Mod_Switch_1A. Essentially, this is the other half of Mod_Switch_1A we're adding in parallel to the **START** switch to keep the AR-40 ac power on when we're in Arduino Mode.

Step 15. Remove the red wire coming from the power transformer that connects to spade lug #5 on the AR-40 board. Connect this wire to the wiper of Mod_Switch_1B.

Step 16. Connect a wire from the normally closed contact of Mod_Switch_1B to spade lug #5 on the AR-40 board.

Step 17. Solder a wire from the relay board rotator Powerpole connector J3 pin 2 to terminal #5 on the AR-40 board.

Step 18. Solder a wire from the relay board rotator Powerpole connector J3 pin 3 to terminal #4 on the AR-40 board.

Step 19. Solder a wire from the relay board rotator Powerpole connector J3 pin 4 to terminal #3 on the AR-40 board.

Step 20. Solder a wire from the relay board rotator Powerpole connector J3 pin 1 to the normally open connector on Mod_Switch_1B.

Mount the relay board to the inside of the AR-40 control unit as shown in **Figure 12.10** and **Figure 12.11**. While the schematic shows the 2.2 µF capacitor (C3) on the Arduino control board, I chose to move it closer to the positioning signal source by soldering it to terminal lug #1 for the signal and terminal lug #2 for the ground as shown in **Figure 12.12**.

The last step in the modification is to mount the connector for the Arduino control unit in the rear of the AR-40 enclosure as shown in Figure 12.9. I used a DB-9 connector in my prototype, but it may be easier to use an 8 pin Mini-DIN connector. Use whatever connector you prefer to get the signals between the AR-40 controller and the Arduino control unit. None of the pins are carrying any significant current, so any gauge of wire is fine. I used a piece of 8-conductor rotator cable to build this cable.

Finally, just to add some cooling because the AR-40 controller is powered on all the time while in Arduino mode, I drilled a series of extra holes in the bottom of the AR-40 enclosure to help keep things cool, just in case.

Figure 12.10 — Mounting the relay board in the AR-40 enclosure.

Figure 12.11 — Another view of the relay board mounted in the AR-40 enclosure.

Figure 12.12 — Installation of the 2.2 µF positioning sensor capacitor (C3 in Figure 12.7). While the schematic shows this capacitor on the Arduino control board, I chose to move it closer to the positioning signal source by soldering it to Terminal Lug #1 for the signal and Terminal Lug #2 for the ground.

The Arduino Board

The Arduino Nano, ADS1115 A/D, and the color TFT display are housed in a Solarbotics Mega SAFE enclosure connected to the AR-40 controller by a short piece of 8-conductor rotator cable as shown in Figure 12.2 at the beginning of this chapter. This project is an excellent example of the Arduino's power and versatility. We'll spend more time doing the modification to the AR-40 controller than wiring up the Arduino board itself. Because this project is designed to interact with a workstation running *Ham Radio Deluxe* software, we'll use the Nano's USB port connection to the workstation to provide the 5 V power for the Arduino enclosure piece of the project.

The Flowchart

The flowchart for the modified AR-40 controller (**Figure 12.13**) is very similar to the one for the AR-40D from the last chapter, except we don't need to handle all the extra pieces we added to the AR-40D. The modified AR-40 controller is designed to add just one new feature to the original control box — communicate and interact with *Ham Radio Deluxe* running on a workstation. That will allow us to re-use some of the functions used in the AR-40D project, but the sketch for this project will be much easier and simpler overall, because we only need to focus on the *HRD* control pieces.

We'll start out by including the Adafruit GFX, Adafruit ST7735, and SPI libraries for the color TFT display, the ADS1115 library for the 16-bit programmable gain A/D converter, and the EEPROM library to allow us to save and load calibration data. Next we'll define all of the I/O pins and constants for the sketch. As mentioned in the previous project, I prefer to use the preprocessor #define statement to assign values such as I/O pin numbers and constants.

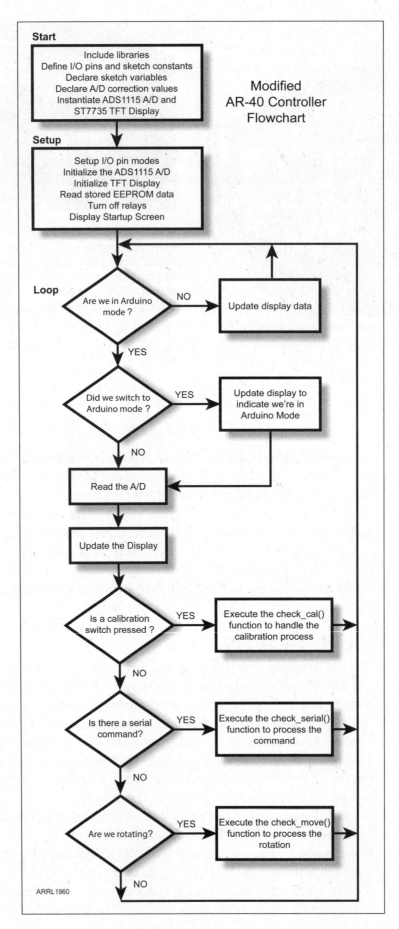

Figure 12.13 — The Modified AR-40 Rotator Controller flowchart.

That way, these values are compiled with the sketch and stored in flash memory, saving valuable SRAM for the variables we'll be using in the sketch.

Note that as in the AR-40D project, we'll need to use three arrays to hold the correction values to apply to the A/D data due to non-linearity and inductive noise on the rotator positioning signal. Finally we'll instantiate (create) the A/D converter and ST7735 color TFT display objects.

In the `setup()` function, we'll set the I/O pin modes and initialize the A/D converter and the TFT display. Next, we'll read any saved calibration data from the EEPROM, and if there is none, we'll use the default calibration settings. Then, we'll make sure the rotation relays are turned off and display the startup screen on the TFT display.

The first thing we'll do in the `loop()` function is to check if the switch on the back of the AR-40 enclosure is in Arduino mode or not. If not, we'll update the display to indicate the Arduino mode is offline, and loop until the switch is moved to the Arduino mode position.

If the switch is Arduino mode, we'll check to see if the Arduino mode was just started, and if so, update the display to show the Arduino mode is now online. Otherwise, we'll read the A/D converter, and update the display if any data has changed.

Next we'll call the `check_cal()` function to see if one of the calibration switches has been pressed. If so, we'll read the calibration data and save it to the Nano's EEPROM.

The next thing we'll do is call the `check_serial()` function that checks the USB/serial port for a command. If there is a valid command in the serial receive buffer, we'll decode the command and execute the desired operation.

Finally, we'll call the `check_move()` function to see if we're currently rotating. If so, we'll continue to process the rotation operation, and de-energize the relays if we've reached the commanded destination.

The Sketch

While similar to the sketch for the AR-40D and even using some of the same functions, the sketch for the Modified AR-40 Rotator Controller is much simpler because the only thing this sketch really needs to do is communicate with *Ham Radio Deluxe* running on a workstation attached to the Arduino's USB/serial port.

The sketch starts out by defining only two debug modes rather than the three modes in the AR-40D. The second debug mode provides detailed debugging information related to reading the A/D converter.

Next, we'll include the Adafruit_GFX and Adafruit_ST7735 libraries, as well as the built-in SPI library for the ST7735 TFT display. Then we'll include the EEPROM library to allow us to read and save the rotator position potentiometer calibration data. Finally, we'll include the ADS1115 library, which will automatically include the I2Cdev and Wire libraries. We will need to be sure that the I2Cdev library is installed, otherwise we'll receive error message from the Arduino IDE relating to this library when we try to compile the sketch.

```
/*
    Modified AR-40 Rotator Controller
    written by Glen Popiel - KW5GP

*/

// #define debug  // Enables Debug Mode - Sends Debug output to Serial Monitor
// #define debug1 // Sends adc debug data to Serial Monitor

#include <Adafruit_GFX.h>       // Core graphics library
#include <Adafruit_ST7735.h>    // Hardware-specific library
#include <SPI.h>
#include <EEPROM.h>             // Include EEPROM Library
#include <ADS1115.h>
```

Next, we'll set the USB/serial communication speed to 19200 baud. This speed was chosen to be compatible with the rotator controller speed settings in the *HRD Rotator* program. Then we'll define the I/O pins we'll need for this project.

```
const int comm_speed = 19200;    // Set the Serial Monitor Baud Rate

#define Sense_24VAC 2   // Assign 24VAC Sense input to Pin 2
#define left  3         // Assign Left (Counter Clockwise) Relay to Pin 3
#define right 4         // Assign Right (Clockwise) Relay to Pin 4
#define cal_zero 9      // Assign Zero Calibrate switch to Pin 9
#define cal_360 5       // Assign 360 Calibrate Switch to Pin 5
#define pos_sense A0    // Assign the position sensor input pin to A0
#define TFT_CS    6     // Assign the TFT CS to Pin 6
#define TFT_RST   7     // Assign the TFT RST to Pin 7
#define TFT_DC    8     // Assign the TFT DC to Pin 8
#define pos_sense_ctl 10  // Assign the Position Sense Control to pin 10
```

Now, we'll need to define the EEPROM data. The Arduino's EEPROM is a sequential read/write type of memory, meaning that you have to process the data one byte at a time until you have read or written the data you need.

To allow us to be sure that we're reading valid calibration data from the EEPROM, the first byte is used to indicate that the EEPROM contains valid data. This can be any number you want it to be. I chose 54 simply because that was how many times I coded and tested the EEPROM read/write function in my various rotator projects through the years. The number itself has no significance other than to indicate the calibration data is valid.

We'll also define the default calibration points we'll use if there is no valid calibration data, as well as setting the Azimuth Tolerance value. The Azimuth Tolerance is set to 1° by default, but because no antenna has a beamwidth that tight, you can set it to a higher tolerance as needed to allow for noise on the rotator positioning signal.

```
#define EEPROM_ID_BYTE 1    // EEPROM ID to validate EEPROM data location
#define EEPROM_ID 54    // EEPROM ID Value
#define EEPROM_AZ_CAL_0 2    // Azimuth Zero Calibration EEPROM location
#define EEPROM_AZ_CAL_MAX 4    // Azimuth Max Calibration Data EEPROM location

#define AZ_CAL_0_DEFAULT 30 // Define the default Cal 0 value
#define AZ_CAL_MAX_DEFAULT 25000 // Define the default Cal Max value

#define AZ_Tolerance 1    // Set the Azimuth Accuracy Tolerance
```

Next, we'll declare and initialize the variables needed by the sketch.

```
int long current_AZ;   // Variable for current Azimuth ADC Value
int AZ_Degrees;    // Variable for Current Azimuth in Degrees
boolean moving = false;   // Variable to let us know if the rotor is moving
int Arduino_mode = -1; // Variable to let us know the Controller Switch is in
                       // Arduino Mode
int Previous_mode = -2; // Variable to track mode switch
int long previous_AZ = -1;   // Variable to track the previous AZ reading
String Azimuth = "    ";
boolean calibrate = false;
boolean first_pass = true;   // Variable used to determine if first pass through
                             // the main loop
```

As with the AR-40D, the modified AR-40 rotator controller needs separate A/D converter correction values for when the rotator is idle, rotating clockwise, or rotating counterclockwise. The correction needed because of ac inductive noise on the rotator positioning signal caused by the rotator cable. Another reason may be that this issue has been in the AR-40 all along, and with the coarseness of the rotator dial potentiometer, we never knew it existed and we were never really concerned with the accuracy of the rotator. Another reason for the correction values is the non-linearity of the rotator position potentiometer caused by the way it is implemented in the rotator bell housing assembly. Now that we've gone digital, why not make things as good as we realistically can? We'll use three arrays as we did with the AR-40, as well as rotation direction dependent correction values for the calibration values.

```
// The Azimuth Correction arrays
const int  AZ_Correction[12] = { -550, -1500, -2100, -1900, -2350, -2400, -2800,
-2500, -2000 , -1850, -1000, 27};

const int  R_correct[12] = { -700, -1600, -2300, -2100, -2500, -2600, -3000,
-2700, -2200 , -2000, -1300, -50};

const int  L_correct[12] = { -550, -1350, -1900, -1700, -2200, -2200, -2600,
-2300, -1800 , -1730, -700, -900};

const int r_zero_adj = 0;
const int l_zero_adj = -210;
const int r_max_adj = 0;
const int l_max_adj = 0;

const int AZ_Max_Correction = 30;
```

Next, we'll declare and initialize the remaining variables we'll need for the sketch.

```
int set_AZ;   // Azimuth set value
int AZ_0;  // Azimuth Zero Value from EEPROM
int AZ_MAX; // Azimuth Max Value from EEPROMs
int zone_size; // Approximate number of A/D counts per zone
int zone = 0; // Current rotation zone

int long update_time = 0;
const int update_delay = 1000;
boolean turn_signal = false;
String previous_direction = "S";

String rotate_direction;

byte inByte = 0;   // incoming serial byte
byte serial_buffer[50];   // incoming serial byte buffer
int serial_buffer_index = 0;   // The index pointer variable for the Serial buffer
String Serial_Send_Data; // Data to send to Serial Port
String Requested_AZ; // RS232 Requested Azimuth - M and short W command
int AZ_To; // Requested AZ Move
int AZ_Distance; // Distance to move AZ
```

And finally, we'll instantiate (create) the ADS1115 A/D and the ST7735 TFT display objects.

```
ADS1115 adc;   // Define the A/D as adc

Adafruit_ST7735 tft = Adafruit_ST7735(TFT_CS,  TFT_DC, TFT_RST);  // Define the
                                                                 // ST7735 TFT as tft
```

In the `setup()` function we'll start by setting the I/O pin modes and making sure that all relays are de-energized.

```
void setup()
{
  // initialize the digital I/O pin modes
  pinMode(right, OUTPUT);
  pinMode(left, OUTPUT);
  pinMode(pos_sense_ctl, OUTPUT);
  pinMode(Sense_24VAC, INPUT_PULLUP);
  pinMode(cal_zero, INPUT_PULLUP);
  pinMode(cal_360, INPUT_PULLUP);

  digitalWrite(pos_sense_ctl, LOW); // Turn off the position sensor control relay
  digitalWrite(right, LOW); // Turn off the rotator control relays
  digitalWrite(left, LOW);
```

Next, we'll initialize the I²C bus for the ADS1115 programmable gain A/D converter and then the ADS1115 itself. Then we'll set the ADS1115 mode to free running, with an input voltage range of 0 to 2.048V and a sample rate of 128 samples per second.

```
  Wire.begin(); // Join the I2C Bus

  adc.initialize(); // Initialize the ADS1115 16 bit A/D module
  Wire.beginTransmission(0x48); // Begin direct communication with ADC
  Wire.write(0x1);   // Connect to the ADC and send two bytes - Set the config
                     // register to all 1's
  Wire.write(0x7F); // MSB
  Wire.write(0xFF); // LSB
  Wire.endTransmission(); // End the direct ADC comms

  adc.setMode(ADS1115_MODE_CONTINUOUS); // free running conversion
  adc.setGain(ADS1115_PGA_2P048);    // set adc gain to 2.048v range,
                                     // .0625 mv/step
  adc.setRate(ADS1115_RATE_128);     // set adc sample rate to 128 samples
                                     // per second
  adc.setMultiplexer(ADS1115_MUX_P0_NG);  // AN0+ Vs ground - Single mode input
                                          // on A0
```

Then we'll initialize the ST7735 color TFT display. Note that there are two initialization options based on the type of ST7735 module you're using. The tab color referred to in the sketch is the small tab on the peel-off screen protector for the display. The majority of the ST7735-type displays these days are the "Black Tab" version, even if the tab itself is a color other than black. If your display does not work properly, try the "Green Tab" option that's commented out in the sketch. Then we'll start the Nano's USB/serial port and then display a brief startup message on the TFT display.

```
// Use this initializer if you're using a 1.8" TFT
tft.initR(INITR_BLACKTAB);    // initialize a ST7735S chip, black tab

// Use this initializer (uncomment) if you're using a 1.44" TFT
//tft.initR(INITR_144GREENTAB);    // initialize a ST7735S chip, green tab

Serial.begin(comm_speed);   // Start the Serial port

tft.fillScreen(ST7735_BLUE); // Clear the display
tft.setRotation(1); // Set the screen rotation
tft.setTextWrap(false);
tft.setTextSize(3);
tft.setTextColor(ST7735_GREEN);
tft.setCursor(40, 10);
tft.print("KW5GP");
tft.setTextSize(2);
tft.setCursor(55, 60);
tft.print("AR-40");
tft.setCursor(45, 80);
tft.print("Rotator" );
tft.setCursor(25, 100);
tft.print("Controller");
```

Next, we'll read the Nano's EEPROM for any saved calibration data. If there is no valid calibration data, the default values we defined earlier are used. We then store the calibration values to the proper variables for use in the sketch.

```
read_eeprom_cal_data();   // Read the EEPROM calibration data

#ifdef debug    // Display the calibration data when in debug mode
  Serial.print("ROM Cal - Zero: ");
  Serial.print(AZ_0);
  Serial.print(" Max: ");
  Serial.print(AZ_MAX);
#endif

  zone_size = AZ_MAX / 12;
  AZ_MAX = AZ_MAX - AZ_Max_Correction;
```

```
#ifdef debug
  Serial.print(" Adj Max: ");
  Serial.print(AZ_MAX);
  Serial.print(" Zone Size: ");
  Serial.println(zone_size);
#endif

  delay(5000);   //Wait 5 seconds then clear the startup message
  tft.fillScreen(ST7735_BLUE); // Clear the display
```

As the final step in the `setup()` function, we'll preset the Azimuth destination variable and then check to see if the AR-40 mode switch is in the Arduino Mode position. If it is, we'll update the display to indicate that we're in Arduino Mode, and wait for commands from *Ham Radio Deluxe* running on the attached workstation.

```
  set_AZ = -1;   // Preset the Azimuth Move Variables

  Arduino_mode = !(digitalRead(Sense_24VAC));
  if (Arduino_mode == 1)
  {
    // We started in Arduino mode
    // Update the display - which also turns on the position sensor control
    // relay and clears the A/D
    Previous_mode = Arduino_mode;
    digitalWrite(pos_sense_ctl, HIGH);
    delay(1000);

#ifdef debug
    Serial.println();
    Serial.println("Start in Arduino Mode");
#endif

    update_display();
    first_pass = false;
  }
}   // End Setup Loop
```

As with the AR-40D, the main `loop()` is relatively short. All of the major work has been moved into functions that are called from the main `loop()`. We start out by checking if the AR-40 Manual/Arduino mode switch has changed position. If so, we'll update the TFT display to show whether Arduino mode is online or offline.

```
void loop()
{
  Arduino_mode = !(digitalRead(Sense_24VAC)); // Check to see if we're in manual
                                              // or Arduino mode

  if (Arduino_mode != Previous_mode)
  {

#ifdef debug
    Serial.print("24VAC Sense: ");
    Serial.println(Arduino_mode);
#endif

    if (Arduino_mode == 1)
    {
      // We just switched into Arduino mode - update the display
      update_display();
    }
  }
```

Next, if we're in Arduino Mode, we'll use the `update_display()` function to update the display with the current rotator position. Then we'll call the `check_serial()` function to get and decode any commands from *Ham Radio Deluxe*. We'll also call the `check_move()` function to process any rotation operations that may be in progress or need to start. If we're not in Arduino mode, we'll display the Arduino Mode offline message.

```
if (Arduino_mode == 1)
  {
    //    read_adc();
    update_display();
    check_cal();   // Check to see if either calibration switch is pressed
    check_serial();
    check_move();
  } else
  {
    offline_display();
  }
  Previous_mode = Arduino_mode;

#ifdef debug1
  delay(250);
```

```
#endif

}    // End Main Loop
```

How's that for a short and simple `loop()`? By offloading the majority of the workload into functions, we've broken the sketch into manageable blocks of code, and the sketch can cycle quickly through the `loop()` while waiting for a command from *Ham Radio Deluxe*. Now, let's get into the various functions that make this whole thing work.

The `check_cal()` function will check to see if one of the calibration pushbutton switches has been pressed. If so, we'll call the appropriate function to save the calibration value to the Nano's EEPROM.

```
void check_cal()   // Checks to see if a calibration switch is pressed and set
                   // calibration accordingly
{
  if (digitalRead(cal_zero) == 0) // Cal Zero button pressed
  {
    set_0_az_cal();
  }
  if (digitalRead(cal_360) == 0) // Cal Zero button pressed
  {
    set_max_az_cal();
  }
}
```

The `all_stop()` function is used to turn the rotation relays off, stop all rotation, and display a braking message for 3 seconds, to allow the rotator to stop turning before accepting new rotation commands.

```
void all_stop() // Relay Off function - stops motion, delays 3 seconds then turns
                // off Brake message
{
  tft.setTextSize(2);
  clear_top_line();
  tft.setCursor(40, 3); // Display message on LCD
  tft.print("Braking");
  digitalWrite(right, LOW);  // Turn off CW Relay
  digitalWrite(left, LOW);   // Turn off CCW Relay
  moving = false;
  rotate_direction = "S";   // Set direction to S (Stop)
  delay(3000);

  clear_top_line();
}
```

The `read_eeprom_cal_data()` function will read the Nano's EEPROM and load any saved calibration data. If there is no saved calibration data, this function will write the default calibration values to the EEPROM by calling the `write_eeprom_cal_data()` function.

```
void read_eeprom_cal_data()   // Read the EEPROM Calibration data
{
  if (EEPROM.read(EEPROM_ID_BYTE) == EEPROM_ID)    // Verify the EEPROM has
                                                   // valid data
  {

#ifdef debug // Display the Calibration data in debug mode
    Serial.print("Read EEPROM Cal Data Valid - AZ_CAL_0: ");
    Serial.print((EEPROM.read(EEPROM_AZ_CAL_0) * 256) + EEPROM.read(EEPROM_AZ_
CAL_0 + 1), DEC);
    Serial.print("   AZ_CAL_MAX: ");
    Serial.println((EEPROM.read(EEPROM_AZ_CAL_MAX) * 256) + EEPROM.read(EEPROM_
AZ_CAL_MAX + 1), DEC);
#endif

    AZ_0 = (EEPROM.read(EEPROM_AZ_CAL_0) * 256) + EEPROM.read(EEPROM_AZ_CAL_0 +
1); // Set the Zero degree Calibration Point
    AZ_MAX = (EEPROM.read(EEPROM_AZ_CAL_MAX) * 256) + EEPROM.read(EEPROM_AZ_CAL_
MAX + 1); // Set the 360 degree Calibration Point

  } else
  { // EEPROM has no Calibration data - initialize eeprom to default values

#ifdef debug
    // Send status message in debug mode
    Serial.println("Read EEPROM Cal Data Invalid - set to defaults");
#endif

    AZ_0 = AZ_CAL_0_DEFAULT;   // Set the Calibration data to default values
    AZ_MAX = AZ_CAL_MAX_DEFAULT;
    write_eeprom_cal_data();   // Write the data to the EEPROM
  }
}
```

The `write_eeprom_cal_data()` function is used to write the calibration data to the Nano's EEPROM.

```
void write_eeprom_cal_data() // Write the Calibration data to the EEPROM
{
#ifdef debug
  Serial.println("Writing EEPROM Cal Data");   // Display status in debug mode
#endif
```

```
    EEPROM.write(EEPROM_ID_BYTE, EEPROM_ID); // Write the EEPROM ID to the EEPROM
    EEPROM.write(EEPROM_AZ_CAL_0, highByte(AZ_0)); // Write Zero Calibration Data
                                                   // High Order Byte
    EEPROM.write(EEPROM_AZ_CAL_0 + 1, lowByte(AZ_0)); // Write Zero Calibration
                                                      // Data Low Order Byte
    EEPROM.write(EEPROM_AZ_CAL_MAX, highByte(AZ_MAX)); // Write 360 Calibration
                                                      // Data High Order Byte
    EEPROM.write(EEPROM_AZ_CAL_MAX + 1, lowByte(AZ_MAX)); // Write 360 Calibration
                                                         // Data Low Order Byte
}
```

The `check_serial()` function is one of the core functions that we call from the main `loop()`. This function is used to check the Nano's USB/serial port for commands from the workstation running *Ham Radio Deluxe*. Also note, that for testing purposes, you can manually test and send Yaesu GS-232A commands to the Arduino controller using the IDE's Serial Monitor. This function is an example of the reusability of Arduino functions. This function is a direct derivative of the function I wrote for the rotator controller projects in my first Arduino book, and has been carried over to every rotator controller project since then, with a few improvements here and there as time went on.

First, we'll check to see if there's a character in the serial buffer, and if so, start building the command one character at time until we receive a carriage return character. Once we receive the complete command, we'll start decoding the command we received.

```
void check_serial() // Function to check for data on the Serial port
{
  if (Serial.available() > 0) // Get the Serial Data if available
  {
    inByte = Serial.read();  // Get the Serial Data

    // You may need to uncomment the following line if your PC software
    // will not communicate properly with the controller

//    Serial.print(char(inByte));  // Echo back to the PC

    if (inByte == 10)  // ignore Line Feeds
    {
      return;
    }
    if (inByte != 13) // Add to buffer if not CR
    {
      serial_buffer[serial_buffer_index] = inByte;

#ifdef debug // Print the Character received if in Debug mode
      Serial.print("RX: ");
      Serial.print(" *** ");
      Serial.print(serial_buffer[serial_buffer_index]);
      Serial.println(" *** ");
#endif
```

```
        serial_buffer_index++;    // Increment the Serial Buffer pointer

   } else
   { // It's a Carriage Return, execute command

     if ((serial_buffer[0] > 96) && (serial_buffer[0] < 123))
     //If first character of command is lowercase, convert to uppercase
     {
       serial_buffer[0] = serial_buffer[0] - 32;
     }
```

Using a Switch...Case series of statements, we'll decode the command we received. The commands are based on the Yaesu GS-232A command set, as this is one of the more straightforward and easier rotator controller command sets to implement, and is supported by nearly every rotator control program.

We'll start by looking at the first character of the command, which indicates the operation to perform. First, we'll decode the "A" command that stops all rotation. When this command is decoded, we'll call the az_rotate_stop() function, which will de-energize the rotation relays and stop all rotation.

```
     switch (serial_buffer[0]) {  // Decode first character of command

       case 65:    // A Command - Stop the Azimuth Rotation

#ifdef debug
         Serial.println("A RX");
#endif

         az_rotate_stop();
         break;
```

Next, we'll decode the "C" command, which will call the send_current_az() function to send the current rotator azimuth position to the workstation running *Ham Radio Deluxe*.

```
       case 67:      // C - return current azimuth

#ifdef debug
         Serial.println("C RX ");
#endif

         send_current_az();   // Return Azimuth if C Command
         break;
```

The next command to decode is the "F" command. This command will call the `set_max_az_cal()` function to set the maximum azimuth rotation calibration value. This command can be used instead of manually pressing the **CAL MAX** pushbutton switch on the Arduino control board.

```
case 70:    // F - Set the Max Calibration

#ifdef debug
        Serial.println("F RX");
        Serial.println(serial_buffer_index);
#endif

        set_max_az_cal();   // F - Set the Max Azimuth Calibration
        break;
```

The "L" command is used to rotate counterclockwise. This function will call the `rotate_az_ccw()` function. Rotation will continue until manually stopped.

```
case 76:    // L - Rotate Azimuth CCW

#ifdef debug
        Serial.println("L RX");
#endif

        rotate_az_ccw();   // Call the Rotate Azimuth CCW Function
        break;
```

Next, we'll decode the "M" command, which is used to rotate to a specified destination. This function will call the `rotate_to()` function that is used to extract the destination from the command string and rotate to the specified bearing.

```
case 77:    // M - Rotate to Set Point

#ifdef debug
        Serial.println("M RX");
#endif

        rotate_to();   // Call the Rotate to Set Point Function
        break;
```

The "O" command is used to set the zero calibration point by calling the `set_0_az_cal()` function, similar to what occurs when you press the zero calibration point pushbutton switch.

```
        case 79:   // O - Set Zero Calibration

#ifdef debug
        Serial.println("O RX");
        Serial.println(serial_buffer_index);
#endif

        set_0_az_cal();   // O - Set the Azimuth Zero Calibration
        break;
```

The "R" command is used to start rotating clockwise. This command will call the `rotate_az_cw()` function, and clockwise rotation will continue until a stop command is received.

```
        case 82:   // R - Rotate Azimuth CW

#ifdef debug
        Serial.println("R RX");
#endif

        rotate_az_cw();   // Call the Rotate Azimuth CW Function
        break;
```

The "S" command does the same as the "A" command and calls the `az_rotate_stop()` function to stop all rotation. The reason for the two commands that do the same thing is to remain consistent with the Yaesu GS-232A rotator command set.

```
        case 83:   // S - Stop All Rotation

#ifdef debug
        Serial.println("S RX");
#endif

        az_rotate_stop();   // Call the Stop Azimuth Rotation Function
        break;

    }
```

If no matching command is decoded, the received command is ignored, the serial input buffer is cleared, and made ready to receive the next command sequence.

```
      serial_buffer_index = 0;   // Clear the Serial Buffer and Reset the Buffer
                                 // Index Pointer
      serial_buffer[0] = 0;
    }
  }
}
```

The `check_move()` function is another core function that is called from the main `loop()`. This function will manage an active rotation command. First, we'll call the `read_adc()` function to read the A/D converter and map the rotator position potentiometer value to degrees. Then, we'll check to see how far we need to continue rotating and we'll call the `fix_180()` function if we're rotating past the 0°/360° point. Because the AR-40 is a North-centered rotator, the 0°/360° transition occurs at the middle of the rotator turning radius. If the rotation is within the azimuth tolerance value, we'll stop the rotation process by calling the `az_rotate_stop()` function. Otherwise we'll call the `rotate_az_cw()` or `rotate_az_ccw()` function to start or continue rotation in the proper direction.

```
void check_move() // Check to see if we've been commanded to move
{
  if (set_AZ != -1)
  { // We're moving - check and stop as needed

    read_adc();

    // Map AZ to degrees

#ifdef debug
    Serial.print(" *** check_move *** AZ_To: ");
    Serial.print(AZ_To);
    Serial.print("  Zone: ");
    Serial.print(zone);
    Serial.print("  Zone Adj: ");
    if (rotate_direction == "R")
    {
      Serial.print(R_correct[zone]);
    } else
    {
      Serial.print(L_correct[zone]);
    }
    Serial.print("  Current_AZ: ");
    Serial.print(current_AZ);
    Serial.print("  AZ_Deg: ");
```

Modified AR-40 Rotator Controller

```
      Serial.print(AZ_Degrees);
#endif

    if (set_AZ != -1)   // If Azimuth is moving
    {
      AZ_Distance = set_AZ - AZ_Degrees;   // Check how far we have to move

#ifdef debug
      Serial.print("  AZ_Distance: ");
      Serial.print(AZ_Distance);
#endif

      fix_180();

#ifdef debug
      Serial.print("  Adj AZ_Distance: ");
      Serial.print(AZ_Distance);
      Serial.print("  Direction: ");
      Serial.println(rotate_direction);
#endif

      if (abs(AZ_Distance) <= AZ_Tolerance)  // No move needed if we're within
                                             // the tolerance range
      {
        az_rotate_stop();   // Stop the Azimuth Rotation
        set_AZ = -1;   // Turn off the Azimuth Move Command
      } else
      { // Move Azimuth - figure out which way
        if (AZ_Distance > 0)    //We need to move CW
        {
          rotate_az_cw();   // Rotate CW if positive
        } else {
          rotate_az_ccw();   // Rotate CCW if negative
        }
      }
    }
  }
}
```

The `send_current_az()` function is used to send the current azimuth position value to the workstation via the USB/serial port. First, the function calls the `read_adc()` function to read the current rotator position potentiometer value, apply all the various correction factors, and return the current position in degrees. This integer value is then converted into a 3 character value padded with leading zeroes to match the Yaesu GS-232A format, and then sent to the workstation running *Ham Radio Deluxe*.

```
void send_current_az() // Send the Current Azimuth Function
{
  read_adc();   // Read the ADC

#ifdef debug
  Serial.println();
  Serial.print("Deg: ");
  Serial.print(AZ_Degrees);
  Serial.print("  Return Value: ");
#endif

  // Send it back via serial
  Serial_Send_Data = "";
  if (AZ_Degrees < 100)  // pad with 0's if needed
  {
    Serial_Send_Data = "0";
  }
  if (AZ_Degrees < 10)
  {
    Serial_Send_Data = "00";
  }
  Serial_Send_Data = "+0" + Serial_Send_Data + String(AZ_Degrees);  // Send the
                                                                   // Azimuth in Degrees
  Serial.println(Serial_Send_Data);   // Return value via RS-232 port
}
```

The `set_max_az_cal()` function is used to set the maximum rotation calibration value and save the calibration value to the Nano's EEPROM. The function starts by calling the `read_adc()` function to read the current rotator position potentiometer, and returns the current raw A/D value. Next, the function will call the `write_eeprom_cal_data()` function to save the calibration data to the Nano's EEPROM.

```
void set_max_az_cal() // Set the Max Azimuth Calibration Function
{
#ifdef debug
  Serial.println("Cal Max AZ Function");
#endif

  calibrate = true;
  read_adc();   // Read the ADC

  // save current az value to EEPROM - Zero Calibration

#ifdef debug
  Serial.println(current_AZ);
#endif

  AZ_MAX = current_AZ;   // Set the Azimuth Maximum Calibration to Current
                         // Azimuth Reading
  write_eeprom_cal_data();   // Write the Calibration Data to EEPROM

#ifdef debug
  Serial.println("Max Azimuth Cal Complete");
#endif

  calibrate = false;
}
```

The `rotate_az_ccw()` function is used to rotate counterclockwise. This function will energize the left rotation relay and display the rotation direction arrow on the TFT display.

```
void rotate_az_ccw() // Function to Rotate Azimuth CCW
{
  digitalWrite(left, HIGH);   // Set the Rotate Left Pin High
  digitalWrite(right, LOW);   // Make sure the Rotate Right Pin is Low
  rotate_direction = "L";   // set the direction flag to "L" (Left - CCW)
  if (!turn_signal || (previous_direction != rotate_direction))
  {
    tft.setTextSize(2);
    clear_top_line();
    tft.setCursor(3, 10); // display the left arrow on the LCD
    tft.print("<==");
```

```
    turn_signal = true;
    previous_direction = rotate_direction;
  }
  moving = true;
}
```

The `rotate_az_cw()` function is used to perform the same operation as the `rotate_az_ccw()` function, except for the clockwise direction.

```
void rotate_az_cw() // Function to Rotate Azimuth CW
{
  digitalWrite(right, HIGH);   // Set the Rotate Right Pin High
  digitalWrite(left, LOW);     // Make sure the Rotate Left Pin Low
  rotate_direction = "R";   // set the direction flag to "R" (Right - CW)
  if (!turn_signal || (previous_direction != rotate_direction))
  {
    tft.setTextSize(2);
    clear_top_line();
    tft.setCursor(120, 10); // display the left arrow on the LCD
    tft.print("==>");
    turn_signal = true;
    previous_direction = rotate_direction;
  }
  moving = true;
}
```

The `az_rotate_stop()` function will de-energize the rotation control relays and display a braking message on the TFT display for three seconds.

```
void az_rotate_stop() // Function to Stop Azimuth Rotation
{
  digitalWrite(right, LOW);   // Turn off the Rotate Right Pin
  digitalWrite(left, LOW);    // Turn off the Rotate Left Pin

  tft.setTextSize(2);
  clear_top_line();
  tft.setCursor(40, 3); // Display Braking message on LCD
  tft.print("Braking");
  set_AZ = -1;
  turn_signal = false;
  delay(3000);

  clear_top_line();

  moving = false;
  rotate_direction = "S";   // Set direction to S (Stop)
  previous_direction = rotate_direction;
}
```

The `rotate_to()` function is used to rotate to a specified direction. This function will first extract the desired bearing from the received command. Then it will call the `read_adc()` function, get the current rotator position, calculate how far we need to rotate, and in which direction. If the destination has us rotate across the 0°/360° position, we'll call the `fix_180()` function to handle the crossover, and then call either the `rotate_az_cw()` or `rotate_az_ccw()` function depending on the direction we need to rotate.

```
void rotate_to() // Function to Rotate to Set Point
{
#ifdef debug
  Serial.println("M Command -  rotate_to Function");
  Serial.print("   Chars RX: ");
  Serial.print(serial_buffer_index);
#endif

  // Decode Command - Format Mxxx - xxx = Degrees to Move to
  if (serial_buffer_index == 4)   // Verify the Command is the proper length
  {
#ifdef debug
    Serial.print("   Value [1] to [3]: ");
#endif

    Requested_AZ = (String(char(serial_buffer[1])) + String(char(serial_buffer[2])) + String(char(serial_buffer[3]))) ;  // Decode the Azimuth Value
    AZ_To = (Requested_AZ.toInt()); // AZ Degrees to Move to as integer


#ifdef debug
    Serial.println(Requested_AZ);
    Serial.print("AZ_To: ");
    Serial.print(AZ_To);
#endif

    read_adc();   // Read the ADC

#ifdef debug
    Serial.print("   Zone: ");
    Serial.print(zone);
    Serial.print("   Current Deg: ");
    Serial.print(AZ_Degrees);
#endif

    AZ_Distance = AZ_To - AZ_Degrees;   // Figure out far we have to move

#ifdef debug
    Serial.print("   AZ_Dist: ");
    Serial.print(AZ_Distance);
```

```
#endif

    fix_180();

    set_AZ = AZ_To;
    moving = true;

#ifdef debug
    Serial.print(" Adj AZ_Dist: ");
    Serial.print(AZ_Distance);
#endif

    if (abs(AZ_Distance) <= AZ_Tolerance)  // No move needed if we're within
                                           // the Tolerance Range
    {
      az_rotate_stop();   // Stop the Azimuth Rotation
      set_AZ = -1;   // Turn off the Move Command
    } else
    { // Move Azimuth - figure out which way
      if (AZ_Distance > 0)    // We need to move CW
      {

#ifdef debug
        Serial.println(" Turn right ");
#endif

        rotate_az_cw();   // If the distance is positive, move CW
      } else
      {

#ifdef debug
        Serial.println(" Turn left ");
#endif
        rotate_az_ccw();  // Otherwise, move counterclockwise
      }
    }
  }
}
```

The `set_0_az_cal()` function is used to set the rotator zero calibration point. This function will call the `read_adc()` function to read the A/D converter, and save the current rotator position value to the zero calibration setting in the Nano's EEPROM.

```
void set_0_az_cal()   // Set Azimuth Zero Calibration
{

#ifdef debug
  Serial.println("Cal Zero Function");
#endif

  calibrate = true;
  read_adc();   // Read the ADC
  // save current Azimuth value to EEPROM - Zero Calibration

#ifdef debug
  Serial.println(current_AZ);
#endif

  AZ_0 = current_AZ;   // Set the Azimuth Zero Calibration to current position
  write_eeprom_cal_data();   // Write the Calibration Data to EEPROM

#ifdef debug
  Serial.println("Zero AZ Cal Complete");
#endif

  calibrate = false;
}
```

The `fix_az_string()` function is a simple function that will convert the integer azimuth degree value to a 3-character string that is padded with leading values to fit the Yaesu GS-232A protocol.

```
void fix_az_string()   // Convert azimuth to string and pad to 3 characters
{
  Azimuth = "00" + String(AZ_Degrees);
  Azimuth = Azimuth.substring(Azimuth.length() - 3);
}
```

The `update_display()` function is another of the core functions called by the main `loop()`. When first switching into the Arduino mode, this function will display an Arduino mode online message.

```
void update_display()
{

  if ((Arduino_mode != Previous_mode) || first_pass)
  {
    // Clear the screen and display normal operation
    tft.fillScreen(ST7735_BLUE); // Clear the display
    tft.setTextSize(1);
    tft.setTextColor(ST7735_GREEN);
    tft.setTextSize(1);
    tft.setCursor(25, 115);
    tft.print("Arduino Mode Online");
    digitalWrite(pos_sense_ctl, HIGH);

#ifdef debug
    Serial.println();
    Serial.println("Switch to Arduino mode - delay for ADC");
#endif

    previous_AZ = -1;  // Reset the previous AZ before starting
    Previous_mode = Arduino_mode;
    first_pass = false;
    delay(1000);

  }
```

Next, the function will call the `read_adc()` function to get the current rotator position, and update the display if the 1-second update timer has expired, and the rotator position has changed.

```
if (abs(millis()) > abs(update_time + update_delay))
  {
    read_adc();
    if (AZ_Degrees != previous_AZ)
    {
      tft.setTextSize(6);
      tft.fillRect(30, 30, 130, 55, ST7735_BLUE);
      tft.setTextColor(ST7735_GREEN);
      tft.setCursor(30, 40);
      tft.print(Azimuth);
      tft.setCursor(140, 30);
      tft.setTextSize(3);
      tft.print("o");
      previous_AZ = AZ_Degrees;
    }
    update_time = millis();
  }


#ifdef debug1
  Serial.print("    ");
  Serial.print(" Deg: ");
  Serial.println(AZ_Degrees); // Send position to Serial Monitor in debug mode
#endif



}
```

The `offline_display()` function is used to display an Arduino mode offline message on the TFT display when the Manual/Arduino mode switch on the AR-40 is switched to the manual operation mode.

```
void offline_display()
{
  if (Arduino_mode != Previous_mode)
  {
    tft.fillScreen(ST7735_BLUE); // Clear the display
    tft.setTextSize(3);
    tft.setTextColor(ST7735_RED);
    tft.setCursor(40, 25);
    tft.print("AR-40");
    tft.setTextSize(2);
    tft.setCursor(25, 55);
    tft.print("Controller");
```

```
    tft.setCursor(45, 75);
    tft.print("Not In" );
    tft.setCursor(10 , 95);
    tft.print("Arduino Mode");
    digitalWrite(pos_sense_ctl, LOW);
  }
}
```

The `read_adc()` function is used to read the ADS1115 16-bit programmable gain A/D converter. If this is the first time this function is called after the Manual/Arduino mode switch is placed in the Arduino Mode position, the function will energize the rotator position potentiometer signal relay to switch the rotator position signal from the AR-40 control board electronics to the input of the ADS1115 A/D converter on the Arduino board.

```
// Read the A/D Converter
void read_adc()
{
  // Read ADC and display position
  if (Arduino_mode != Previous_mode)   // First ADC read, we need to allow
                                       // settle time
  {
    digitalWrite(pos_sense_ctl, HIGH);
    delay(1000);
  }
```

Next, we'll set the A/D gain to a range of 0 to 2.048 V with a sample rate of 128 samples per second, and then read the rotator position potentiometer input on A/D input 0. If the `read_adc()` function was called as part of a calibration operation, the function exits and returns the raw A/D count value without corrections applied.

```
  adc.setGain(ADS1115_PGA_2P048);   // set adc gain to 2.048v range,
                                    // .0625 mv/step
  adc.setRate(ADS1115_RATE_128);    // set adc sample rate to 128 samples
                                    // per second

  // Read adc
  current_AZ = adc.getDiff0();   // Read ADC channel 0


#ifdef debug1
  Serial.print("   Read ADC: "); // display ADC read status in debug mode
  Serial.print(current_AZ);   // Display ADC value in debug mode
#endif

  if (calibrate)
  {
    return;
  }
```

Finally, we'll make sure the sure the calculated rotator bearing is not below zero or above 360°, and then apply the rotation zone corrections by calling the `zone_correct()` function. The function will then map the A/D value to the rotator position bearing, and covert the rotator bearing to a 3-digit character string padded with leading zeroes as per the Yaesu GS-232A format.

```
  if (current_AZ < 0)
  {
     current_AZ = 0;
  }

  if (current_AZ > AZ_MAX)
  {
     current_AZ = AZ_MAX;
  }

  zone_correct(); // Adjust A/D based on rotation zone

  map_az();

  fix_az_string();   // Convert azimuth to string then pad to 3 characters and
                     // save in Azimuth

#ifdef debug1
  Serial.print(" Moving: "); Serial.print(moving);
  Serial.print(" set_AZ: "); Serial.print(set_AZ);
#endif

}
```

The `clear_top_line()` function is used to clear the top line of the TFT display that is used to display the rotation direction arrow.

```
void clear_top_line()
{
  tft.fillRect(0, 0, 160, 30, ST7735_BLUE);
}
```

The `zone_correct()` function is used to adjust the A/D count value based on the rotation "zone." Due to the non-linearity of the AR-40 rotation position potentiometer, we need to compensate differently based on the position of the rotator position potentiometer in the rotator motor bell housing assembly. Also, due to the ac rotator motor power, there is significant inductive noise on the rotator position sensor that is different, based on the direction of rotation. The 360° rotation angle is broken out into 12 "zones" to provide the best solution to apply the necessary correction values. These correction values are saved in three sets of arrays. One set is used when the rotator is idle, and the other two are applied based on the direction of rotation. I know it sounds complex, and it was a royal pain to figure out, but the important thing is that it works, and when it comes down to working with older rotators that were never designed for computer-control, you don't always have the luxury of a complete redesign to correct the issue you're having to work around. So, you make do with what you have, and realize that in the end, it actually works out rather well.

First, the function will determine which rotation zone we're currently in. If we're in zone 0 or zone 11, we have to take into account that we're at either extreme on the rotator position and we have to manually plug in the values of the starting point for zone 0 and the ending point for zone 11 respectively.

```
void zone_correct() // Adjusts A/D value based on rotation zone
{
  int start_zone;
  int end_zone;
  int correction;
  int start_pos;   // Start of current zone
  int end_pos;   // End of current zone
  int current_pos;   // current position in zone
  int calculated_AZ;   // Used to adjust current AZ at zero point

  // Figure out what rotation zone we're in

  zone = current_AZ / zone_size;

  if (zone < 0)
  {
    zone = 0;
  }
  if (zone > 11)
  {
#ifdef debug1
    Serial.println(" Zone forced to 11");
#endif
    zone = 11;
  }

  start_pos = (zone * zone_size);
  end_pos = start_pos + zone_size;

#ifdef debug1
  Serial.print("  Zone: ");
  Serial.print(zone);
#endif

  if (zone == 0)
  {
    start_zone = 0;
    if (!moving)
```

```
    {
      start_pos = AZ_0;
      end_zone = AZ_Correction[0];
    } else
    {
      // we're moving - use moving correction factors
      if (rotate_direction == "R")
      {
        // We're turning right
        start_pos = AZ_0;
        end_zone = R_correct[0];
        current_AZ = current_AZ - r_zero_adj ;
      } else
      {
        // We're turning left
        start_pos = AZ_0;
        end_zone = L_correct[0];
        current_AZ = current_AZ - l_zero_adj ;
      }
    }
  }

  if (zone == 11)
  {
    if (moving)
    {
      // we're moving - use moving correction factors
      if (rotate_direction == "R")
      {
        start_zone = R_correct[zone - 1];
        end_zone = R_correct[zone];
        current_AZ = current_AZ  - r_max_adj;
      } else
      {
        // We're turning left - use left corrections
        end_pos = AZ_MAX;
        start_zone = L_correct[11];
        end_zone = l_max_adj;
        current_AZ = current_AZ  - l_max_adj;
      }
    } else
    {
      start_zone = AZ_Correction[zone - 1];
      end_zone = AZ_Correction[zone];
      calculated_AZ = current_AZ - AZ_0;
    }
  }
```

If we're currently in zone 1 thru 10, we can get the starting and ending A/D value for the zone from the correction array, and then apply the proper set of correction values.

```
  calculated_AZ = current_AZ;

if (zone > 0 && zone < 11)
  if (moving)
  {
    // Add the moving correction value when moving
    if (rotate_direction == "R")
    {
      start_zone = AZ_Correction[zone - 1];
      end_zone = R_correct[zone];
    } else
    {
      // We're turning left - use left corrections
      start_zone = AZ_Correction[zone - 1];
      end_zone =  L_correct[zone];
    }
  } else
  {
    start_zone = AZ_Correction[zone - 1];
    end_zone = AZ_Correction[zone];
    calculated_AZ = current_AZ - AZ_0;
  }

#ifdef debug1
  Serial.print("  Zone data: ");
  Serial.print(start_zone);
  Serial.print(" ");
  Serial.print(end_zone);
  Serial.print("  Zone range: ");
  Serial.print(start_pos);
  Serial.print(" ");
  Serial.print(end_pos);
#endif
```

Next, we'll need to figure out where in the zone we're at, in order to properly use the slope-intercept formula to apply the correct amount of correction. We'll then add the calculated correction value to the current rotator position, which should give us a fairly accurate rotator position value.

```
// Map the correction factor based on position in zone

correction = map(calculated_AZ, start_pos, end_pos, start_zone, end_zone);
// Map the correction
```

```
  current_AZ = current_AZ + correction;

  if (current_AZ < 0)
  {
    current_AZ = 0;
  }

#ifdef debug1
  Serial.print(" Adj: ");
  Serial.print(correction);
  Serial.print(" Final AZ: ");
  Serial.print(current_AZ);

#endif


}
```

The `map_az()` function will map the current A/D value to a range of zero to 360°, and correct for the North-centering of the rotator position.

```
void map_az()
{
  AZ_Degrees = map(current_AZ, AZ_0, AZ_MAX, 0, 360); // Map the current_AZ to
                                                     // calibrated degrees

  if (AZ_Degrees < 0) // Limit Zero point to 0
  {
    AZ_Degrees = 0;
  }

  if (AZ_Degrees > 360) // Limit max point to 360
  {
    AZ_Degrees = 360;
  }

  if (AZ_Degrees < 180)  // Adjust for North-centered controller
  {
    AZ_Degrees = 180 + AZ_Degrees;
  } else
  {
    AZ_Degrees = AZ_Degrees - 180;
  }

}
```

The fix_180() function is used to adjust for the North-centering position reporting of the rotator position potentiometer.

```
void fix_180()
{
  if (AZ_Degrees >= 180 && AZ_Degrees <= 360  && zone != 11) // Make sure we
                  // don't rotate left beyond 180 and we're not in zone 11
  {
    if ((AZ_Degrees + AZ_Distance < 180) && (AZ_Distance < 0))
    {
      AZ_Distance = 360 + AZ_Distance;
    }
  } else {
    // We're in zone 4 thru 12 - Make sure we don't rotate right beyond 180

#ifdef debug
    Serial.print(" In Zone 4 thru 12");
#endif

    if ((AZ_Degrees + AZ_Distance > 180) && (AZ_Distance > 0))
    {
      AZ_Distance = -360 + AZ_Distance;
    }
  }
}
```

Connecting to *Ham Radio Deluxe*

To connect to a workstation running *Ham Radio Deluxe* as shown in **Figure 12.14**, connect a USB cable from the workstation to the Nano's USB port. Set the Manual/Arduino mode switch on the AR-40 control box to

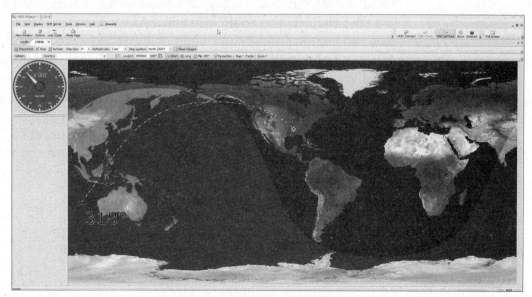

Figure 12.14 — An *HRD Rotator* screen showing the AR-40 pointed to 318 degrees using the Arduino-powered modified rotator control box.

Modified AR-40 Rotator Controller   12-43

Figure 12.15 — The TFT display on the Arduino control unit while the AR-40 bell housing is rotating. This display shows a heading of 11 degrees and the arrow at the upper right indicates clockwise rotation.

Arduino mode. When in the Arduino Mode, the TFT display will show that the Arduino Mode is online, along with the current rotator position as shown in **Figure 12.15**.

On the *HRD Rotator* program, set the rotator type to Yaesu GS-232A Az, set the Port to the COM port the Arduino is connected to, and the speed to 19200. Because the AR-40 is a North-centered rotator, you'll want to set the Stop position setting to South (180°). Then, click the **CONNECT** button on the screen, and *Ham Radio Deluxe* should connect to the Arduino controller.

If you're having problems getting *HRD Rotator* to connect, verify the settings above, and be sure that the debug modes in the sketch are commented out, as the debug data will confuse *Ham Radio Deluxe*. You should now be able to control the rotator from *HRD*. When connected, you'll see an option on the screen to change the step size. The default of 1° should be fine, but you can change this to a higher step size if the rotator doesn't position and stop accurately due to the position sensor nonlinearity, rotator position sensor noise, and the length of the rotator cable.

Enhancement Ideas

Because the goal of the modified AR-40 project was to just get the AR-40 to communicate and interact with *Ham Radio Deluxe*, there's plenty of room for improvement. For example, you can have the TFT display emulate the LED ring used in the Rotator Turn Indicator project while the rotator is turning. Or, you could add the direct-entry keypad control as we do with the CDE/HyGain rotator in a later chapter. You could also add a rotary encoder to the Arduino control box, and use the pushbutton switch on the rotary encoder to simulate the AR-40's dial potentiometer and Start switch when in Arduino mode. Finally, you could add the left/right paddle switch as we did with the AR-40D in the previous chapter, to give you even more flexibility with rotator control while in the Arduino mode.

Sketches, libraries, Fritzing diagrams, and other useful files for the projects in this book can be found at **www.sunriseinnovators.com** and **www.kw5gp.com**.

13 USB CW Keyboard

One of the most popular projects in my first Arduino book was a CW Keyboard that used a PS/2-type computer keyboard. That was one of my very first Arduino projects ever, and I still use it in my own station whenever I work CW. I keep saying I'll go back to using a straight key or a set of paddles, but it hasn't happened yet. It turns out I'm much better at typing than I am at keying. However, as computer technology advances, the 1980s-era PS/2-style keyboard has been replaced by USB or wireless keyboards.

For this book, I wanted to include some USB and Bluetooth projects, so I researched USB and Bluetooth technologies on the Arduino, hoping to come up with some next-generation Arduino projects. As it turns out, the Arduino doesn't directly support the Human Interface Device (HID) role very well at all. Yes, virtually all Arduinos can connect to a workstation via USB, but that's as a device, not as a USB host. Because you can't connect a USB device directly to another USB device, you have to have a USB host to communicate with the device. This means I couldn't just take a USB keyboard and connect it to an Arduino. So much for the easy way.

As for the Bluetooth projects, it's pretty much the same story. The standard Bluetooth modules for the Arduino don't support the HID role. I wasn't able to do a Bluetooth project in time for this book, but rest assured, the lab workbench is piled high with Bluetooth stuff for future experiments.

To solve the USB issue, early in the Arduino's life several vendors created a USB host shield to allow the Arduino to be a USB host. This book is not about

Figure 13.1 — The USB CW Keyboard.

Figure 13.2 — The mini USB host module.

plugging in shields and writing a few lines of code to create a ham radio-related Arduino project, and those shields were relatively expensive at the time, so I kept looking for other solutions.

Recently, I came across the mini USB host shield/module with an SPI interface. Okay, technically it's still a shield in a general sense of the word, but now I could fit it in an enclosure with an Arduino Nano, and give it a much better and larger 2.2-inch color TFT display than the 16-character by 2-line LCD used in the original PS/2 Keyboard project (see **Figure 13.1**).

As you can see from the photo in **Figure 13.2**, it is not a shield in the traditional sense, but more like a module. As it turns out, this module was designed as a shield for the Arduino Pro-Mini. Because I prefer to use Arduino Nanos, for this project we'll just treat it like any other module and go from there.

A Brief Look at USB

USB is short for Universal Serial Bus. Created in 1996, USB was designed to be a standard for cables, connectors, and protocols to interface between computers and peripherals. As part of the standard, USB also provides power to the attached devices. USB was designed as a far simpler method of connecting devices to a computer than the large multiconductor parallel and RS-232 serial cables used back then. A major advantage of USB at the time was the higher data transfer speeds of 12 Mb/s with USB 1.0. USB 2.0 gave us speeds up to 480 Mb/s, and now with USB 3.0, the speeds are up to 5 Gb/s and are headed even higher with newer versions.

One nice thing about USB is that the cables are the same for USB 1.0, and 2.0. USB 3.0 introduced a new device-end connector, but for all intents, a USB host port is compatible with older USB devices running a lesser version of the USB protocol. USB cables are also hot-swappable, meaning they can be plugged and unplugged without the need to turn off either the host or device.

USB is also highly versatile. The USB interface on a computer is self-configuring, eliminating the need for manually configuring device settings. Most computers will automatically install any needed drivers for a new USB device without any action needed on the part of the user. However, herein lies the issue with the Arduino when acting as a USB host. There are literally thousands of USB devices that can be connected to a USB host port, and they all have their own specific drivers. These devices can include keyboards, mice, headphones, microphones, webcams, external hard drives, thumb drives, and even many modern transceivers. The Arduino would need a driver for each of these for it to be able to properly communicate with every USB device.

The USB protocol breaks the functionality of various devices into classes. Each USB device will send its class code to the USB host when it is first connected, so that the host can determine what type of device it is. The host will then query the new device and gather more information in order to load the correct device drivers and complete the auto-configuration process. Because the basic Arduino project does not have enough memory or storage for every possible USB device configuration, nor does it have the ability to automatically download and install any missing drivers, the Arduino can only support a subset of the USB devices available.

Fortunately for us, peripherals such as keyboards and mice are well supported in the Arduino USB Host library. I was even able to use a wireless keyboard with a USB dongle with the Arduino USB Host library. Even better, this library is fully compatible with the mini USB host module that we'll be using in this project. For those of you who wish to know more about the USB protocol, I found an excellent explanation at **www.beyondlogic.org/usbnutshell/usb1.shtml**.

Block Diagram

So now that we have the biggest piece of the Arduino and USB keyboard connection puzzle solved, we can go on with the process of designing the USB CW Keyboard project. **Figure 13.3** shows the block diagram.

The goal for this project is to duplicate the functionality of the original PS/2 CW Keyboard, but also give it a bit up an upgrade. We'll use an Arduino Nano and the mini USB host shield/module we talked about earlier. In addition, we'll be upgrading from the original 16-character by 2-line LCD display to a 2.2-inch ILI9341-type 240×320 pixel color graphic TFT display (**Figure 13.4**). Due to some differences between the original PS/2 Keyboard library and the USB Host library which we'll discuss in detail later, we won't be able to use the original Morse library that was used in the PS/2 version.

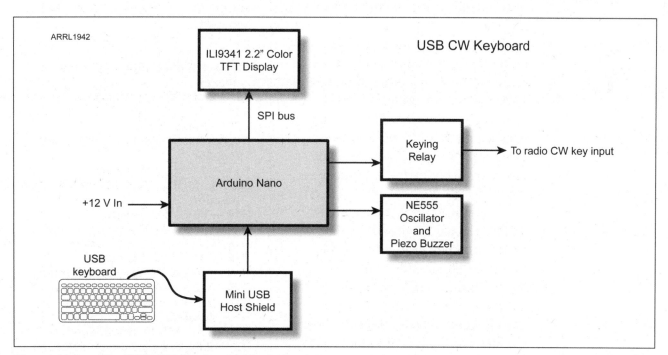

Figure 13.3 — The USB CW Keyboard block diagram.

Figure 13.4 — The USB CW Keyboard 2.2-inch TFT display.

The new interrupt-driven Morse code library does not have the ability to send a CW tone, so we'll need to add an external NE555 timer-based oscillator driving a piezo buzzer to provide the CW tone. As with the PS/2 version, this version will also drive a reed relay used to key the transceiver. You can switch between keying the CW tone circuit or the transceiver relay.

Building the USB CW Keyboard

Figure 13.5 shows the schematic for the USB CW Keyboard project. The project is built on a piece of copper-clad perfboard cut to fit inside a Solarbotics Mega SAFE enclosure. This will allow us to house the electronics inside and mount the 2.2-inch TFT display on the inside of the top cover using 2.2 mm hardware as shown in **Figure 13.6**. I've found that 2.2 mm screws, washers, and nuts are nearly a perfect fit for the mounting holes in Arduino boards and most of the modules. I prefer the hex-head version of the screws, and I recommend getting them in various lengths.

The USB host modules are available from a number of online sources. They all are based on the same chipset, regardless of module size, and the USB Host library works with all of them. However, there are a few things to be aware of when wiring up the USB host module.

The mini USB host module is a 3.3 V device and it will be damaged if you power it with 5 V, but most USB keyboards don't run at 3.3 V. So, in order to power the attached devices with 5 V, we'll need to cut a trace on the USB host module and apply 5 V to the VBUS pin as shown in **Figure 13.7**. Cutting the trace will isolate the USB power pin from the rest of the USB

Figure 13.5 — The USB CW Keyboard schematic.
C1 — 0.01 µF ceramic capacitor
D1, D2 — 1N4004 diode
J1, J3 — 2-pin DuPont male header socket
J2 — 2-pin DuPont female header socket
K1 — SPST DIP reed relay with 5 V coil
LS1 — Piezo buzzer
R1, R3, R5, R7, R9, R12 — 1 kΩ, ⅛ W resistor
R11 — 1.2 kΩ, ⅛ W resistor
R2, R4, R6, R8, R10, R13 — 1.5 kΩ, ⅛ W resistor
R14 — 56 Ω, ½ W resistor
R15 — 100 Ω, ⅛ W resistor
R16, R17 — 10 kΩ, ⅛ W resistor
R18 — 47 kΩ, ⅛ W resistor
R19 — 100 kΩ 10-turn potentiometer
R20 — 10 kΩ 10-turn potentiometer
S1 — SPST toggle switch
U1 — Arduino Nano
U2 — Mini USB host shield/module
U3 — ILI9341 type 2.2-inch color TFT display
U4 — NE555 timer IC
Enclosure — Solarbotics Mega SAFE

Figure 13.6 — The USB CW Keyboard in the enclosure.

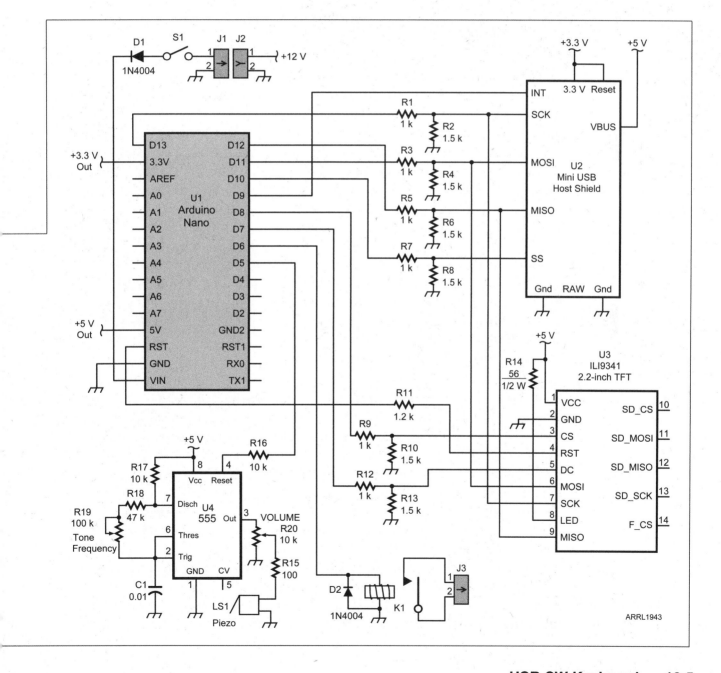

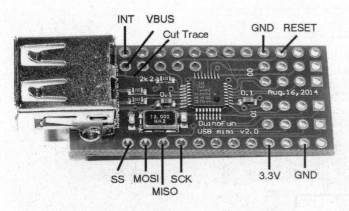

Figure 13.7 — The Mini USB host module pinout and trace cut location.

module and only supply 5 V to the attached USB device. We'll use voltage dividers between the Nano and the USB host module for the SPI bus signals to reduce those voltages to a safe level as well (for example, R1/R2). You could also replace the mini USB host module with a regular Arduino Uno shield and use an Uno in place of the Nano. The full-size USB host shield for the Uno is completely 5 V powered and some even have a 3.3 V/5 V option.

Some of the inexpensive mini USB host modules you get from places such as eBay have an error on the silkscreen labeling on the board. These boards have the MOSI and SCK pins mislabeled. The correct labeling is shown in Figure 13.7.

The ILI9341-type 2.2 inch color TFT display has somewhat similar power issues, except you can supply 5 V to the display for power without damaging the device. However, the signal lines must all be connected to the Nano using voltage dividers. I have found the inexpensive Chinese versions of the TFT display are a bit fussy on the voltage divider resistor values, which is why I use the 1 kΩ/1.5 kΩ voltage divider resistors in this project (for example, R12/R13). One special note about the ILI9341-type TFT displays is that you do need to have the reset (RST) pin on the TFT connected to the Arduino's reset pin through a 1.2 kΩ resistor for the TFT to reset and start properly.

The NE555 timer circuit for the CW tone is a simple astable oscillator that is triggered from the Arduino Nano. You can control both the tone frequency (R19) and the volume (R20) of the piezo buzzer on the output of the NE555 circuit. For the NE555 oscillator circuit to trigger properly from the Arduino I/O pin, we'll need to add a 10 kΩ pull down resistor (R16) on the I/O pin.

The Flowchart

Figure 13.8 shows the flowchart for the USB CW Keyboard. What makes this sketch work is the power in all of the libraries we'll be using. For the ILI9341 color TFT display, we'll need to include the Adafruit GFX, Adafruit ILI9341, and SPI libraries. The USB Host library has several sub-libraries, but for this project, we'll include the hidboot library. This library is for Human Interface Devices (HID) in boot protocol mode, a simple USB HID-device communication method that is supported by all USB keyboards. We'll also include the Lewis Morse Code library.

Because of the way that the USB Host library handles communication with the keyboard, we'll need to use an interrupt-driven library to send the Morse code, as the USB keyboard implementation doesn't work well with anything that spends a lot of time in delays such as the CW timing between elements. Without this interrupt-driven Morse code library, I'm not sure this project would have been viable at all. And finally, we'll need to include the TimerOne library to set timer interrupts for the Morse library to check for outbound CW data.

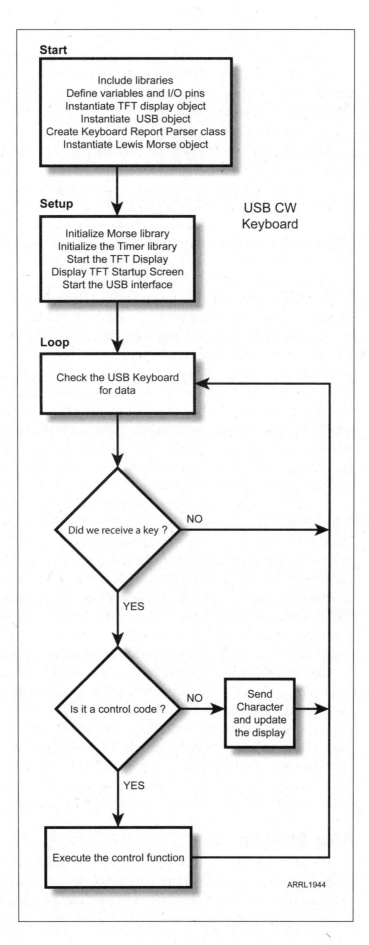

Figure 13.8 — The USB CW Keyboard flowchart.

Next, we'll define the variables and I/O pins we'll need in the sketch. It is important to note that the SS and INT pins on the USB host module must be connected to digital I/O pins 10 and 9 respectively. These I/O pins are hard-coded in the USB Host library, and while theoretically they can be changed by editing the pin definitions in the library, why go through all that trouble? Just use the pins they are configured for.

We'll need to instantiate (create) the USB host and TFT display objects, as well as the Lewis library Morse object. One special initialization task for this project is that we'll need to define a Keyboard Report Parser class and object for the USB keyboard data. Unlike a PS/2 keyboard that will send one character at a time, the USB keyboard will follow the standard USB protocol for transferring data and send the data over in blocks known as "reports." The USB host will need to decode these reports that are based on the USB device type, and extract the keyboard data that way.

Because of the way the keyboard data is transferred, and the way it is handled by the USB Host library, the typeahead buffer for the USB Keyboard is only 10 characters long. A fast typist could easily overrun the keyboard buffer, so you will need to be careful and not type too fast. This will be especially noticeable at slower CW speeds. This could easily be corrected with the implementation of a keyboard buffer routine in the sketch that could add new characters to the tail end of a larger buffer to keep from overrunning the CW sending side of things. But remember, one of the goals of the projects in this book is to provide you with a working and functional platform that you can improve on and add your own personal touches, so I'll leave the larger typeahead buffer to you as an enhancement challenge.

In the `setup()` function, we'll initialize the Lewis Morse Code, and the TimerOne libraries. We'll also start the TFT display and the USB host interface. Then we'll display a brief startup message on the TFT display.

The `loop()` function for this project is relatively straightforward. Unlike the previous two projects where we had the sketch broken out into multiple functions with a smaller `loop()` function, in this project we'll have the majority of the sketch in the main `loop()`. The reason for this is that this sketch performs basically a single operation, wait for a keystroke on the keyboard, and either send it or perform the selected command operation.

The first thing we'll do is check the keyboard for data. Once we receive a keystroke, we'll check to see if it is either a valid Morse code character or a command code. The command codes in this sketch will allow you to adjust the keying speed from 5 to 30 words per minute (WPM) and enable either the CW tone or transmitter keying modes. If the received keystroke is a Morse code character, we'll send it to either the NE555 tone circuit or the keying relay. If it's a command code, we'll execute the specified command. We'll then go back to the top of the loop and wait for the next keystroke.

The Sketch

As with the AR-40 sketch, the USB CW Keyboard has multiple levels of debugging information. Uncommenting the debug `#define` statements as needed when troubleshooting the sketch will provide a wealth of information on the

Arduino IDE's Serial Monitor as the sketch runs. Next, we'll include all of the libraries needed for the sketch. The USB Host library has a number of individual USB host-related sub-libraries to support a number of USB host functions. For this sketch, we'll be using the hidboot library to support the USB keyboard in boot protocol mode. For the TFT display, we'll include the Adafruit GFX and Adafruit ILI9341 libraries, and the SPI library. Finally, we'll include the Lewis Morse Code library, along with the TimerOne library to support time-based interrupt operations.

```
/*
  USB CW Keyboard

  By Glen Popiel - KW5GP

  Uses USB Mini Host Shield module, 2.2" ILI9341 TFT Display, and standard USB
Keyboard

  Keyboard can be wireless with USB dongle - detected automatically


  Uses USB_Host_Shield_Library_2.0
  Copyright (C) 2011 Circuits At Home, LTD. All rights reserved.

  This software may be distributed and modified under the terms of the GNU
  General Public License version 2 (GPL2) as published by the Free Software
  Foundation and appearing in the file GPL2.TXT included in the packaging of
  this file. Please note that GPL2 Section 2[b] requires that all works based
  on this software must also be made publicly available under the terms of
  the GPL2 ("Copyleft").

  Contact information
  -------------------

  Circuits At Home, LTD
  Web      :  http://www.circuitsathome.com
  e-mail   :  support@circuitsathome.com

  Uses Lewis Interrupt-driven Morse Code Library

  Uses TimeOne Library to generate timer interrupts for Lewis Library

*/

//#define debug  // Uncomment this to enable debugging information
//#define usb_debug  // Uncomment this line to enable USB Host Library debug
                    // information - requires #define debug also uncommented
//#define debug1  // Uncomment this line for extra debug information
```

```
#include <hidboot.h>  // USB Host shield library for Human Interface Devices in
                      // boot protocol mode
#include <SPI.h>  // Library to drive SPI devices
#include <Adafruit_GFX.h> // The Adafruit Graphics Library
#include <Adafruit_ILI9341.h> // The ILI9341 TFT Display Library
#include <Lewis.h>  // The Lewis Interrupt driven Morse Code Library
#include <TimerOne.h> // Library to set timer interrupts
```

Next, we'll define the digital I/O pins and the various messages and parameters needed in the sketch.

```
#define TFT_DC 7  // Define the TFT DC pin
#define TFT_CS 8  // Define the TFT CS pin
#define beep_pin 5  // Define the pin for the CW tone
#define key_pin 6 // Define the relay keying pin

#define cw_start_speed 10 // Set the keyer starting speed to 10 wpm

#define osc_fail "USB OSC Fail" // Define the USB oscillator failure error text
#define usb_ok "USB OK" // Define the USB oscillator ok status text
#define repeat_delay 100  // Set repeating key delay to 100ms

#define cw_font_size 4  // The CW text display font size
#define status_font_size 2  // The font size for the status line

String usb_status_msg;  // Variable for the current USB status message
                       // to display
String cw_data = "";  // String variable to hold cw output text data
String key_ascii = "";  // Single character string variable to hold converted
                       // key value
bool system_status = true;  // Flag indicating current system status -
                           // true = ok, false = bad
bool repeat_enabled = false;  // Flag indicating if we've met the requirements
                             // to repeat keys
int first_pass_delay = 10000; // Number of milliseconds to wait for USB Host to
                             // identify and begin communication with HID Keyboard
int repeat_timeout = 500; // Number of milliseconds a key must remain pressed
                         // to initiate repeat
int key_code; // Variable for the USB non-ASCII key code
int ascii_key_code; // Variable for the ASCII key code
int prev_key_code;  // Variable for the previous USB non-ASCII key code
int prev_ascii_key_code;  // Variable for the previous USB ASCII key code
unsigned long repeat_timer; // Timeout timer variable
int key_speed = cw_start_speed; // Define the initial keyer speed and set it
                               // to the defined start_speed
int last_speed = 0; // Variable for the previous keyer speed
bool key_mode = false;  // Used to select beep or key 0=beep 1=key
bool last_mode = true; // Set the previous mode opposite to the starting key
                      // mode so it will update the tft
```

Then we'll create the ILI9341 TFT display instance and then do something we haven't done before, create a new C++ class that will be used to parse the keyboard report data.

```
// Use hardware SPI (on Uno, #13, #12, #11) and the above for CS/DC
Adafruit_ILI9341 tft = Adafruit_ILI9341(TFT_CS, TFT_DC);  // Create the
                                                          // TFT instance

class KbdRptParser : public KeyboardReportParser     // Define a class to parse
                                                     // the incoming keyboard data
{
    void PrintKey(uint8_t mod, uint8_t key);

  protected:
    void OnControlKeysChanged(uint8_t before, uint8_t after);

    void OnKeyDown   (uint8_t mod, uint8_t key);
    void OnKeyUp     (uint8_t mod, uint8_t key);
    void OnKeyPressed(uint8_t key);
};
```

Finally, we'll create a USB instance and start the USB Keyboard in HID boot protocol mode. Along with that, we'll create a Keyboard Report Parser instance and a Lewis Morse Code instance.

```
USB     Usb;  // Create a USB instance

HIDBoot<USB_HID_PROTOCOL_KEYBOARD>    HidKeyboard(&Usb);  // Start keyboard in
                                                          // HID Boot protocol mode

KbdRptParser Prs; // Create a Keyboard Report Parser Instance

Lewis Morse;  // Create a Morse Library Instance
```

In the `setup()` function, we'll start the Morse code instance, in beep (CW tone) mode, at 10 words per minute, and use interrupts. The ability to use interrupts is key for this sketch and project to work properly. This allows the sketch to process other operations such as receiving the next keystroke and implementing command code functions while it is also sending the CW characters. Next, we'll initialize the `Timer` interrupt to have the Morse process check every 10 milliseconds for a character to send, and attach the `myISR()` interrupt service function that will handle the actual interrupt processing.

```
void setup()
{

  Morse.begin(-1, beep_pin, 10, true);  // Morse.begin(rx_pin, tx_pin,
                                        // words_per_minute, use_interrupts)

  Timer1.initialize(10000); // Initialize the timer interrupt to 10 milliseconds
  Timer1.attachInterrupt(myISR);   // Attach the Interrupt Service Routine
```

Next, we'll start the TFT display, clear it, then display a brief startup message.

```
  tft.begin();   // Start the TFT display

#ifdef debug
  Serial.begin( 115200 );
  Serial.println("ILI9341 TFT Start");   // Verifies that the TFT is operational
  Serial.print("Display Power Mode: 0x");
  Serial.println(tft.readcommand8(ILI9341_RDMODE), HEX);
  Serial.println("");
  Serial.println("USB Start");
#endif


  clear_display();   // Clear the TFT display
  tft.setTextColor(ILI9341_CYAN, ILI9341_BLACK);   // Set the display text to
                   // Cyan. Black background color used to "clear" previous text
  tft.setRotation(3); // horizontal display // change rotation as needed/desired
  tft.setTextSize(4);   // large font
  tft.setCursor(90, 30);  //Set the Cursor and display the startup screen
  tft.print("KW5GP");
  tft.setCursor(80, 100);
  tft.print("USB CW");
  tft.setCursor(55, 160);
  tft.print("Keyboard");

  delay(5000);   //Wait 5 seconds then clear the startup message
```

```
  tft.setTextColor(ILI9341_GREEN, ILI9341_BLACK);  // Set the display text to
                 // Cyan. Black background color used to "clear" previous text
  clear_display();  // Clear the TFT display
```

Then we'll start the USB host port. Note that if the USB port fails to start for some reason, the sketch will loop endlessly at this point until the issue is resolved and the Nano is reset. Then we'll set up the HID Keyboard Report Parser that will be used to extract the keystrokes from the keyboard. Finally, we'll get things ready for receiving keystrokes, move the TFT display cursor to the top line, and set the CW keying mode to beep.

```
if (Usb.Init() == -1) // Check for USB initialization failure
  {
    usb_status_msg = osc_fail;  // Set the status message to oscillator fail
    system_status = false;  // Indicate a system error
    update_status();  // Update the status line on the bottom of the TFT display

#ifdef debug
    Serial.println(osc_fail);
#endif

    do  // Loop endlessly, no point in continuing
    {
      // do nothing further until reset
    } while (true);
  } else
  {
    usb_status_msg = usb_ok;  // Indicate USB started ok
    update_status();  // Update the status line on the bottom of the TFT display
  }

  delay( 200 ); // Wait for USB to come online

  HidKeyboard.SetReportParser(0, &Prs); // Set up the HID report parser

#ifdef debug
  Serial.println("USB Ready");
  Serial.println("");
#endif

  key_code = 0; // Clear the incoming key code
  ascii_key_code = 0; // Clear the incoming ASCII key code
  repeat_timer = millis() + first_pass_delay; // Set the repeat key timer to wait
                                  // a bit on startup

  tft.setCursor(0, 10); // Set the TFT cursor to the top line

  mode_set();  // Set the CW keying mode (beep or key)
}
```

In the main `loop()`, we'll start by clearing the incoming key variables and set up to allow us to repeat keystrokes when the key is held down. That's another issue with the USB Host library as it relates to keyboards; there's no repeat function. So, we'll implement one manually in the sketch. Then, we'll check the USB host port for keyboard data.

```
     void loop()
{
  key_code = 0; // Clear the incoming key code
  ascii_key_code = 0; // Clear the incoming ASCII key code

  if (!repeat_enabled && (millis() > first_pass_delay)) // Check to see if we can
                       // repeat the key and it's past the startup delay time
  {
    repeat_timer = millis() + repeat_timeout; // Set the repeat timer
                                              // timeout time
  }

  Usb.Task(); // Check the USB keyboard for data
```

Next, we'll check to see if the key has been held to long enough to trigger the repeat key operation and prepare to repeat the keystroke. Then we'll check the received key code and verify that it's either a valid International Morse code character, or one of the command codes we use for setting the speed or changing the beep/key mode. First, we'll check for invalid ASCII values in the received key codes and ignore them. If it is a valid ASCII key code, we'll update the TFT display with the new character, and send it using the Lewis Morse Code library.

```
  if (millis() >= repeat_timer) // if the key is still down and we've exceeded
                                // the repeat timeout
  {
    delay(repeat_delay);   // Delay the repeat delay
    key_code = prev_key_code; // Set the key code to last received key code
    ascii_key_code = prev_ascii_key_code; // Set the ASCII key code to last
                                          // received ASCII key code
  }

  if (key_code != 0)  // If the key code is not zero, we have received a key
  {
#ifdef debug1
    Serial.print("Key code: ");
    Serial.print(key_code);
#endif
    if (ascii_key_code != 0)   // If the ASCII key code is not zero, the key
                               // is an ASCII key
    {
      // It's an ASCII character
```

```
#ifdef debug1
    Serial.print("   ASCII code: ");
    Serial.print(ascii_key_code);
    Serial.print("   ASCII Character: ");
    Serial.println(char(ascii_key_code));
#endif

    prev_key_code = key_code; // Set the previous key code to current received
                              // key code
    prev_ascii_key_code = ascii_key_code; // Set the previous ASCII key code
                                          // to current ASCII received key code

    // Verify the ASCII character is a valid International Morse Code
    // character, display and send it

    switch (ascii_key_code) // Check for invalid keys and ignore
    {
      case 19:   // <Enter>
        break;

      case 35:   // #
        break;

      case 37:   // %
        break;

      case 42:   // *
        break;

      case 91:   // - [
        break;

      case 92:   // - \
        break;

      case 93:   // - ]
        break;

      case 94:   // ^
        break;

      case 96:   // `
        break;

      case 123:  // {
        break;

      case 124:  // |
        break;

      case 125:  // - }
        break;

      case 126:  // ~
        break;

      default:   // It's a valid
                 // character - send it
        update_cw_text(); // Update the
            // TFT display with the new key

        send_cw();   // Send the CW
                     // character
        break;
    }
  } else
```

If the received key code is not a valid ASCII code, we'll next check to see if it's a command code. For this project, we use the Up and Down arrow keys on the keyboard to increase and decrease the CW sending speed. We'll use the Left arrow key to enable the CW tone beep mode, and the Right arrow key to enable the transmitter key mode. We'll also configure the Del (Delete) key to send 3 dits to indicate we made a typo and let the receiving station know it's an error. You could change it and have it send all 8 dits, but sending 3 dits, spaced like individual letter E characters, is often used to indicate an error, for the sake of convenience.

```
      {
        //It's a control code
#ifdef debug
        Serial.print(" - Control Code");
#endif
        switch (key_code)
        {
          case 82:
            // Up Arrow - Increase speed
#ifdef debug
            Serial.println("  Up Arrow");
#endif

            if (key_speed < 30) // Limit max speed to 30 wpm
            {
              key_speed = key_speed + 1;


            }
            mode_set();
            update_status();
            break;

          case 81:
            // Down Arrow - Decrease speed
#ifdef debug
            Serial.println("  Down Arrow");
#endif

            if (key_speed > 5)
            {
              key_speed = key_speed - 1;
            }
            mode_set();
            update_status();
            break;

          case 80:
            // Left Arrow - Set beep mode
```

```
#ifdef debug
          Serial.println(" Left Arrow");
#endif
          key_mode = false;   // Enable Beep mode
          mode_set();
          update_status();
          break;

        case 79:
          // Right Arrow - Set Key Mode
#ifdef debug
          Serial.println(" Right Arrow");
#endif

          key_mode = true; // Enable key mode
          mode_set();
          update_status();
          break;

        case 76:
          // Delete Key - Send 3 dits (EEE)
#ifdef debug
          Serial.println(" Delete/Error");
#endif

          for (int i = 1; i <= 3; i++)
          {
            ascii_key_code = 69;   // The ASCII code for "E"
            key_ascii = "E";

            tft.setTextColor(ILI9341_RED, ILI9341_BLACK);   // Red on black -
                      // background color required to "clear" previous text
            update_cw_text();
            tft.setTextColor(ILI9341_GREEN, ILI9341_BLACK);  // Red on black -
                      // background color required to "clear" previous text

            send_cw();   // Send the CW character
          }
          break;

        default:
          // If no match - ignore the key
#ifdef debug
          Serial.println();
#endif
          break;
      }
    }
  }
}
```

That completes the main `loop()` for this project. As you can see, it's relatively straightforward. Now, let's dig into the functions and see what they do. The first function is the `clear_display()` function, which is used to fill the TFT display with the background color. Filling the TFT display with the background color is the way we "erase" data on a TFT display. There is no way to type over existing information on the screen and have it erase the previous information, unless you first fill that area with the background color.

```
// clear_display function
void clear_display()  // Clears the TFT display
{
  tft.fillScreen(ILI9341_BLACK);   // Fill the screen with the background color
}
```

The next function is the `update_status()` function. The bottom line of the TFT display is used to display status information, such as the status of the USB host port, the CW mode (CW tone beep or transmitter keying), and CW sending speed.

```
// update_status Function
void update_status() // Prints status messages on bottom line of TFT display
                    // in green
{
  if (system_status)  // True if good, false if bad
  {
    //We're all good - update status in green
    tft.setTextColor(ILI9341_GREEN, ILI9341_BLACK);   // Green on black
                          // background color required to "clear" previous text
  } else
  {
    // There is a system error
    tft.setTextColor(ILI9341_RED, ILI9341_BLACK);   // Red on black background
                          // color required to "clear" previous text
  }
  tft.setTextSize(status_font_size);   // Set the font size to the status line
                                       // font size
  tft.setCursor(10, 220); // Move to the bottom line of the display

  tft.print(usb_status_msg);   // Display the USB status

  if (key_mode != last_mode)   // Check to see if the keyer mode has changed -
                               // and if so, update the display
  {
    tft.setCursor(130, 220);
    if (key_mode)
    {
      tft.print("Key ");
```

```
  } else
  {
    tft.print("Beep");
  }
  last_mode = key_mode;
}

if (key_speed != last_speed)   // Check to see if the keyer speed has changed -
                               // and if so, update the display
{
  tft.setCursor(230, 220);
  tft.print("WPM: ");
  tft.print(key_speed);
  tft.print(" "); // Clear the second digit if less than 10 wpm
  last_speed = key_speed;
}
tft.setTextSize(cw_font_size);   // Set the display font size back to the CW
                                 // text font size

// Return the cursor to where we were
// Set the cursor to line 0 and reprint the cw_data string - neat trick..it
// overwrites what is there without blinking and sets
// the cursor back to where we were
tft.setCursor(0, 10);
tft.print(cw_data);
}
```

The `update_cw_text()` function is used to update the text on the TFT display. We use a larger font to display the CW characters that are being sent, and we'll also scroll a line up at a time when we reach the bottom of the text area on the display. Because there is no scrolling function for the TFT display, we'll have to handle that manually. For this reason, you'll need to use the Adafruit_ILI9341 TFT library, or a library that's as fast as that one, as several of the other TFT libraries I tested were too slow to handle the scrolling operation without causing the display to flicker (which I find irritating).

```
// update_cw_text function
void update_cw_text() // Updates the cw text on the display
{
  tft.setTextSize(cw_font_size);  // Set the cw text font size

  key_ascii = String(char(ascii_key_code)); // Convert the ASCII key code
                                            // to a string
  key_ascii.toUpperCase();  // Make all the characters uppercase

  // Scroll bottom line up a line if at end of line 4
  if (cw_data.length() >= 52)
  {
    cw_data = cw_data.substring(13);  // Remove the first 13 characters
    tft.setCursor(0, 10); // Move the cursor to the top line
    tft.print(cw_data); // Print the cw data string variable
    for (int i = 1; i <= 13; i++) // Print a space 13 times to scroll the data
                                  // up a line
    {
      tft.print(" ");
    }
    tft.setCursor(0, 107);  // Set the cursor to the start of line 4
  }

  cw_data = cw_data + key_ascii;  // Add the character to the cw data string

  tft.print(key_ascii); // display the received ASCII character

#ifdef debug
  Serial.print("cw_data string length: ");
  Serial.print(cw_data.length());
#endif

}
```

The `send_cw()` function is used to send the CW keystroke using the currently selected CW tone or transmitter key mode.

```
// send_cw function
void send_cw()   // Sends the CW character
{
#ifdef debug
  Serial.print(" Sending: ");
  Serial.println(key_ascii);
#endif

  Morse.print(key_ascii); // Send the character to the Morse library
}
```

The `mode_set()` function is called when a Left or Right arrow keystroke is detected. In key mode, the Lewis Morse library will send the CW to the relay on digital I/O pin 6 that is used to key the transmitter. In beep mode, the library will send the CW using digital I/O pin 5 that is used to trigger the NE555 tone oscillator. The function will also update the TFT display to indicate the new status information.

```
//mode_set function
void mode_set()    // Function to Set the mode to beep or keying and/or
                   // set keying speed
{
  if (key_mode) // Keying mode True = Key False = Beep
  {
    Morse.begin(-1, key_pin, key_speed, true); // Key on pin 6
  } else
  {
    Morse.begin(-1, beep_pin, key_speed, true); // Key on pin 5
  }

#ifdef debug
  Serial.print("Key Mode: ");
  Serial.print(key_mode);
  Serial.print("  WPM: ");
  Serial.println(key_speed);
#endif

  update_status();  // Update the information on the status line

}
```

USB CW Keyboard   13-21

The myISR function comes from the Morse library example code to process the interrupts for the Lewis Morse library. This allows the sketch to do other operations while sending the CW data simultaneously.

```
// myISR function
void myISR()  // Function to handle interrupts for the Morse library
{
  Morse.timerISR();
}
```

The USB Host library functions come directly from the hidboot keyboard example sketch. Because USB devices don't send data on a per-character basis, and instead use the USB data report mechanism, we'll need these functions to parse the received keyboard data for us. This sketch is a perfect example of how to use the example code that is often provided with Arduino libraries. Even with all the research I have done regarding USB operations, I doubt very seriously if I would even know where to start with getting the data from the keyboard. Fortunately, the example code that came with the USB Host library has everything I need to get the job done. I may not fully understand how and what these functions are doing, but when it comes to things like the Arduino, sometimes you just trust the library, and copy and paste the pieces you need from the example code. That's what the Arduino and the Open Source community are all about.

The first USB host function is the KbdRptParser::PrintKey() function, which is used to display information about the pressed key on the Arduino IDE's Serial Monitor if the usb_debug information is enabled at the beginning of the sketch.

```
// USB Host Library Functions -----------------------------------------

void KbdRptParser::PrintKey(uint8_t m, uint8_t key)
{
  MODIFIERKEYS mod;
  *((uint8_t*)&mod) = m;

#ifdef usb_debug
  Serial.print((mod.bmLeftCtrl   == 1) ? "C" : " ");
  Serial.print((mod.bmLeftShift  == 1) ? "S" : " ");
  Serial.print((mod.bmLeftAlt    == 1) ? "A" : " ");
  Serial.print((mod.bmLeftGUI    == 1) ? "G" : " ");

  Serial.print(" > ");
  PrintHex<uint8_t>(key, 0x80);
  Serial.print(" < ");
#endif

#ifdef usb_debug
  Serial.print("Key = ");
  Serial.println(key);
#endif
```

```
#ifdef usb_debug
  Serial.print((mod.bmRightCtrl   == 1) ? "C" : " ");
  Serial.print((mod.bmRightShift  == 1) ? "S" : " ");
  Serial.print((mod.bmRightAlt    == 1) ? "A" : " ");
  Serial.println((mod.bmRightGUI  == 1) ? "G" : " ");
#endif
};

void KbdRptParser::OnKeyDown(uint8_t mod, uint8_t key)
{
#ifdef usb_debug
  Serial.print("DN ");
#endif

  repeat_enabled = true;
  PrintKey(mod, key);

  uint8_t c = OemToAscii(mod, key);

  if (c)
    OnKeyPressed(c);

  key_code = key;
  ascii_key_code = c;

#ifdef usb_debug
  Serial.print("Before OEM to ASCII: ");
  Serial.print(key, HEX);
  Serial.print("   After OEM to ASCII: ");
  Serial.println(c, HEX);
#endif
}
```

The `KbdRptParser::OnKeyDown()` function is used to indicate to the USB Host library that a key has been pressed down. Because of the way USB keyboards function, they will send information to the USB host that a key has been pressed, and again when the key is released.

```
void KbdRptParser::OnKeyDown(uint8_t mod, uint8_t key)
{
#ifdef usb_debug
  Serial.print("DN ");
#endif

  repeat_enabled = true;
  PrintKey(mod, key);

  uint8_t c = OemToAscii(mod, key);

  if (c)
    OnKeyPressed(c);

  key_code = key;
  ascii_key_code = c;

#ifdef usb_debug
  Serial.print("Before OEM to ASCII: ");
  Serial.print(key, HEX);
  Serial.print("   After OEM to ASCII: ");
  Serial.println(c, HEX);
#endif
}
```

The `KbdRptParser::OnControlKeysChanged()` function is used by the USB Host library to get the key code information when a control key on the keyboard has been pressed or released.

```
void KbdRptParser::OnControlKeysChanged(uint8_t before, uint8_t after)
{
  MODIFIERKEYS beforeMod;
  *((uint8_t*)&beforeMod) = before;

  MODIFIERKEYS afterMod;
  *((uint8_t*)&afterMod) = after;

#ifdef usb_debug

  if (beforeMod.bmLeftCtrl != afterMod.bmLeftCtrl)
  {
    Serial.println("LeftCtrl changed");
```

```
  }
  if (beforeMod.bmLeftShift != afterMod.bmLeftShift)
  {
    Serial.println("LeftShift changed");
  }
  if (beforeMod.bmLeftAlt != afterMod.bmLeftAlt)
  {
    Serial.println("LeftAlt changed");
  }
  if (beforeMod.bmLeftGUI != afterMod.bmLeftGUI)
  {
    Serial.println("LeftGUI changed");
  }

  if (beforeMod.bmRightCtrl != afterMod.bmRightCtrl)
  {
    Serial.println("RightCtrl changed");
  }
  if (beforeMod.bmRightShift != afterMod.bmRightShift)
  {
    Serial.println("RightShift changed");
  }
  if (beforeMod.bmRightAlt != afterMod.bmRightAlt)
  {
    Serial.println("RightAlt changed");
  }
  if (beforeMod.bmRightGUI != afterMod.bmRightGUI)
  {
    Serial.println("RightGUI changed");
  }

#endif

}
```

The `KbtRptParser::OnKeyUp()` function is used to indicate to the USB Host library that a key has been released.

```
void KbdRptParser::OnKeyUp(uint8_t mod, uint8_t key)
{

#ifdef usb_debug
  Serial.print("UP ");
  PrintKey(mod, key);
#endif

  prev_key_code = 0;
  prev_ascii_key_code = 0;
  repeat_enabled = false;
}
```

And finally, the `KbtRptParser::OnKeyPressed()` function will return the ASCII value of the key that is pressed.

```
void KbdRptParser::OnKeyPressed(uint8_t key)
{
#ifdef usb_debug
  Serial.print("ASCII value: ");
  Serial.print(key);
  Serial.print(" Character: ");
  Serial.println((char)key);
#endif
  // ascii_key_code = key;
};
```

Enhancement Ideas

As mentioned earlier, the USB keyboard library only has a 10 character typeahead buffer. At slower CW speeds, it would be easy for a fast typist to overrun the buffer. Expanding the buffer by as few as 10 characters could add a great deal of flexibility to this project.

In the original PS/2 CW Keyboard, there were five user-programmable memories. It should not be difficult to add programmable memory capability to this project and even save the memory information to the Nano's EEPROM.

While we're saving things to the Nano's EEPROM, why not save the speed and CW mode (beep tone or transmitter key) settings to EEPROM as well?

And finally, rather than use the keyboard arrow keys, what about using a rotary encoder with a pushbutton switch to set the sending speed, mode, and select/program memories?

Sketches, libraries, Fritzing diagrams, and other useful files for the projects in this book can be found at **www.sunriseinnovators.com** and **www.kw5gp.com**.

Yaesu CAT Display

Many of today's transceivers have the ability to be controlled via computer, a feature also known as CAT control. (CAT stands for computer-aided transceiver.). Most transceiver CAT interfaces are now based on RS-232, although some use another serial communication method such as Icom's CI-V interface. Over time, the CAT control functions have expanded to the point that you can do practically anything over the CAT interface that you can with the radio's physical knobs and buttons. Advancements in radio technology have resulted in more features than can be viewed or managed using the radio's front panel display and controls. With more and more features moving into menu systems, no matter how well the menus are done, it is always a challenge to remember which menu has the feature you want to view or change.

One of the first things I do when I get a radio that is new to me is to go through the manual to make myself familiar with the menu system. Then, I'll skim through the CAT commands, and realize that I'd probably need a workstation connected to the radio with some sort of CAT control program to access some of the functions on the radio.

But maybe I don't want to carry a workstation with me out in the field on a portable operation, but would still like to have access to some of the CAT functions. Here's where that "wouldn't it be cool if..." thing that always gets me in trouble kicks in. Wouldn't it be cool if I could have an Arduino access some of the CAT functions on the rig and display them on a color TFT graphic display?

For this project, we'll build a basic Arduino-powered CAT interface and display, and set it up to use with my Yaesu FT-450D go-box that I use for portable events such as ARRL Field Day and the Mississippi QSO

Figure 14.1 — The Yaesu CAT Display.

Figure 14.2 — The Yaesu FT-450D.

Party, two of my favorite events. The plan is to remotely display some of the basic operating parameters so that visitors and other operators can see, for example, what frequency we're operating on, without having to lean over and see the transceiver display. Because the Yaesu FT-450D (**Figure 14.2**) shares the same RS-232 interface and many of the same CAT commands with the Yaesu FT-950 and FTDX1200, this project can be used with those transceivers as well. With minor changes to the sketch, this project could be used with virtually any transceiver that has CAT control capability and a serial interface.

The parameters I have chosen to display for this project are the VFO-A and VFO-B frequencies, an indicator as to which VFO is currently active, and the current operating mode. Dozens of choices for what to display are available through the CAT menu, so feel free to pick the ones you would like. The Yaesu CAT command structure is very straightforward and easy to work with as you'll see when we get to the sketch.

Block Diagram

Figure 14.3 shows the block diagram for the Yaesu CAT display project. The hardware is little more than the Arduino Uno, TFT display, and a MAX232 RS-232 driver/receiver chip used to interface to the transceiver. We'll use an ST7735-type 1.8-inch color graphic 160×128 pixel TFT display, although there's no reason you couldn't use a larger display to have room for more information.

Because we'll need to communicate with the transceiver using RS-232, we'll use the AltSoftSerial library. That library allows us to use digital I/O pins to communicate with the transceiver and not have to deal with the complexities of using the Arduino's USB serial port to handle the CAT communications along with programming and debugging.

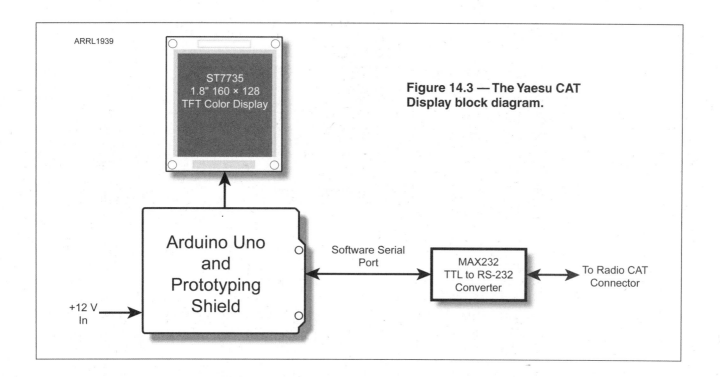

Figure 14.3 — The Yaesu CAT Display block diagram.

Construction

Figure 14.4 shows the schematic diagram for the Yaesu CAT Display project. This is yet another example of how you can use only a handful of parts and wires to create a simple, functional project with the Arduino.

The MAX232 RS-232 driver/receiver chip is the key to this project. The MAX232 is my first choice when it comes to translating between 5 V serial communication and RS-232, which uses ±12 V for the signal levels. This voltage is far too high to connect directly to the Arduino, nor can the Arduino directly generate the ±12 V needed to transmit RS-232 commands. The cool thing about the MAX232 is that it does not need a plus and minus power supply to generate the signal voltages. Instead, the MAX232 is powered by +5 V and an internal charge pump generates the ±12 V signal levels needed. The receiver side of the MAX232 will convert the ±12 V signal levels into standard 5 V TTL logic levels that can be connected directly to the Arduino I/O pins.

To connect to the Yaesu FT-450D/FT-950/FTDX1200 transceivers, you will need to build a cable to connect from the MAX232 to the DB-9 CAT interface connector on the transceiver. The connector on the radio has male pins, so you will need a cable-mounted DB-9 female to match. Connect the MAX232 **T1OUT** to pin 3 (**TX DATA**) on the DB-9 connector, the MAX232 **R1IN** pin to DB-9 pin 2 (**RX DATA**), and **GROUND** to DB-9 pin 5. You will also need to connect pin 7 (**RTS**) to pin 8 (**CTS**) on the DB-9 connector,

It is important to note that the AltSoftSerial library we use to communicate with the transceiver is hardcoded to use digital I/O pin 9 to transmit data to the transceiver, and digital I/O pin 8 to receive data from the transceiver. We use the AltSoftSerial library mainly because it is more efficient and easier to use than the built-in SoftwareSerial library. The AltSoftSerial library also handles data

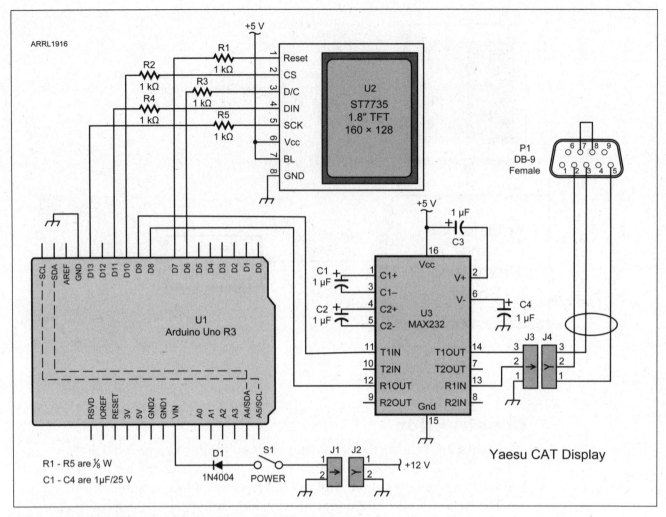

Figure 14.4 — The Yaesu CAT Display schematic diagram.
C1-C4 — 1 µF, 25 V capacitor
D1 — 1N4004 diode
J1 — 2-pin DuPont male header
J2 — 2-pin DuPont female header socket
J3 — 3-pin DuPont male header
J4 — 3-pin DuPont female header socket
P1 — DB-9 female on cable to radio
R1-R5 — 1 kΩ, ⅛ W resistor
S1 — SPST toggle switch
U1 — Arduino Uno
U2 — ST7735-type 1.8-inch color TFT display
U3 — MAX232 RS-232 driver/receiver
Enclosure — Solarbotics Mega SAFE

transfer in full-duplex mode, while the built-in SoftwareSerial library can only handle data in half-duplex mode. For projects such as this one that use serial communication, I prefer to use the digital I/O pin software serial method rather than having to add jumpers to the project and other workarounds to prevent signal conflicts on the Arduino's USB/serial port on I/O pins 0 and 1. The AltSoftSerial library can handle data rates up to 57,600 baud, so it should be more than adequate for most serial CAT applications.

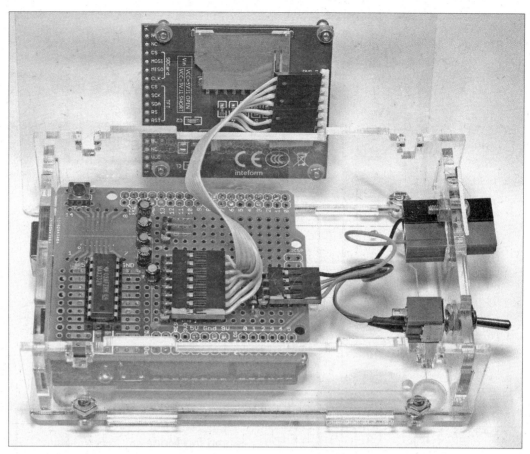

Figure 14.5 — The Yaesu CAT Display prototyping shield and construction.

Figure 14.5 shows the construction of the prototyping shield for the Yaesu CAT display. This photo shows one of the main reasons why I like to use prototyping shields along with the 2.54 mm DuPont-style headers and sockets. This arrangement allows me to quickly disconnect everything and remove the prototyping shield to correct wiring errors or make modifications to the project. This is also a good example of the usage of the 2 mm hardware to mount the TFT display to the top cover of the enclosure. You'll also notice a pair of Anderson Powerpole connectors is used in a chassis mount to supply the 12 V dc for the project. Because this project is made to work in conjunction with a 12 V dc-powered transceiver, you can easily add a Powerpole cable from the station power supply to the project.

Flowchart

Figure 14.6 shows the flowchart for the Yaesu CAT Display. For this sketch, we'll need to include the Adafruit GFX, Adafruit ST7735, and SPI libraries for the TFT display. We'll also need to include the AltSoftSerial library to handle the serial communication with the transceiver's CAT interface. We're using the AltSoftSerial library instead of the built-in SoftwareSerial library due to the fact the AltSoftSerial library has better performance and operates in full-duplex mode.

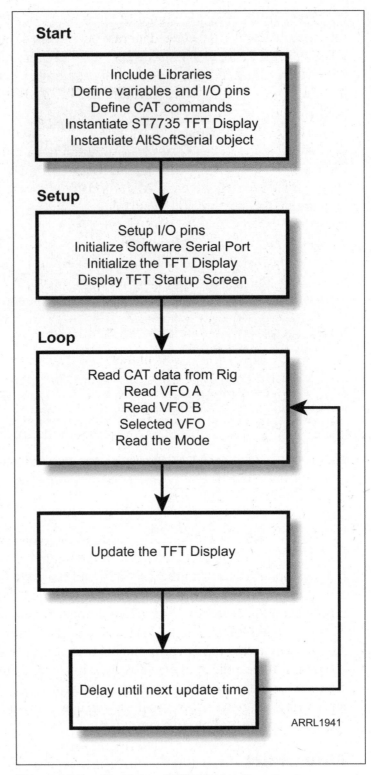

Figure 14.6 — The Yaesu CAT Display flowchart.

Next, we'll define the digital I/O pins for the TFT and data rate for the CAT interface. The CAT interface data rate defaults to 4800 baud for the FT-450D, FT-950, and FTDX1200 transceivers. We'll also define the CAT commands used to have the transceiver send us the information we want to display.

Finally, we'll instantiate (create) the ST7735 TFT display and the AltSoftSerial objects.

In the `setup()` function, we'll start the Arduino's Serial Monitor port if debugging is enabled, start the AltSoftSerial port, and initialize the TFT display. Then we'll display a brief startup message on the TFT display, and get the TFT display ready to display the CAT information by displaying a template of the data that doesn't change on the display to reduce flickering during display updates.

In the main `loop()` function, we'll read the CAT data from the transceiver. First we'll read VFO-A, then VFO-B, then determine which VFO is currently selected, and then read the operating mode. We'll update the TFT with the data and repeat the display process every 5 seconds.

The Sketch

As with all of the projects in this book, the sketch and libraries for this project can be found at **www.sunriseinnovators.com/Arduino3** and **www.kw5gp.com/Arduino3**. We'll start the sketch by including the Adafruit GFX and Adafruit ST7735 libraries, and the SPI libraries for the ST7735 TFT display. Then we'll include the AltSoftSerial library that we'll use to communicate with the transceiver via RS-232.

```
/*
  Yaesu CAT Display
  for the Yaesu FT-450/FT-950/FT-1200 and others
  written by Glen Popiel - KW5GP

  Released under the GPLv3 license
*/

#include <Adafruit_GFX.h>     // Core graphics library
#include <Adafruit_ST7735.h>  // Hardware-specific library
#include <SPI.h>   // SPI library so we can communicate with the TFT display
#include <AltSoftSerial.h>   // AltSoft Serial Library - better than the
                             // SoftwareSerial library
```

Next, we'll define the CAT serial rate at 4800 baud, the default for the Yaesu transceiver. We'll also define some basic operating parameters such as the TFT display update delay and the I/O pins needed for the TFT display. The `tft_delay` is needed to allow the TFT display time to process commands, and a 10 ms delay seems to work just fine.

```
//#define debug // Uncomment to send debug information to the Serial Monitor

#define CAT_serial_rate 4800   // Define the radio CAT baud rate -
                               // Default is 4800 baud
#define comm_speed 9600  // Define the Arduino Serial Monitor baud rate
#define timeout_delay 5000   // The delay to signify a data timeout from the radio
#define update_delay 5000  // The delay between Arduino TFT display updates

#define TFT_CS       10   // Assign the TFT CS to pin 10
#define TFT_RST       7   // Assign the TFT RST to pin 7
#define TFT_DC        6   // Assign the TFT DC to pin 6

#define tft_delay 10   // set the TFT command delay to 10ms
```

Next, we'll define the CAT command strings we'll need to send to the transceiver. For this project, from the transceiver we'll get the VFO-A and VFO-B frequencies, an indication of which VFO is currently active, and the current operating mode. You can customize this display and get different CAT information simply by changing or adding CAT commands, and modifying the displayed information.

```
#define get_VFO_A "FA;" // Get VFO-A command
#define get_VFO_B "FB;" // Get VFO-B command
#define get_vfo "VS;" // Get which VFO is selected
#define get_mode "MD0;"    // Get Mode command
```

Next, we'll declare and initialize the remaining variables needed in the sketch. Then, we'll instantiate (create) the ST7735 TFT display and AltSoftSerial objects.

```
boolean first_pass = true;    // Variable used to determine if first pass through
                              // the main loop
char rx_char; // Variable for each character received from the radio
unsigned long timeout, current_millis;  // Variables to handle communication
                                        // timeout errors
bool receiving, timeout_error;  // Variables for the receive radio data process
String CAT_buffer;   // String to hold the received CAT data
String VFO_A, VFO_B, vfo, mode; // Variables to hold the parsed
                                // and formatted CAT data

Adafruit_ST7735 tft = Adafruit_ST7735(TFT_CS,  TFT_DC, TFT_RST);
   // Initialize the TFT display instance

AltSoftSerial altSerial;  // Initialize the software serial instance
```

In the setup() function, we'll first start the Arduino IDE Serial Monitor port if debugging is enabled, then we'll start the software serial port on digital I/O pins 8 and 9 at 4800 baud, which is the default for the CAT interface on the Yaesu FT-450D, FT-950, and FTDX1200. Next, we'll initialize the ST7735 TFT display and show a brief startup message.

```
void setup()
{
#ifdef debug
  Serial.begin(comm_speed); // Set the Arduino Serial Monitor baud rate
  while (!Serial) ; // wait for Arduino Serial Monitor to open
  Serial.println("FT-450/900/1200 CAT Display Begin");
#endif

  //Start the Software Serial port
  altSerial.begin(CAT_serial_rate); // Start the software serial port to
                                    // the transceiver

  tft.initR(INITR_18BLACKTAB);   // initialize a 1.8" TFT with ST7735S chip,
                                 // black tab
  delay(tft_delay);
  tft.fillScreen(ST7735_BLACK); // Clear the display - fill with BLACK background
  delay(tft_delay);
  tft.setRotation(1); // Set the screen rotation
  delay(tft_delay);
  tft.setTextWrap(false); // Turn off Text Wrap
  delay(tft_delay);
  tft.setTextSize(3); // Set the Font Size
  delay(tft_delay);
  tft.setTextColor(ST7735_BLUE); //Set the Text Color
  delay(tft_delay);
  tft.setCursor(40, 10);  //Set the Cursor and display the startup screen
  delay(tft_delay);
  tft.print("KW5GP");
  delay(tft_delay);
  tft.setTextSize(2);
  delay(tft_delay);
  tft.setCursor(50, 60);
  delay(tft_delay);
  tft.print("Yaesu");
  delay(tft_delay);
  tft.setCursor(15, 80);
  delay(tft_delay);
  tft.print("Transceiver");
  delay(tft_delay);
  tft.setCursor(15, 100);
  delay(tft_delay);
  tft.print("CAT Display" );
```

Finally, we'll clear the TFT display, and then we'll display a template of the TFT data that doesn't change between updates to save time and reduce display flickering during updates.

```
  delay(5000);   //Wait 5 seconds then clear the startup message
  tft.fillScreen(ST7735_BLACK); // Clear the display
  delay(tft_delay);

  update_display(); // update the display
  first_pass = false; // turn off the update display first pass flag
}
```

In the main `loop()`, we'll start by calling the `send_command()` function to get the frequency of VFO-A. We'll then call the `get_response()` function to receive the CAT response from the transceiver, and then we'll call the `format_string()` function to format the CAT response for the TFT display. This is the basic sequence needed to request and receive CAT data for all of the Yaesu CAT commands.

```
void loop()
{
  send_command(get_VFO_A);   // Send the Get VFO-A Frequency Command

  get_response();   // Receive the VFO-A frequency response

  format_string();   // Format the VFO Response data for display
  VFO_A = CAT_buffer;

#ifdef debug
  Serial.print("VFO-A: ");
  Serial.println(VFO_A);
#endif
```

We'll then repeat the sequence for VFO-B.

```
  send_command(get_VFO_B);   // Send the Get VFO-B Frequency Command

  get_response();   // Receive the VFO-B frequency response

  format_string();   // Format the VFO Response data for display
  VFO_B = CAT_buffer;

#ifdef debug
  Serial.print("VFO-B: ");
  Serial.println(VFO_B);
#endif
```

Next, we'll get the CAT information for which VFO is currently selected. Because we're only extracting a single character from the response, we don't need to format the string for the display, as this information will be formatted by the `display_update()` function.

```
  send_command(get_vfo);     // Send the VFO Selected command

  get_response();    // Receive the VFO Selected response
  vfo = CAT_buffer.substring(2);

#ifdef debug
  Serial.print("VFO: ");
  Serial.println(vfo);
#endif
```

Finally, we'll get the CAT information for the current operating mode. The CAT response is a single number or letter corresponding to each mode. We'll then extract the mode character from the CAT response and then call the `convert_mode()` function to translate the CAT response into text for the display. For example, we'll convert 1 to LSB, 2 to USB and so on. Then we'll call the `update_display()` function to update the information on the TFT display, delay five seconds, then repeat the process.

```
  send_command(get_mode);     // Send the Get Mode Command

  get_response();    // Receive the Get Mode response
  mode = CAT_buffer.substring(3);

  convert_mode(); // Convert the CAT Mode response into the corresponding text

#ifdef debug
  Serial.print("Mode: ");
  Serial.println(mode);
#endif

  update_display(); // Update the TFT display

  delay(update_delay);   // Delay until the next display update
}
```

The `get_response()` function is used to receive the CAT command response from the transceiver. The first thing this function does is set up a timeout timer so that the sketch doesn't get hung up waiting for a response from the transceiver that for some reason failed or was garbled.

```
void get_response() // Receives the CAT command response from the radio
{

#ifdef debug
   Serial.print("    Start RX  ");
#endif

   // Set a timeout value
   current_millis = millis();  // Get the current time
   timeout = current_millis + timeout_delay; // Calculate the timeout time
   // Check for millis() rollover condition - the Arduino millis() counter rolls
   // over about every 49.7 days

   if (timeout < current_millis) // We've calculated the timeout during
                                 // a millis() rollover event
   {
      timeout = timeout_delay; // Go ahead and calculate as if we've rolled over
                               // already (adds a few millis to the timeout delay)
   }
```

You'll note that the timeout calculation is a bit more complex than you would think it should be, and you are correct. The millis() timer in the Arduino returns an unsigned long integer, which ranges from zero to 4,294,867,295. This means that the millis() timer will roll over every 49.7 days, and we need to take this rollover into account when dealing with timeout calculations. Because it is possible for this project to be left running for extended periods of time, we need to take into account that the millis() timer may roll over. That's why the timeout calculation formula used in this sketch is a bit more involved than usual.

Next, we'll get ready to receive a CAT response from the radio. This of course assumes that we've already sent a CAT command to the radio and are expecting a CAT response. We'll keep looping waiting for received characters until the loop either times out or receives a semicolon (;) indicating an end of CAT message from the radio.

```
  // Get ready to receive CAT response from the radio
  receiving = true;
  timeout_error = false;
  CAT_buffer = "";
  do
  {
    if (millis() > timeout) // We've exceeded the timeout delay
    {
      // We timed out - exit
      receiving = false;
      timeout_error = true;

      CAT_buffer = "";
      break;
    }
    if (altSerial.available() && receiving) // If there's a character in the rx
                                            // buffer and we're ok to receive
    {
      rx_char = altSerial.read(); // Get the incoming character
      if (rx_char == ';') // ";" indicates the end of the response
                         // from the radio
      {
        receiving = false;  // Turn off the ok to receive flag
      } else
      {
        CAT_buffer = CAT_buffer + rx_char;  // Add the received character to the
                                            // CAT rx string
      }

    }
  } while (receiving);  // Keep looping while we're ok to receive data from
                       // the radio

#ifdef debug
  if (timeout_error)
  {
    Serial.println("   Timeout ");
  } else
  {
    Serial.print("   Received complete: ");
    Serial.println(CAT_buffer);

  }
#endif
}
```

Yaesu CAT Display   14-13

The `send_command()` function is used to send a CAT command string to the radio. The CAT command to send is passed as a parameter to the function, so that this function can be re-used to send any desired CAT command.

```
void send_command(String CAT_command)   // Sends the CAT command string to
                                        // the radio
{
  // Send the CAT Command

#ifdef debug
  Serial.print("Sending: ");
  Serial.print(CAT_command);
#endif

  altSerial.print(CAT_command); // Sends the CAT command to the radio

#ifdef debug
  Serial.print("  Send Complete  ");
#endif
}
```

The `convert_mode()` function is used to convert the single character operating mode CAT response to a text string that can be displayed on the TFT display. We'll use a `Switch...Case` statement to handle the mode character to text conversion.

14-14   Chapter 14

```c
void convert_mode()   // Convert the mode to the mode name
{
  // Mode is a single character string

  switch (mode[0])// convert the single character mode string to a char variable
  {
    case '1': // LSB
      mode = "LSB";
      break;

    case '2': // USB
      mode = "USB";
      break;

    case '3': // CW
      mode = "CW";
      break;

    case '4': // FM
      mode = "FM";
      break;

    case '5': // AM
      mode = "AM";
      break;

    case '6': // DATA (RTTY-LSB)
      mode = "DATA (RTTY-LSB)";
      break;

    case '7': // CW-R
      mode = "CW-R";
      break;

    case '8': // USER-L
      mode = "USER-L";
      break;

    case '9': // DATA (RTTY-USB)
      mode = "DATA (RTTY-USB)";
      break;

    case 'B': // FM-N
      mode = "FM-N";
      break;

    case 'C': // USER-U
      mode = "USER-U";
      break;

    default: // Unknown mode
      mode = " ";
      break;

#ifdef debug
      Serial.print("Mode String: ");
      Serial.println(mode);
#endif
  }
}
```

The `update_display()` function will update the CAT information displayed on the TFT display. The first time this function is called, it will display a template of the display information that doesn't change. This is done to reduce flickering from clearing and redisplaying the template information on every update. Because this information doesn't change, there's no need to waste the time to update it every five seconds.

```
void update_display() // Updates the TFT display
{
  if (first_pass) // Only do this part the first time the function is called
  {
    // Clear the screen and display normal operation
    tft.fillScreen(ST7735_BLACK); // Clear the display
    delay(tft_delay);
    tft.setTextSize(1); // Set the text size to 1
    delay(tft_delay);
    tft.setTextColor(ST7735_BLUE); // Set the text color to BLUE
    delay(tft_delay);
    tft.setTextSize(1); // Set the text size to 2
    delay(tft_delay);
    tft.setCursor(30, 5);
    delay(tft_delay);
    tft.print("Yaesu CAT Display");  // Display screen title
    tft.setCursor(5, 40);
    delay(tft_delay);
    tft.print("VFO A :");   // Display VFO A
    tft.setCursor(5, 60);
    delay(tft_delay);
    tft.print("VFO B :");   // Display VFO B
    delay(tft_delay);
    tft.setCursor(5, 80);
    delay(tft_delay);
    tft.print("Mode  :");   // Display Mode
    delay(tft_delay);
```

If this is not the first time through the operation, we'll update the CAT information on the display.

```
} else
  {
    clear_data();  // Clear the value display area

    tft.setCursor(50, 40);
    delay(tft_delay);
    tft.print(VFO_A); // Display VFO-A Value

    tft.setCursor(50, 60);
    delay(tft_delay);
    tft.print(VFO_B); // Display VFO-B Value

    tft.setCursor(50, 80);
    delay(tft_delay);
    tft.print(mode);   // Display mode

    if (vfo == "0")
```

```
  {
    // VFO-A is active
    tft.setCursor(115, 40); // Select the active VFO line
  }

  if (vfo == "1")
  {
    // VFO-B is active
    tft.setCursor(115, 60); // Select the active VFO line
  }

  tft.print("Active"); // Indicate the active VFO

  }
}
```

The `clear_data()` function is used to clear the data area of the TFT display. With a TFT display, we can't just type over the existing data — we have to erase it first by filling the display with the background color.

```
void clear_data() //Clears the data area of the display
{
  tft.fillRect(50, 40, 110, 50, ST7735_BLACK); // Clear the CAT data area
  delay(tft_delay);
}
```

And finally, the `format_string()` function is used to format the received VFO frequency into a string that can be displayed on the TFT display.

```
void format_string()    // Formats the VFO data RX string data for display
{
  CAT_buffer = CAT_buffer.substring(2);  // First, strip off the CAT command echo
  CAT_buffer = CAT_buffer.substring(0, 2) + "." + CAT_buffer.substring(2, 5) +
". " + CAT_buffer.substring(5); // Add periods to separate digits
  if (CAT_buffer.startsWith("0"))
  {
    // It's Below 10MHZ - strip off the leading zero
    CAT_buffer = CAT_buffer.substring(1);
  }
}
```

Enhancement Ideas

This project only touches on the things you can do using the CAT interface on a transceiver. For example, we only read a couple of things and displayed them. With CAT, you can also control the various transceiver settings. With all of the Arduino controls available such as switches, keypads, rotary encoders and the like, you can build your own CAT-based control panel with all of your favorite functions and menu settings at your fingertips.

You could also increase the size of the color TFT display, add gridlines, and create other graphics to really add to the flash of your external CAT display/control unit. By replacing the Arduino Uno with an Arduino Mega or similar processor, you can even add touchscreen capability to your CAT project.

Finally, because we're using serial communications, you could easily replace the RS-232 portion of this project with an Arduino Bluetooth link unit attached to the transceiver, and the actual Arduino display/control unit remotely connected via Bluetooth. You could even take this all the way cool, and add internet-based CAT control/display capabilities just by adding an ethernet module to the project. This is one project that at the surface is fundamentally simple and easy to build, but has tremendous possibilities for you to enhance and improve upon.

CDE/HyGain Keypad Entry Rotator Controller

The 4×4 matrix keypad has been used with the Arduino since the early days, and one is included in just about every Arduino starter kit. But, I'm not a huge fan of these thin, flexible keypads, and they tie up eight digital I/O lines to decode the key presses. Recently, several new types of 4×4 keypads have become available. Adafruit has the Neo-Trellis keypad (**Figure 15.2**), which is a nice elastomer 4×4 keypad with addressable RGB or monochrome LED backlighting that uses the I^{2}C bus. In addition, RobotDyn offers an inexpensive analog version that uses resistors to select the output voltage based on the key pressed, similar to the Yaesu FH-2 keypad described in an earlier chapter. The nice thing about the RobotDyn keypad (**Figure 15.3**) is that it has removable, clear keycaps so you can create your own key legends.

A recurring issue with computer-controlled rotators is that you need an attached workstation running software such as *Ham Radio Deluxe* to manage the rotation. Also, how many times have you looked at the meter on the CDE/HyGain rotator and mentally converted that meter reading to a bearing in your head, just so you could point the antennas in the right direction?

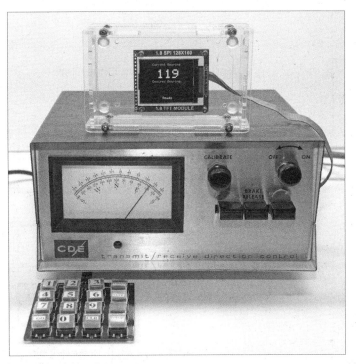

Figure 15.1 — The CDE/HyGain Keypad Entry Rotator Controller.

Figure 15.2 — The Adafruit NeoTrellis keypad.

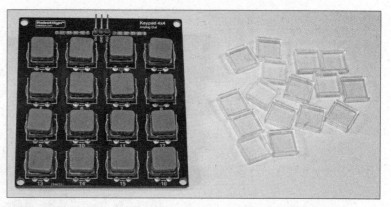

Figure 15.3 — The RobotDyn analog keypad with its clear caps removed.

Thanks to another one of those "Wouldn't it be cool if…" moments, I thought it might be cool to use a 4×4 keypad to directly enter the desired antenna bearing, press "Go," and have the Arduino take care of rotating the antenna and showing the bearing on a TFT display. This project does just that.

Block Diagram

Based on the rotator projects described in the earlier chapters, we have all the pieces we need to make this project work. **Figure 15.4** shows the block diagram. We'll be using many of our favorite components from previous projects, including a 1.8-inch ST7735-type color graphic TFT display, an ADS1115 16-bit programmable gain A/D converter, and an analog 4×4 keypad. I chose the RobotDyn analog keypad because it only requires one analog input pin on the Arduino, and it has removable keycaps so we can use our own key labels

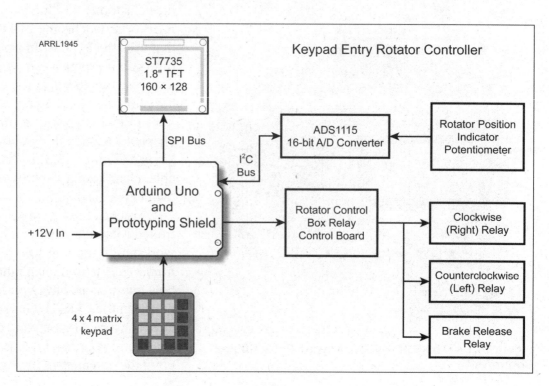

Figure 15.4 — The Keypad Entry Rotator Controller block diagram.

15-2   Chapter 15

customized for this project. We'll also use a small external relay board to control the brake release and clockwise/counterclockwise rotation. The relay contacts are wired in parallel with the three switches on the front panel of the control box.

Construction

Figure 15.5 shows the schematic diagram for the Keypad Entry Rotator Controller project. The plan is to mount the Arduino, a prototyping shield, and a separate relay board to control the rotation and brake operations inside the CDE/HyGain rotator control unit. The keypad and a 1.8-inch ST7735-type color TFT display will go in separate enclosures outside the control box.

We'll mount the Arduino Uno and a prototyping shield on the underside of the CDE/HyGain control unit, up against the rear panel, as shown in **Figure 15.6**. By mounting the Uno as shown in the photo, you can cut a small hole for the USB cable in the rear of the controller. This will allow you to access the Uno's USB port and modify the sketch or use the Serial Monitor without having to open the controller unit. We'll mount the relay board on the underside as well. Note that the relay board was cut and drilled to mount on the lower two screws for the 28 V ac transformer mount.

A small circular hole with a strain relief is used to pass the +12 V dc power and keypad cables as shown. You will want to keep the TFT display cable as short as possible. That cable runs from the connector (J9) on the prototyping shield, exiting the enclosure between the rear lip of the enclosure and cover to keep it short. (The cable is not shown in the photo but connects to the 8-pin header strip on the prototyping shield, near resistors R1-R5.)

The TFT itself is mounted in a Solarbotics Mega SAFE as shown in Figure 15.1 at the beginning of this chapter, but you can use whatever enclosure you like, because the only thing in the enclosure is the display itself. The RobotDyn 4×4 analog keypad is mounted inside a matching enclosure available from RobotDyn and connected to the Arduino board using a short piece of ribbon cable.

One of the reasons I chose the RobotDyn keypad for this project is the ability to use your own key labels. The original set of plain inkjet-printed labels I made is shown in **Figure 15.7**. After doing that and realizing that the keypad looked more like a ransom note than a nice project keypad, I replaced the white paper labels with laser-printed decal labels. You can get decal paper for inkjets from craft stores such as Hobby Lobby. You can also buy a laser-printer version of the decal paper online. I like using the laser-printed labels because you don't have to spray them with the waterproofing fixer spray after you print them. **Figure 15.8** shows the finished keypad in its case after switching to the printed decal labels.

We'll mount the ADS1115 16-bit programmable A/D converter and other components on the prototyping shield. You may recognize that this is a different style of ADS1115 module than I have been using. The module shown in the photo was the most recent ADS1115 module available in late 2020. The ADS1115 A/D is used to read the CDE/HyGain's rotator position potentiometer. For this project, we'll be using a 100 kΩ /33 kΩ resistor divider (R9/R10)

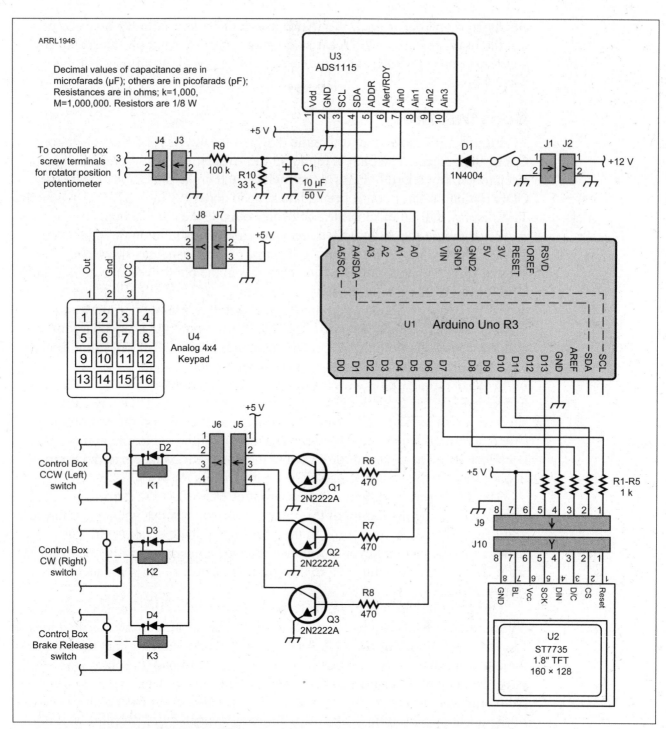

Figure 15.5 — The Keypad Entry Rotator Controller schematic.

C1 — 10 µF, 50 V electrolytic capacitor
D1-D4 — 1N4004 diode
J1, J3 — 2-pin DuPont-style male header
J2, J4 — 2-pin DuPont-style female header socket
J5 — 4-pin DuPont-style male header
J6 — 4-pin DuPont-style female header socket
J7 — 3-pin DuPont-style male header
J8 — 3-pin DuPont-style female header socket
J9 — 8-pin DuPont-style male header
J10 — 8-pin DuPont-style female header socket
K1-K3 — SPST relay with 5 V coil

Q1-Q3 — 2N2222A NPN transistor
R1-R5 — 1 kΩ, 1/8 W resistor
R6-R8 — 470 Ω, 1/8 W resistor
R9 — 100 kΩ, 1/8 W resistor
R10 — 33 kΩ, 1/8 W resistor
S1 — SPST toggle switch
U1 — Arduino Uno
U2 — ST7735-type 1.8-inch color TFT display
U3 — ADS1115 A/D converter
U4 — Analog 4×4 keypad (see text)
Enclosure — Solarbotics Mega SAFE

Figure 15.6 — The Keypad Entry Rotator Controller boards mounted in the bottom of the CDE/HyGain rotator control unit. The Arduino Uno is mounted under the prototyping shield.

Figure 15.7 — The "ransom note" keypad labels.

Figure 15.8 — The finished RobotDyn keypad in its optional enclosure with laser-printed decal keypad labels.

to reduce position voltage to a safe level. The reason for these high resistor values is so that we don't interact with the rotator controller's analog meter. The basic rule with resistors is that if the value you're using is 10 times or greater than the resistance of the circuit (known as the 10-to-1 rule), the added resistance has virtually no effect. By using a total of 133 kΩ worth of resistance, the position signal seen by the A/D converter is reduced to a safe value for the ADS1115 A/D, but also has virtually no impact on the existing meter in the controller.

While we're talking about the CDE/HyGain direction indication meter, I found out something that I think we all knew, but never really took into account. During my testing, I discovered that the accuracy can vary by as much as 5°. Adding the 10 µF capacitor (C1) on the rotator position potentiometer signal smoothed out the signal and helped to improve the accuracy. As you'll find with this project and the computer-controller rotator projects in earlier chapters that use the 16-bit ADS1115 A/D converter, accuracy down to 1° is possible. At the end of the day, though, there's no reason to obsess over a few degrees, because no antenna that we'll be using has a 1° beamwidth.

Calibrating the Keypad

Because we're using an inexpensive analog keypad comprised of different resistor values to indicate the key pressed, and real resistors vary slightly from the marked value, it's more than likely that no two keypads will have the exact same analog values for the various keys. Theoretically, the resistance spacing between the keys is enough that small variations are not a concern, but it never hurts to calibrate the keypad used in your project just to be sure. To help with this, I found a library for the analog keypad that includes a calibration routine in its examples. While that library wasn't suitable for use in this project, I was able to adapt the calibration sketch to provide the values we'll need to calibrate the keypad for the main project.

Odds are that you can just build the project and use the project sketch without making any changes, as the analog keypad values are not that critical for proper key decoding. However, on the off chance you're having key decode errors, you can use the Keypad Calibration sketch to get the analog values for your keypad to insert into the project sketch. **Figure 15.9** shows the flowchart for the Keypad Calibration sketch.

To use the Keypad Calibration sketch, you press every key on the pad in order. The sketch will then read the analog value for each key and save the values in an array. At the end of the sketch, the array is printed on the IDE's Serial Monitor for you to cut and paste the information into the main project sketch.

We'll start the Keypad Calibration sketch by defining the analog input pin for the keypad and declaring the sketch variables, including the 16 element array we'll use to hold the keypress analog values.

The `setup()` function for this sketch is about as simple as it gets. We'll start the IDE Serial Monitor port and read the keypad's analog input pin to get the "no key pressed" value. We'll print the basic instructions for the calibration process to the Serial Monitor, and then start the actual calibration process.

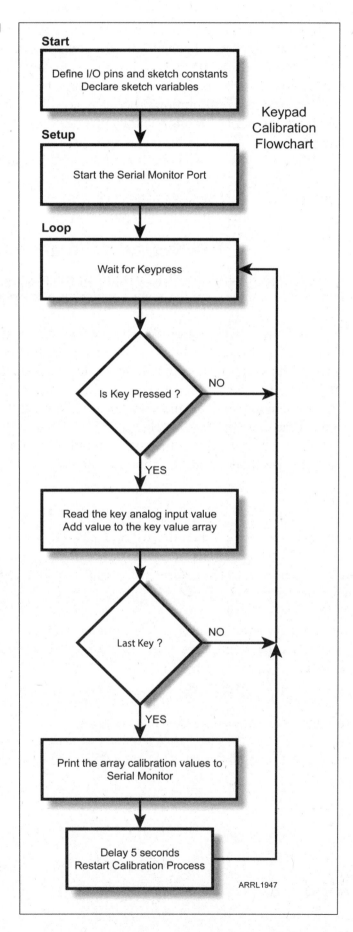

Figure 15.9 — Keypad Calibration sketch flowchart.

In the main `loop()`, we'll wait for a key to be pressed. Note that if you press the wrong key, whatever key you pressed will be assigned the value the A/D converter reads. It's best to just press the keys sequentially as they are numbered on the keypad board's silkscreen lettering. If you do make an error, you can repeat the calibration process as no values are actually saved, just printed to the Serial Monitor so they can be copied and pasted into the main project sketch.

When a key is pressed, we'll read the analog value of the key and add that value to the key value array. Once we read the last key on the keypad, we'll print the full text of the command statement and the array values so you can just copy and paste this into the main project sketch.

The Keypad Calibration Sketch

Sketches, libraries, Fritzing diagrams, and other useful files for the projects in this book can be found at **www.sunriseinnovators.com** and **www.kw5gp.com**.

The Keypad Calibration sketch will read the analog input pin the keypad is connected to and create an array of the key values for use by the main project sketch. The sketch will output an array declaration statement that you can copy from the IDE's Serial Monitor display and paste it into the main project sketch. The initial comments for the sketch show the default values used for the keypad used in the main project sketch.

We'll start by defining the analog input pin and the basic parameters and variables we'll need for the calibration sketch.

```
// Analog Keypad Test
//
// This sketch will monitor the analog pin and create a map array that can be
// copied from
// the serial monitor into your sketch; allowing for easier setup of arbitrary
// analog keypads
//
// Just run it and follow the instructions in the serial monitor
// The line to copy will be very similiar to the following.  Review the
// AnalogKeypad_Simple
// example sketch on how to apply it.
//
// The default for the RobotDyn 4x4 Analog Keypad
// keypad_map[] =
// {1023,931,853,787,675,633,596,563,503,480,458,438,401,322,269,231};
//
//

//#define debug

#define keypad_pin    A0
#define no_key_pressed 0
```

```
#define samples 5  // The number of times to read the key
#define read_delay 10   // Delay between data samples
#define variance 10

int key_value;
int keypad_map[16]; // The number of keys on the keypad
int key;
int zero_value;
int analog_value;
```

In the `setup()` function, we'll start the IDE's Serial Monitor and read the analog value when no key is pressed. You can transfer this value into the main sketch for the `no_key_pressed` variable as needed. Then you'll be instructed to press one key at a time until all of the keys have been read and their values stored in the array. You will notice on the keypad circuit board silkscreen that the key switches are numbered from 1 to 16. You'll need to press the keys in the order of their silkscreen label on the keypad circuit board to complete the calibration sequence.

```
void setup()
{
  Serial.begin(115200);

  Serial.print("No key pressed value defined as: ");
  Serial.print(no_key_pressed);

  read_key(); // Read key 5 times and average - put value in key_value
  Serial.print("   Actual Value:");
  Serial.println(key_value);
  zero_value = key_value;
  Serial.println("Keypad Calibration Sequence Initialized");
  Serial.println();
  Serial.println("Press one button at a time, hold it until instructed to
release");
  Serial.println();
}
```

In the main `loop()`, we'll set up a For... and a Do... loop to read the analog value for each keypress. We'll then save the keypress value in an array.

```
void loop()
{
  for (key = 0 ; key <= 15; key++)
  {
    Serial.print("Press Key ");
    Serial.println(key + 1);

    do
    {
      read_key();
    } while (key_value < (zero_value + variance));

    read_key();
    Serial.print("Release Key ");
    Serial.println(key + 1);

#ifdef debug
    Serial.print("  Key Value = ");
    Serial.println(key_value);
#endif

    keypad_map[key] = key_value;

    do
    {
      read_key();
    } while (key_value > (zero_value + variance));

    if (key != 16)
    {
      Serial.println("Delaying for next key");
      delay(read_delay);
    }
  }
```

Once we have all of the key values in the array, we'll format it to look like the array declaration statement you'll need to plug into the main project sketch. We'll then print the generated command, and you can use the copy and paste function in the IDE to copy this into the main project sketch.

```
// All key data acquired - Print the keypad map array

Serial.println("Copy the following line into your sketch");
Serial.println();

Serial.print("const int keypad_map[] = {");
for (key = 0; key <= 15; key++)
{

  Serial.print(keypad_map[key]);
  if (key != 15)
  {
    Serial.print(",");
  } else
  {
    Serial.println("};");
    Serial.println();
  }
}

Serial.println("Keypad calibration complete");
delay(10000);
Serial.println("Restarting calibration sequence");
delay(5000);

}
```

CDE/HyGain Keypad Entry Rotator Controller

The `read_key()` function is used to read the analog value of the pressed key. When a key is pressed, we'll read the Arduino's A/D converter several times and use the average of the key values as the calibration value for that key.

```
void read_key()
{

#ifdef debug
  Serial.println();
  Serial.print("Averaging ");
#endif

  int average = 0;
  int count = samples;
  for (count = 0; count < samples + 1 ; count++)
  {
    analog_value = analogRead(keypad_pin);
    delay (read_delay);
    if (count != 0)
    {
      average = average + analog_value;
    }


#ifdef debug
      Serial.print(" ");
      Serial.print(count);
      Serial.print(": ");
      Serial.print(average / count);
#endif

  }
  key_value = average / samples;
 }
}
```

Keypad Entry Rotator Controller Flowchart

Figure 15.10 shows the Keypad Entry Rotator Controller flowchart for the main project. This sketch is very similar to the computer-controlled rotator control sketches, except the initial rotation is commanded by the keypad and stopped when either the destination is reached or the Stop key is pressed.

We'll start by including the Adafruit GFX, Adafruit ST7735, and SPI libraries for the TFT display. We'll also include the Adafruit ADS1015 and Wire libraries for the ADS1115 A/D converter on the I²C bus. While the Adafruit ADS1015 library works fine, I prefer to use the ADS1115_WE library by Wolfgang Ewald, as it's a more full-featured library with access to more of the ADS1115 A/D's features. For this project however, the Adafruit library is easier to use and does what we need it to do. You'll note that no library is used

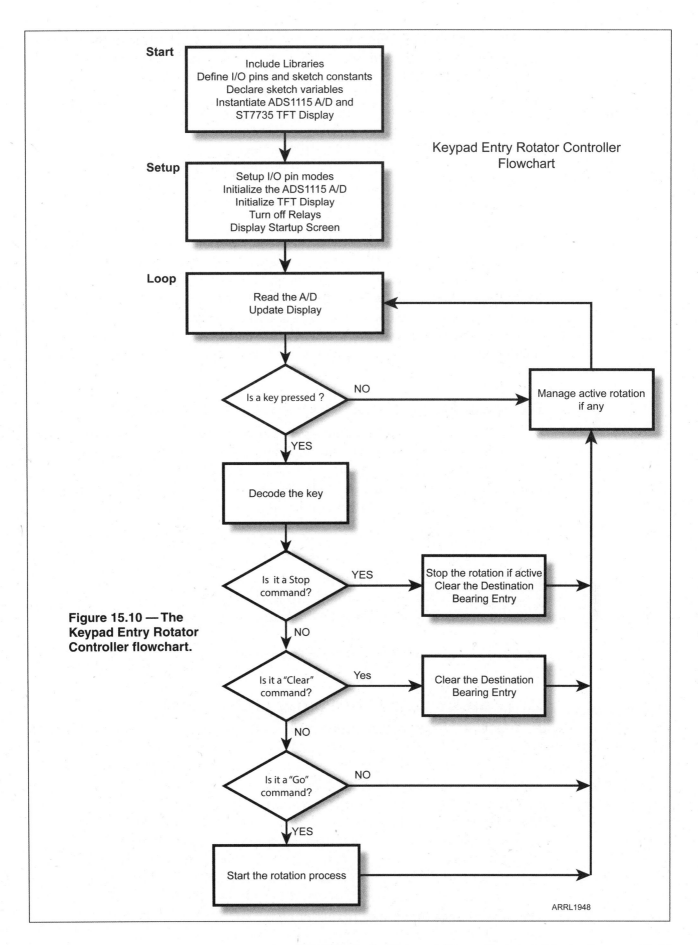

Figure 15.10 — The Keypad Entry Rotator Controller flowchart.

for the analog keypad. While there is a library available for the analog keypad, I was not happy with the implementation and its features weren't really beneficial for this particular project, so we just handled the analog keypad output manually.

Next, we'll define the I/O pins and constants used in the sketch, along with the keypad configuration array. Then we'll be instantiating (creating) the ADS1115 A/D and ST7735 TFT display objects.

In the `setup()` function, we'll set the digital I/O pin modes, initialize the ST7735 TFT display, and initialize the ADS1115 A/D converter. We'll also make sure the brake release and rotation relays are turned off, and display a brief startup message on the TFT display.

In the main `loop()`, we'll check for a keypress on the keypad and update the rotator bearing on the TFT if it has changed from the previous display update. If a key is pressed, we'll decode it, and if it's a number, we'll add it to the bearing variable and wait for the Go key to be pressed. If it's the Clear key, we'll erase the bearing entry and start the process over.

Once the Go key is pressed with a valid bearing entry, we'll process the rotation until the specified destination has been reached or the Stop key is pressed.

Keypad Entry Rotator Controller Sketch

We'll start the sketch by defining several debug modes, each providing additional debugging information about specific portions of the sketch. Next we'll include the Adafruit GFX, Adafruit ST7735, and SPI libraries for the ST7735 color TFT display. We'll also include the Adafruit ADS1015 library and the Wire library for the ADS1115 A/D converter.

```
/*

   CDE/HyGain Keypad Entry Rotator Controller

   The following table is for the RobotDyn 4x4 Analog Keypad

   Run the Analog Keypad Calibration sketch to create the Keypad Mapping table if
needed

   The default mapping for the RobotDyn 4x4 Analog Keypad
    const int keypad_map[] = {1023,931,853,787,675,633,596,563,503,480,458,438,401
,322,269,231};

   Uses Adafruit ADS1015 Analog to Digital Converter library
   Uses Adafruit ST7735 and GFX TFT display libraries
   Uses Wire library to communicate with I2C A/D module
   Uses SPI library to communicate with SPI TFT display

*/


//#define debug     // Enables diagnostic information
```

```
//#define debug1  // Keypad related diagnostic information
//#define debug2  // Rotator Positioning and A/D diagnostic information
//#define debug3  // Individual A/D read values
//#define debug4  // Rotation calculation and movement diagnostic information

#include <Wire.h> //I2C Library
#include <Adafruit_ADS1015.h> // ADS1x15 A/D Library
#include <Adafruit_GFX.h>     // Core graphics library
#include <Adafruit_ST7735.h>  // Hardware-specific library
#include <SPI.h>   // SPI library so we can communicate with the TFT display
```

Next, we'll define the calibration settings for the rotator position potentiometer. Note that unlike the other rotator control projects, this project does not have a means to dynamically calibrate the rotator position potentiometer and save the calibration settings to EEPROM. You can easily add this functionality back in yourself as an enhancement to the project.

We'll also define the keypad calibration values and map the keys on the keypad to the text that we will use to match with them. This is the part of the sketch where you would replace the keypad_map[] array with the values from the Keypad Calibration sketch if you need to. We'll also create a string with the numbers 0-9 that we will use to determine if a keypress is a number or a command key.

```
#define default_cal_0 150   // The default ADC Zero (-180 degrees) value
#define default_cal_max 26895   // The default ADC Max (+180 degrees) value


// Map the calibrated analog keypad values to an array
const int keypad_map[] = {1023, 931, 853, 787, 675, 633, 596, 563, 503, 480,
458, 438, 401, 322, 269, 231};
// Map the keypad keys to their matching text
const String key_text[] = {"1", "2", "3", "A", "4", "5", "6", "B", "7", "8",
"9", "C", "D", "0", "E", "F"};
const String numbers = "1234567890";   // Creates a string used to compare if
                                       // a pressed key is a number
```

Next, we'll define the I/O pins and other items we'll need in the sketch. The CDE/HyGain Ham-series rotators do not have as much of a positioning error issue as the AR-40, but we'll still want the ability to trim the position just a bit. The trim settings are in this section as well. Note: When adjusting the trim settings, be sure to turn off all debug information so that the sketch runs at its normal operating speed to track properly with the rotation speed.

```
#define comm_speed 115200 // Define the Arduino Serial Monitor baud rate
#define keypad_pin  A0   // Define the keypad analog input pin
#define no_key_pressed 0  // The A/D value for when no key is pressed
#define samples 5 // The number of times to read the key when pressed
#define read_delay 10  // Delay between keypad data samples
#define variance 10 // The A/D count tolerance for the calibrated key value
                   // versus the actual value
#define update_time 1000  // The display update time in milliseconds
#define degree_tolerance 1  // The rotator error tolerance in degrees
#define brake_time 3000 // Define the Brake delay in milliseconds
#define release_delay 500 // Define the delay between the brake release and
                          // rotation start
#define left_rotate_trim -3 // Define the rotate error fine tuning value for left
                            // (counterclockwise) rotation
#define right_rotate_trim 3 // Define the rotate error fine tuning value for right
                            // (clockwise) rotation

#define TFT_CS     10  // Assign the TFT CS to pin 10
#define TFT_RST    7   // Assign the TFT RST to pin 7
#define TFT_DC     8   // Assign the TFT DC to pin 6
#define tft_delay 10   // set the TFT command delay to 10ms

#define left_relay 4   // Define the I/O pin for the rotate left relay
#define right_relay 5  // Define the I/O pin for the rotate right relay
#define brake_relay 6  // Define the I/O pin for the brake release relay
```

Then, we'll declare the remainder of the variables we'll need for the sketch and instantiate (create) the ADS1115 A/D and ST7735 TFT display objects.

```
int key_value;  // Variable to hold the averaged A/D value of the pressed key
int analog_value; // Variable to hold the converted analog value of the
                  // pressed key
int zero_point = no_key_pressed + variance;
int key_index = -1;  // The index point to the location of the pressed
                     // key in the keypad_map array
String test_key;  // Variable to hold the text value of the pressed key
String move_to = "";  // Variable for the text of the entered bearing
bool moving = false;  // Flag to indicate if the rotator is turning
bool data_changed = true; // Flag to indicate that the display data has changed
int entered_bearing = -1; // Variable to hold the entered destination
int current_bearing;   // Variable to hold the current bearing
```

```
int previous_bearing = -1;   // Variable to hold the previous bearing
int actual_bearing;  // Variable to hold the raw bearing value (-180 to +180)
boolean first_pass = true;   // Variable used to determine if first pass through
                             // the main loop
int adc_value;  // Variable to hold the ADS1115 A/D Value
int adc_cal_0 = default_cal_0;   // Variable to hold the ADS1115 ADC zero
                                 // calibration point
int adc_cal_max = default_cal_max;   // Variable to hold the ADS1115 ADC max
                                     // calibration point
int destination_bearing;   // Variable to hold the destination bearing
String rotate_direction;   // Variable to indicate if the rotation is to the left
                           // (counterclockwise) or right (clockwise)
unsigned long next_update_time = 0; // Variable to hold the next display
                                    // update time in millis()

Adafruit_ADS1115 ads;   // Create an instance for the ADS1115 16-bit A/D

Adafruit_ST7735 tft = Adafruit_ST7735(TFT_CS,  TFT_DC, TFT_RST);
  // Create the TFT display instance
```

In the setup() function, we'll set the digital I/O pin modes and verify that the brake release and rotation control relays are turned off. We'll also start the Serial Monitor port if debugging is enabled.

```
void setup()
{
  pinMode(left_relay, OUTPUT);   // Set the rotate left relay pin to output
  pinMode(right_relay, OUTPUT);  // Set the rotate right relay pin to output
  pinMode(brake_relay, OUTPUT);  // Set the brake release relay pin to output

  digitalWrite(left_relay, LOW);   // Turn off the rotate left relay
  digitalWrite(right_relay, LOW);  // Turn off the rotate right relay
  digitalWrite(brake_relay, LOW);  // Turn off the brake release relay

#ifdef debug
  Serial.begin(comm_speed); // Set the Serial Monitor port speed
  Serial.println("Starting");
#endif
```

Next, we'll set the ADS1115 A/D gain for a range of 0 to 4.096 V and start the A/D converter. Then we'll start the ST7735 TFT display and show a brief startup message. After the startup message, we'll clear the display, read the A/D converter, and update the TFT display with the current rotator position information.

```
ads.setGain(GAIN_ONE);          // Set the ADC Gain to 1x gain  +/- 4.096V
                                // 1 bit = 0.125mV

ads.begin();// Start the A/D Converter

tft.initR(INITR_18BLACKTAB);    // initialize a 1.8" TFT with ST7735S chip,
                                // black tab
delay(tft_delay);
tft.fillScreen(ST7735_BLACK); // Clear the display - fill with BLACK background
delay(tft_delay);
tft.setRotation(1); // Set the screen rotation
delay(tft_delay);
tft.setTextWrap(false); // Turn off Text Wrap
delay(tft_delay);
tft.setTextSize(3); // Set the Font Size
delay(tft_delay);
tft.setTextColor(ST7735_BLUE); //Set the Text Color
delay(tft_delay);
tft.setCursor(40, 10);  //Set the Cursor and display the startup screen
delay(tft_delay);
tft.print("KW5GP");
delay(tft_delay);
tft.setTextSize(2);
delay(tft_delay);
tft.setCursor(10, 60);
delay(tft_delay);
tft.print("Keypad Entry");
delay(tft_delay);
tft.setCursor(45, 80);
delay(tft_delay);
tft.print("Rotator");
delay(tft_delay);
tft.setCursor(25, 100);
delay(tft_delay);
tft.print("Controller" );

delay(5000);   //Wait 5 seconds then clear the startup message
tft.fillScreen(ST7735_BLACK); // Clear the display - set to use a black
                              // background
delay(tft_delay);
```

```
  read_adc(); // Read the ADC to get the initial values

  update_display(); // update the display

} // End of Setup Loop
```

In the main `loop()`, we'll first call the `read_key()` function to read the current value of the keypad analog input pin. Next, we'll call the `update_display()` function that will update the bearing on the display if the update timer has timed out and if the bearing has changed.

```
void loop()
{
  read_key(); // Check the keypad to see if a key is pressed
  if ((millis() > next_update_time && data_changed) || first_pass)  // Check to
                                    // see if we need to update the TFT display
  {
    update_display(); // Update the data on the TFT display
  }
```

Then we'll check to see if a valid key has been pressed. If a key has been pressed, we'll get the text associated with the pressed key.

```
  if (key_index != -1)  // Check to see if there is a valid key pressed
  {
    test_key = key_text[key_index]; // Get the text for the pressed key
    // We have a valid key

#ifdef debug1
    Serial.print("Valid Key: ");
    Serial.println(key_index);
#endif

    // Key has been decoded - determine action
    //If it's a number, check that's it's a valid number
    // Key codes 0-2, 4-6, 8-10 and 13 are digit entries

#ifdef debug1
    Serial.print("Key Text: ");
    Serial.print(test_key);
    Serial.print("   Checking numbers string: ");
    Serial.println(numbers.indexOf(test_key));
#endif
```

Next, we'll check to see if the pressed key is a number. If the rotator is not currently rotating, this number will be added to the bearing input string.

```
    if (numbers.indexOf(test_key) != -1)  // Check to see if the test key is in
                                          // the numbers string (it's a number key)
    {
      if (!moving)   // Allow bearing entry when not moving
      {
        // It's a valid number - add it to direction string

        move_to = move_to + key_text[key_index];   // Add the key entry to the
                                                   // desired bearing
        entered_bearing = move_to.toInt();   // Convert the desired bearing
                                             // to a string

#ifdef debug
        Serial.print("Valid Number: ");
        Serial.println(key_text[key_index]);
        Serial.print("Bearing: ");
        Serial.println(entered_bearing);
#endif
```

Now, we'll check to be sure that the entered bearing hasn't gone over 359 degrees. If it has, we'll clear the bearing entry and restart the `loop()`.

```
        if (entered_bearing >= 360)   // Clear the entered bearing if it's over 359
        {
#ifdef debug
          Serial.println("Entry Error - Bearing Cleared");
#endif
          move_to = "";
          entered_bearing = -1;
        }
        data_changed = true;   // Flag to indicate the display needs to
                               // be updated
      }
```

If the key pressed is not a number, we'll check to see if it is a command key. First, we'll check to see if it is an "E" key, indicating that we want to clear the destination bearing entry.

```
    } else
    {
      // It's a command key
#ifdef debug
      Serial.println("Command Key Pressed");
#endif
```

15-20   Chapter 15

```
        if (test_key == "E")   //Clear Bearing Entry
        {
          if (!moving)   // Don't execute if moving
          {
            move_to = ""; // Reset the bearing entry string
            entered_bearing = -1; // Reset the entered bearing
            data_changed = true;// Flag to indicate the display needs to
                                // be updated

#ifdef debug
            Serial.println("Clear Key Pressed - Bearing cleared");
#endif
          }
```

Next, we'll check to see if it is the "D" key, which performs the role of the Go key. If a valid bearing has been entered and we're not already rotating, we'll call the start_rotate() function to begin the rotation process.

```
        } else
        {
          if (test_key == "D")   // Go Key
          {
            if (entered_bearing != -1 && !moving) // Only execute if not moving
                                    // and a valid destination has been entered
            {
#ifdef debug
              Serial.print("Go Key Pressed - Destination Bearing: ");
              Serial.println(entered_bearing);
#endif

              start_rotate(); // Process the rotation

            }
```

CDE/HyGain Keypad Entry Rotator Controller

Lastly, we'll check to see if it is the "F" key, indicating that it's the Stop key. The Stop key will immediately stop all rotation, clear the destination bearing entry, and update the display.

```
      } else
      {
        if (test_key == "F")   // Stop Key
        {

#ifdef debug
          Serial.print("Stop Key Pressed");
#endif

          if (moving) // If rotating, stop rotation
          {
#ifdef debug
            Serial.println(" - Rotation stopped");
#endif

            stop_all(); // Stop the rotation
          }
          if (entered_bearing != -1)   // Clear the bearing entry if it's a
                                       // valid destination
          {
#ifdef debug
            Serial.println(" - Bearing cleared");
#endif
          }
          move_to = ""; //   Reset the bearing string
          entered_bearing = -1; // Reset the bearing entry
          data_changed = true;  // Indicate the display needs updating
        }
      }
    }
  }
}
```

We'll then call the `read_adc()` function to read the ADS1115 16-bit programmable gain A/D converter. If we're rotating, we'll call the `check_move()` function to manage any active rotation process.

```
  read_adc(); // Read the ADS1115 A/D

  if (moving) // If we're moving - check where we are and stop if we're at
              // the destination
  {
    check_move(); // Calls the function to process the rotation
  }
} // End of Main Loop
```

The `read_key()` function will read the analog input pin for the 4×4 analog keypad. The keypad signal will be read five times and the results averaged to get us the analog value of the pressed key.

```
void read_key() // read_key function - Reads the 4x4 analog keypad
{
  int average = 0;  // Local variable to hold the average analog value
  int count = samples;  // Local variable to hold the number of data samples
                        // to get
  key_value = 0;  // Clear the incoming key value
  key_index = -1; // Clear the keypad_map array pointer

  if (analogRead(keypad_pin) > zero_point)  // Valid Key Pressed -
                                            // otherwise exit
  {
    for (count = 0; count < samples + 1 ; count++)  // Get the desired number of
                                                    // keypad data samples
    {
      analog_value = analogRead(keypad_pin);  // Read the keypad analog
                                              // input pin
      delay (read_delay); // Delay between samplings
      if (count != 0) // Only if the number of requested samples is not zero
      {
        average = average + analog_value; // Add the analog value to the average
      }
      key_value = average / samples;  // Calcluate the average value
    }

#ifdef debug
    if (key_value > zero_point) // If the key analog value is not at the zero
                                // voltage point
    {
#ifdef debug1
      Serial.print("Key Value: ");
      Serial.println(key_value);
#endif

    }

#endif
```

Then, we'll wait for the key to be released and then call the `decode_key()` function.

```
  do
  {
    // Wait until the key is released
  } while (analogRead(keypad_pin) > zero_point);   // Stay in the loop as long
                                                   // as the keypad is pressed

    decode_key(); // Decode the key press
  }
}
```

The `decode_key()` function will take the value of the pressed key and scan the contents of the `keypad_map[]` array, looking for a matching value.

```
void decode_key() // Function to translate key code to its position number
{
  key_index = 1;   // Local variable for the pointer to the keypad_map array

#ifdef debug1
  Serial.println("Scanning Array");
#endif

  for (int x = 0; x <= 15; x++)   // Scan the keypad_map array looking for
                                  // closest valid match
   {
#ifdef debug1
     Serial.print("Index: ");
     Serial.print (x);
#endif

    if (abs(key_value - keypad_map[x]) < variance)   // If the data matches
                                                     // within the specified tolerance
    {

#ifdef debug1
      Serial.print("    Match");
      Serial.print("    Key Char: ");
      Serial.println(key_text[x]);
#endif

      key_index = x;   // Set the key pointer to the matching key

      break;   // Exit as soon as we have a valid match
    }

#ifdef debug1
```

```
    Serial.println();
#endif

  }
#ifdef debug1
  Serial.print("Key Index: ");
  Serial.println(key_index);
#endif
}
```

The `read_adc()` function will first read the rotator position potentiometer on input 0 of the ADS1115 A/D converter.

```
void read_adc() // Function to read the ADS 1115 A/D Converter
{
  adc_value = ads.readADC_SingleEnded(0); // Get the raw ADC value

#ifdef debug3
  Serial.print("Current ADC Value: ");
  Serial.println(adc_value);
#endif
```

If the value has changed since the last A/D read, we'll then convert and map the analog value to a degree value from −180 to +180 degrees, and flag the TFT display to update the rotation bearing value.

```
  // Now map the value to degrees (-180 to +180)
  actual_bearing = map(adc_value, adc_cal_0, adc_cal_max, -180, 180);
  // This is the true bearing (-180 to + 180 degrees)
  // Only process the data if it's outside of the degree tolerance value or
  // if it's the first time through this code
  if ((abs(actual_bearing - previous_bearing) > degree_tolerance) || previous_bearing == -1)
  {
  // Convert the bearing from -180 to +180 into two zones of 180 - 359 and
  // 0 - 180 for proper calculation and display
    if (actual_bearing < -180)  // Set the lower limit to -180
    {
      actual_bearing = -180;
    }

    if (actual_bearing > 180) // Set the upper limit to +180
    {
      actual_bearing = 180;
    }
```

CDE/HyGain Keypad Entry Rotator Controller

```
#ifdef debug2
    Serial.print("Actual Bearing: ");
    Serial.print(actual_bearing);
#endif

    if (actual_bearing < 0)    // Now map it to the display value
    {
      current_bearing = actual_bearing + 360;
    } else {
      current_bearing = actual_bearing;
    }

#ifdef debug2
    Serial.print("   Current Bearing: ");
    Serial.println(current_bearing);
#endif

    previous_bearing = actual_bearing;  // Update the previous bearing value
    data_changed = true;   // Indicate the display needs to be updated
  }
}
```

The first time the `update_display()` function is called, it will clear the TFT screen and display a basic template of data that doesn't change during sketch operation. This is done to reduce the amount of information that has to be erased and rewritten on every display update that can give the appearance of flickering on the TFT.

```
void update_display() // Updates the TFT display
{
  if (first_pass) // Only do this part the first time the function is called
  {
    // Clear the screen and display normal operation
    tft.fillScreen(ST7735_BLACK); // Clear the display
    delay(tft_delay);
    tft.setTextSize(1); // Set the text size to 1
    delay(tft_delay);
    tft.setTextColor(ST7735_BLUE); // Set the text color to BLUE
    delay(tft_delay);
    tft.setCursor(35, 5);
    delay(tft_delay);
    tft.print("Current Bearing");  // Display Current Rotator Bearing
    tft.setCursor(35, 60);
    delay(tft_delay);
    tft.print("Desired Bearing");  // Display Entered Rotator Bearing
    delay(tft_delay);
    tft.setCursor(67, 115);
    delay(tft_delay);
```

```
      tft.setTextColor(ST7735_GREEN); // Set the text color to GREEN
      delay(tft_delay);
      tft.print("Ready");
      first_pass = false; // Turn off the first pass flag
```

If this is not the first time the function has been called, we'll call the `clear_data()` function to clear the data display areas and display the current bearing. If the rotator is rotating, we'll display this in yellow, otherwise it will be displayed in green.

```
} else
  {
    clear_data();  // Clear the value display area

    tft.setTextSize(4); // Set the text size to 4
    delay(tft_delay);
    if (moving) // If the rotator is turning set the current bearing color to
                // yellow - otherwise set it to green
    {
      tft.setTextColor(ST7735_YELLOW); // Set the text color to GREEN
      delay(tft_delay);
    } else
    {
      tft.setTextColor(ST7735_GREEN); // Set the text color to GREEN
      delay(tft_delay);
    }
    tft.setCursor(50, 25);
    delay(tft_delay);

    convert_bearing(current_bearing); // convert the current bearing to a
                                      // 3 digit formatted string
    tft.setTextColor(ST7735_GREEN); // Set the text color to GREEN
    delay(tft_delay);
```

CDE/HyGain Keypad Entry Rotator Controller

If there is a valid destination bearing entered, the display will be updated with the currently entered destination bearing and the next display update time is calculated.

```
   if (entered_bearing != -1)  // If the entered bearing is a valid entry
   {
     tft.setCursor(50, 80);
     delay(tft_delay);
     convert_bearing(entered_bearing);// Convert the entered bearing to a
                                      // 3 digit formatted string
   }
   data_changed = false; // The display is updated - Turn off the display
                         // update flag
 }

 next_update_time = millis() + update_time;  // Calculate the time for the
                                             // next update

}
```

The `clear_data()` function is used to clear specific areas of the display. This is done to reduce the amount erasing and rewriting of display data, which results in flickering on the display. We'll only clear the areas where the current bearing and the entered destination bearing are displayed.

```
   if (entered_bearing != -1)  // If the entered bearing is a valid entry
   {
     tft.setCursor(50, 80);
     delay(tft_delay);
     convert_bearing(entered_bearing);// Convert the entered bearing to a
                                      // 3 digit formatted string
   }
   data_changed = false; // The display is updated - Turn off the display
                         // update flag
 }

 next_update_time = millis() + update_time;  // Calculate the time for the
                                             // next update

}

void clear_data() //Clears the data area of the display
{
  tft.fillRect(45, 23, 78, 33, ST7735_BLACK); // Clear the display data areas
  delay(tft_delay);
  tft.fillRect(45, 78, 78, 33, ST7735_BLACK); // Clear the display data areas
  delay(tft_delay);
}
```

The `clear_status()` function is similar to the `clear_data()` function, except it will clear the bottom line of the TFT display that is used to provide status information.

```
void clear_status() // Clears the status line of the display
{
  tft.fillRect(0, 112, 159, 15, ST7735_BLACK); // Clear the display data areas
  delay(tft_delay);
}
```

The `start_rotate()` function is used to set up and start the rotation process. It will use the difference between the current bearing and the destination bearing to determine the proper direction to rotate.

```
void start_rotate() // Handles the rotate process
{
  // sets up and starts the rotation process

  // Check to be sure it's a valid move (greater than degree tolerance) Move has
  // already been checked for being a valid entry

  // We use actual bearing value since we may cross the 359 to 0 degree boundary
  // We already have actual - we need to calculate it for the desired
  // destination

  if (entered_bearing > 180)  // Convert the destination to the -180 to +180
                              // actual degree format
  {
    // We're in 181 to 359 area - subtract 180 degrees
    destination_bearing = entered_bearing - 360;
  } else
  {
    // We're in 0 to 180 - don't do anything with it
    destination_bearing = entered_bearing;
  }

#ifdef debug4
  Serial.print("Current = ");
  Serial.print(actual_bearing);
  Serial.print("   Start Moving - Destination = ");
  Serial.print(entered_bearing);
  Serial.print("   Actual Destination Bearing = ");
  Serial.print(destination_bearing);
#endif
```

Note that rotation will not occur if the entered destination bearing is within the tolerance range of the current rotator position. Next, we'll check to see if the entered value is 180°. On the standard CDE/HyGain Ham-series rotators, you have two 180° points on the meter, far left, and far right. If the rotator is commanded to go to 180°, we'll rotate to the closest one.

```
// Only start if the desired destination is outside of the degree tolerance
if (abs(actual_bearing - destination_bearing) > degree_tolerance)
{
   if (destination_bearing == 180) // If desired bearing is 180 degrees, turn
                                   // towards the closest 180 degree point
   {
    // Find the nearest one and go to it
    if (actual_bearing < 0) // Convert the destination bearing to the actual
                            // bearing format
    {
      // Turn Left
      destination_bearing = -180;

#ifdef debug4
      Serial.println(" Turn left");
#endif

      rotate_direction = "L"; // The rotation direction to the nearest 180
                              // should be Left
      rotate(); // Start the rotation
    } else
    {
      // Turn Right

#ifdef debug4
      Serial.println(" Turn right");
#endif

      rotate_direction = "R"; // The rotation direction to the nearest 180
                              // should be Right
      rotate(); // Start the rotation
    }
```

If the desired bearing is not 180°, we'll determine which direction to rotate and call the `rotate()` function to begin the rotation process.

```
      } else
      {
        // Do we turn right (clockwise) or left (counterclockwise)?
        if (actual_bearing > destination_bearing) //  If the current bearing is
                                     // greater than the destination - turn left
        {
          // Turn left

#ifdef debug4
        Serial.println(" Turn left");
#endif

          rotate_direction = "L"; //  The rotation direction should be Left
          rotate(); // Start the rotation

        } else
        {
          // Turn right
#ifdef debug4
        Serial.println(" Turn right");
#endif

          rotate_direction = "R"; //  The rotation direction should be Right
          rotate(); // Start the rotation
        }

        moving = true;  // Set the rotating flag
      }
```

If the desired rotation is within the tolerance range, we won't execute the move and will exit the start rotation process.

```
    } else
    {
      // Don't execute the move
      moving = false; // Turn off the moving flag
      move_to = ""; // Reset the desired bearing string
      entered_bearing = -1; // Reset the entered bearing value
      data_changed = true;   // Indicate that the display needs to be updated
    }
  }
```

The `convert_bearing()` function will receive an integer bearing value, convert it to a three character string, and display it on the TFT.

```
void convert_bearing(int calc_bearing)   // Function to convert int bearing
                                         // to String
{
  // Convert the bearing to a 3 character string and pad with zeroes
  String display_bearing = String(calc_bearing);
  while (display_bearing.length() < 3)    // Do this while the string length is
                                          // less than 3 digits
  {
    // While the string length is less than 3 digits
    display_bearing = "0" + display_bearing;  // Pad with leading zeroes until
                                              // we have a 3 digit value
  }
  tft.print(display_bearing); // Display the padded bearing value
}
```

The `stop_all()` function is used to stop all rotation. It will first clear the TFT status line, update it with a red braking message, and de-energize the brake release and rotation control relays. After a 3 second delay to allow rotation to stop, the braking message will be cleared.

```
void stop_all() // Function to stop all rotation
{
  // Stop all rotation and trigger braking delay
#ifdef debug4
  Serial.println("Stop all Rotation - Initiate Braking delay");
#endif

  clear_status(); // Clear the display status line
  tft.setTextSize(1); // Set the text size to 1
  tft.setTextColor(ST7735_RED); // Set the text color to RED
  delay(tft_delay);
  tft.setCursor(65, 115);
  delay(tft_delay);
  tft.print("Braking"); // Indicate we are braking
  digitalWrite(left_relay, LOW);
  digitalWrite(right_relay, LOW);
  delay(brake_time);
  digitalWrite(brake_relay, LOW);
  clear_status();
  tft.setTextColor(ST7735_GREEN); // Set the text color to RED
  delay(tft_delay);
  tft.setCursor(67, 115);
  delay(tft_delay);
  tft.print("Ready");
```

```
  moving = false; // Turn off the moving flag
  move_to = ""; // Reset the desired bearing string
  entered_bearing = -1; // Reset the entered bearing value
  data_changed = true;  // Indicate that the display needs to be updated
}
```

The `rotate()` function is used to start the rotation operation. It will first change the color of the current bearing display to yellow to indication rotation is occurring, then it will energize the brake release relay along with the proper rotation direction control relay.

```
void rotate() // Function to perform rotation
{
  clear_status(); // Clear the display status line
  tft.setTextSize(1); // Set the text size to 1
  delay(tft_delay);
  tft.setTextColor(ST7735_YELLOW); // Set the text color to Yellow
  delay(tft_delay);
  tft.setCursor(60, 115);
  delay(tft_delay);
  tft.print("Rotating"); // Indicate we are rotating
  if (rotate_direction == "L")  // Turn on the left rotation relay and release
                                // the brake
  {
    digitalWrite(brake_relay, HIGH);  // Release the brake
    delay(release_delay);
    digitalWrite(left_relay, HIGH); // Enable the left relay
  } else
  {
    // Turn on the left rotation relay and release the brake
    digitalWrite(brake_relay, HIGH);  // Release the brake
    delay(release_delay);
    digitalWrite(right_relay, HIGH); // Enable the right relay
  }
}
```

The `check_move()` function is used to manage an active rotation process. This function will check to see if we've reached the destination bearing. If so, we'll call the `stop_all()` function to end the rotation process.

```
void check_move()
{
  // We only get here if it's already moving
  // Checks to see if we've reached the destination bearing
  // We've just read the ADC so no need to read it again
  if (rotate_direction == "L")   // destination is less than current position
  {
    if (actual_bearing <= (destination_bearing + left_rotate_trim))   // Stop
       // when we get there - include left rotate trim
    {
      stop_all(); // Stop all rotation
    }
  } else
  {
    // We're turning right
    if (actual_bearing >= (destination_bearing + right_rotate_trim))   // Stop
       // when we get there - include right rotate trim
    {
      stop_all(); // Stop all rotation
    }
  }
}
```

Enhancement Ideas

As mentioned earlier, this project does not include the rotator position potentiometer calibration switches. Because no two rotator controllers are alike, you might want to consider adding these back in and saving the calibration settings to EEPROM as we do with some of the other rotator controller projects.

This would also be an ideal project to implement a Bluetooth link between the keypad and the rotator control box, and maybe even make the TFT display remote as well. You could expand this to add extra keypad controller and give any station at an event such as ARRL Field Day the ability to rotate an antenna using a common Bluetooth link on the control unit.

And of course, you could always add the interface to allow rotator control from a workstation running *Ham Radio Deluxe*. You could even add a Bluetooth link between the workstation and the rotator control unit so you could have wireless rotator control and position display capabilities.

16 RTTY Reader

Radioteletype, more commonly known as RTTY, was one of the main reasons for upgrading my Novice license to General back when I first got into ham radio. My local ham buddies had a 2-meter RTTY net, and when I first saw that noisy, clanking, Model 15 mechanical teletype machine and smelled the machine oil, I was hooked. I'm absolutely certain that my parents wished that I had chosen a different hobby, especially one that did not include having that noisy beast in my bedroom going off at all hours. Yes, I was enjoying the heck out of operating CW as a Novice, but there was something about the whole RTTY thing that was fascinating to me, and I just had to upgrade as soon as possible.

If you don't count CW, RTTY was the first digital mode used by hams. Instead of the standard 8-bit ASCII character set, amateur radio RTTY uses a 5-bit Baudot-coded character set, transmitted at 45.45 baud. That may seem like an unusual speed at first, but this actually works out to 60 words per minute. RTTY is considered to be a keyboard-to-keyboard mode, meaning that operators essentially type to each other over the air.

On-air RTTY operation is similar to the more modern PSK31 digital mode. RTTY has no error correction, handshaking, or time synchronization that is needed with modes such as JT65, FT8, and FT4. For "canned" messages, some of the

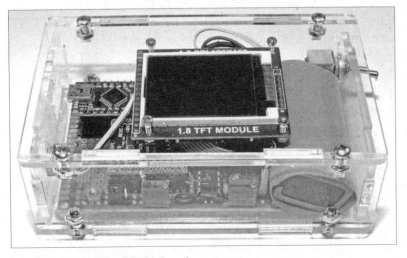

Figure 16.1 — The RTTY Reader.

old RTTY machines had paper tape readers and punches instead of the computer software and memories used today. In the old days, hams enjoyed collecting and sharing typewritten "pictures" via RTTY that were stored on paper tape. (An online search for "RTTY art" will turn up numerous examples. Some are quite intricate.)

The RTTY Signal

RTTY consists of a frequency shift keyed (FSK) signal that is modulated between two carrier frequencies, called the *mark* frequency and the *space* frequency. The difference between the mark and space frequencies is called the *FSK shift*, usually 170 Hz for an amateur RTTY signal. Typically, these frequencies are 2125 Hz for mark and 2295 Hz for space, a frequency shift of 170 Hz. These are called *high tones*, and you may see references to *low tones*, which are typically 1275 Hz for mark and 1445 Hz for space with a 170 Hz shift. If your transceiver does not have a dedicated RTTY (FSK) mode, it will use audio frequency shift keying (AFSK) with the transceiver set to LSB (lower sideband). On the air, FSK and AFSK signals are indistinguishable. AFSK RTTY can also be sent on VHF/UHF using an FM transceiver.

Back in the day, a piece of equipment known as a *terminal unit* was used to convert the mark and space tones into the current drive for the "loop" circuit that powered the signal relay inside the teletype machine. This signal relay was used to control all the gears and synchronous motor timing to select and type the individual characters. In today's RTTY environment, the tone decoding is done via software, and the terminal unit and current loop are long gone unless you are still using a mechanical teletype machine.

Because the Baudot RTTY data word is only 5 bits, you can only have 32 different characters. To get around this limitation, Baudot incorporates the letters (LTRS) and figures (FIGS) shift functions, allowing access to a second set of 32 characters. The LTRS shift consists mainly of alphabetic characters, and the FIGS shift consists mainly of decimal numerals and punctuation marks. Two unique Baudot characters called LTRS and FIGS are used by the sender to command the decoder to switch between these two character sets.

The old mechanical teletypewriter keyboards have two keys to send the LTRS and FIGS characters. If you're going to have a computer translate between 5-bit Baudot and 8-bit ASCII, you'll need to take the FIGS and LTRS shift operation into account, and handle it within your software. If you miss a FIGS or LTRS shift due to noise or fading while receiving, the output will appear to be garbled because you're receiving on the opposite character set from what you should be. For example, 599 will be received as TOO. Most RTTY programs perform what is known as *unshift on space* (USOS) to restore the LTRS shift every time a space character is received.

Back in my early ham years, I was also tinkering around with the RCA 1802 COSMAC Elf CPU, the world's first CMOS microprocessor. While all my friends and co-workers were playing with the new Apple, RadioShack, and 8080/Z-80 based systems, I was playing with the RCA 1802. I liked the 1802 because it was about the closest thing you could get to a microcontroller back then and I was doing a lot of microcontroller-style projects at that time. The

Figure 16.2 — The SC16IS750 UART module.

1802 found a niche in the early embedded computer environment, and was widely used in the industrial and aerospace worlds. The 1802 was used in NASA's Galileo space probe to Jupiter in 1989 and other space projects, including the Hubble Space Telescope. The 1802 also found a home in ham radio satellites, as the onboard internal housekeeping unit (IHU) for AMSAT's Phase 3D, as well as OSCAR 10 and 13. On a more personal note, the very first magazine article I ever had published was about building an ASCII-to-Baudot RTTY converter for an RS-232 video display terminal, so that I could do RTTY without all the noise and motor vibration of the big clunky machines.

Since those early days, RTTY has always been one of my favorite operating modes. Today, virtually all RTTY operation on the ham bands uses PC-based RTTY programs such as *Ham Radio Deluxe DM780*, *MMTTY*, and *FLDIGI*. While thinking up projects for this book, I came across a new Arduino module that really caught my eye, and my first thought was to use an Arduino to create a handheld RTTY decoder and display as shown in Figure 16.1 on the first page of this chapter. While I've never been a big fan of using the Arduino's software serial library to handle all the serial bit timing, this new module, the NXP SC16IS750 universal asynchronous receiver transmitter (UART) module shown in **Figure 16.2**, would be the perfect thing to receive the 5-bit RTTY serial data and convert it to the 8-bit ASCII used by the Arduino and most other modern computers.

What is a UART?

A UART is an IC that is used to send and receive serial data. A typical UART can handle serial communication in either half-duplex or full-duplex mode. In its original design, the UART would convert the parallel data from a computer into a serial RS-232-compatible data stream. The UART manages all of the functions necessary to transmit and receive the serial data, including handshaking, bit timing, data buffering, and error detection. A UART can be configured to handle a data word from 5 to 8 bits long, at a programmable baud rate based on the input clock frequency. Typically, the input clock would be 16 times the baud rate, and either a crystal or NE555 oscillator circuit would provide this clock input.

The UART also supports even, odd, or no parity, and 1, 1½, or 2 stop bits. Handshaking and data flow control could be done using the RTS (Request to Send), CTS (Clear to Send), DTR (Data Ready), and other standard serial hardware flow control functions. Many UARTs also incorporated software flow control using the Xon/Xoff characters in the data stream.

Because the UART is designed to handle virtually all of the serial communication functions, usually very little extra work is needed to add hardware serial communication functions to your program. Simply put, the UART is the one-stop-shop for all things serial communications-wise.

The NXP SC16IS750 UART

The NXP SC16IS750 UART is an update to the original bi-directional parallel-to-serial-data UART design. Instead of a discrete 8-bit digital parallel interface, the SC16IS750 can be configured to use either the I²C or SPI bus to transfer the UART data to the Arduino. This frees up a number of I/O pins for the UART, and in fact, the UART design in this project only requires eight wires, including the power and ground connections.

With the exception of the I²C or SPI bus selection and the I²C device address, all of the UART configuration functions are fully programmable and managed via software. Additionally, the SC16IS750 incorporates a 64-bit transmit and receive FIFO (first in, first out) buffer. For baud rate selection, the UART module uses an external 14.7456 MHz crystal on the board, in conjunction with a programmable divider inside the UART, to create the internal clock rate of 16 times the desired baud rate. The SC16IS750 has a number of internal configuration registers. The main one we're interested in is the line control register (LCR), which controls the data word length, parity, and the number of stop bits.

The SC16IS750 UART chip itself is designed for 2.5 or 3.3 V operation, but all of the I/O pins are 5 V tolerant, and the module itself is powered by 5 V, with a 3.3 V power regulator onboard. I²C or SPI operation is selected by an input pin on the module, and one of 16 different I²C bus addresses can be selected using the A0 and A1 pins. Additionally, the SC16IS750 also has eight programmable I/O pins available for your use.

While this feature is not used in this project, the SC16IS750 can be configured to use interrupts to provide status information. For example, it could signal that a character has been received, the transmit buffer is empty, a communication error was detected, or the state of one of the onboard I/O pins has changed.

The RTTY Reader Project

Figure 16.3 shows the block diagram for the RTTY Reader project. We'll use an Arduino Nano, the SC16IS750 UART, a 1.8-inch color graphic TFT display, and a tone decoder circuit to convert the RTTY audio tones from the receiver to logic-level signals for the UART Rx input. The finished project will be mounted in a Solarbotics Mega SAFE along with a 9 V battery for portable operation. We'll be using the SPI bus to communicate with the TFT display, and the I²C bus for the UART. The I²C bus configuration for the UART was chosen because although the SC16IS750 library supports the SPI configuration, the I²C configuration within the sketch is easier to understand and use.

For the tone decoder portion of the cir-

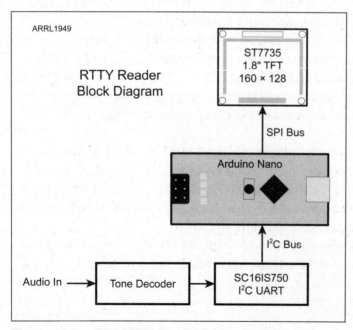

Figure 16.3 — The RTTY Reader block diagram.

cuit, we'll create two different decoder circuits to highlight the differences between two of the most popular tone decoder chips, the Exar XR-2211 FSK demodulator/tone decoder, and the ever-popular LM567 tone decoder. Back in the 1970s and 80s, the XR-2211, along with its counterpart, the XR-2206 tone generator, were my "go-to" chips for any RTTY or FSK audio decoding projects. The XR-2211 is an inexpensive and very good decoder. The hardest part about using the XR-2211 is figuring out the values of the various tuning and filtering components, but once you get those values worked out, the actual construction is very simple.

The LM567 is a single-tone decoder, but is also very easy to work with, and at 100 mA, has a much higher open-collector output sinking current that the XR-2211's 5 mA limit. Both decoder designs were kept simple and basic, although there is a lot of extra filtering and tuning functionality left in both chips for you to explore in your own designs. Because both decoders have their own specific set of advantages and disadvantages, the decision was made to design both decoder options, and let you choose which one you prefer.

For this project, the XR-2211 version is designed for 1000 Hz mark and 1170 Hz space frequencies. (Note that these are not the typical frequencies for high tones or low tones, as described earlier, but any two frequencies within the normal SSB audio passband will work as long as they are 170 Hz apart.) These frequencies are in the middle of the normal HF SSB audio range, without having to use the RTTY receive setting on your transceiver (if it has one). I find that the lower tones are easier to tune in and are easier on my ears.

By changing just three components, you can modify the XR-2211 version to operate at the standard 2125/2295 Hz mark and space tone frequencies if you prefer to do it that way. The LM567 version was designed to only use the 1000/1170 Hz frequencies, although it too could be switched to the 2125/2295 Hz frequencies by changing just a component or two.

The XR-2211 FSK Decoder Circuit

Figure 16.4 shows the schematic for the XR-2211 FSK decoder circuit. While the XR-2211 can also be used as a single-frequency tone decoder, for this circuit, we'll be using it as an FSK decoder. Potentiometer R8 is used to adjust the input level. The XR-2211 input signal level can be anything between 10 mV and 3 V RMS. The reason for this wide range is that the XR-2211 has an input preamplifier that also functions as a voltage limiter, providing a constant signal level at the output.

While this project really only uses the FSK decoder to provide a logic-level low signal on the output pin, the XR-2211 is actually tuned to a frequency between the two tones and can output a signal indicating whether the received frequency is higher or lower than the center frequency of the circuit. Because we're just using the XR-2211 as a basic FSK decoder, we're really only interested in identifying the 1000 Hz mark frequency. You can add some additional components to this circuit to improve the signal filtering and decoding response, but for this project, we'll just be using a basic FSK decoder without any additional filtering.

The basic calculations for determining the component values that determine

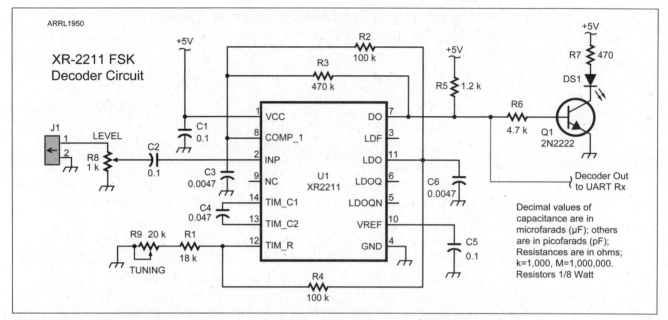

Figure 16.4 — The XR-2211 FSK decoder circuit for 1000/1170 Hz mark/space tones. See text for discussion of modifications for operation with the standard 2125/2295 Hz tone pair.

C1, C2, C5 — 0.1 μF ceramic capacitor
C3, C6 — 0.0047 μF ceramic capacitor
C4 — 0.047 μF ceramic capacitor (for 1000 Hz mark tone; use 0.022 μF for 2125 Hz)
DS1 — 5 mm red LED
J1 — 2-pin DuPont male header
Q1 — 2N2222 NPN Transistor
R1 — 18 kΩ, ⅛ W resistor (for 1000 Hz mark tone; use 15 kΩ for 2125 Hz)
R2 — 100 kΩ, ⅛ W resistor
R3 — 470 kΩ, ⅛ W resistor
R4 — 100 kΩ, ⅛ W resistor (for 1000 Hz mark tone; use 470 kΩ for 2125 Hz)
R5 — 1.2 kΩ, ⅛ W resistor
R6 — 4.7 kΩ, ⅛ W resistor
R7 — 470 Ω, ⅛ W resistor
R8 — 1 kΩ 10-turn potentiometer
R9 — 20 kΩ 10-turn potentiometer
U1 — XR2211 FSK demodulator/tone decoder

the XR-2211's center frequency and bandwidth are relatively straightforward and can be found in the XR-2211 datasheet. For 1000 Hz, the component values work out to 18 kΩ for R1, 100 kΩ for resistor R4, and 0.047 μF for C4. You can easily modify the XR-2211 decoder circuit for the standard mark and space tones of 2125 and 2295 Hz by changing the values to 15 kΩ for R1, 470 kΩ for R4, and 0.022 μF for C4.

The pullup resistor (R5) on the output is a much lower value than is typically used for a pullup resistor to drive a 2N2222 transistor. This is because the XR-2211 can only sink 5 mA of current, and without a lower value of pullup resistor, the output can't drive the UART receive input and the transistor used to light the LED on the board. Because the transistor stage inverts the output signal, you'd need a second 2N2222 transistor to invert the output signal again for the UART if you wanted to provide more drive to the UART Rx input. When idle, the output of the XR-2211 is at logic-level high, which is also what the UART is expecting to see when no characters are being received.

The LM567 Decoder Circuit

Figure 16.5 shows the LM567 tone decoder circuit. The LM567 is a simple, easy-to-use general-purpose single-tone decoder chip. With modern RTTY devices, we no longer need to decode both the mark and space frequencies unless we want to implement a higher level of eliminating false decoder outputs. A simple way to do this would be to implement two 567 tone decoders, and not allow the receive input to the UART to change unless one or the other decoder was active. This is a bit of overkill, because fading or noise that would cause the loss of the space tone would most likely affect more than a single character anyway. So, in this case, a minimal RTTY tone decoder is adequate for the job.

The LM567 is very easy to design with and use. The tone frequency to decode is determined by a single resistor (R2) and capacitor (C5) attached to pins 5 and 6 of the decoder chip. As with the XR-2211, you can add levels of phase-locked loop and output filtering as desired, simply by the addition of a component or two. Because we're using a minimal tone decoder for this project, we haven't added any filtering to the design, but an online search will yield some ideas.

In addition to the tone frequency, the bandwidth is also selectable through the use of a single capacitor (C3) and/or changing the input signal level. Unlike the XR-2211 decoder, the LM567 is much more sensitive to different signal levels, with a maximum design specification of 200 mV RMS on the audio

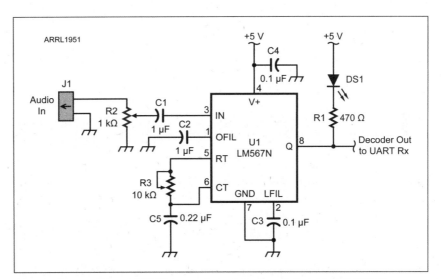

Figure 16.5 — The LM567 tone decoder circuit for 1000/1170 Hz mark/space tones.
C1, C2 — 1 µF, 16 V electrolytic capacitor
C3, C4 — 0.1 µF ceramic capacitor
C5 — 0.22 µF ceramic capacitor
DS1 — 5 mm red LED
J1 — 2-pin DuPont male header
R1 — 470 Ω, ⅛ W resistor
R2 — 1 kΩ 10-turn potentiometer
R3 — 10 kΩ 10-turn potentiometer
U1 — LM567 tone decoder

input pin. For this project, I chose to use the 1000 Hz mark frequency for the LM567 decoder to make it easier to tune in signals on the radio. Information on determining the component values for tone frequency and bandwidth may be found in the LM567 datasheet.

Building the Circuit Board

Figure 16.6 shows the completed XR-2211 decoder version of the RTTY Reader circuit board. The Nano and UART module are mounted in "header strip" sockets on a piece of perfboard that is cut and drilled to fit the Uno mounting holes in the Solarbotics Mega SAFE enclosure. This allows the entire project to fit in the enclosure, along with a 9 V battery for truly portable operation.

Figure 16.7 shows the schematic for the RTTY Reader circuit board. Once again, this circuit is an example of how functional and interesting projects with the Arduino can be created simply. The RTTY Reader project could easy be constructed in a single afternoon.

In the schematic you will notice that the UART's I^{2}C/SPI pin (pin 1) is connected to +5 V, causing the module to be in the I^{2}C operating mode. UART addressing pins A0 and A1 (pins 2 and 3) are connected to ground, selecting I^{2}C bus address BB. All of the remaining UART configuration and operation occurs using the I^{2}C bus.

You will also see J2 in the schematic. This is a three pin header jumper that allows you to either select normal operation (connecting the decoder output to the UART Rx input) or loopback, connecting the UART Tx output pin to the UART Rx input pin. This was added as part of the design and initial testing process, and left in the design as a troubleshooting aid. This allows you to perform complete end-to-end testing of the project using the Serial Monitor to verify that your project is functioning correctly.

Figure 16.6 — The RTTY Reader circuit board.

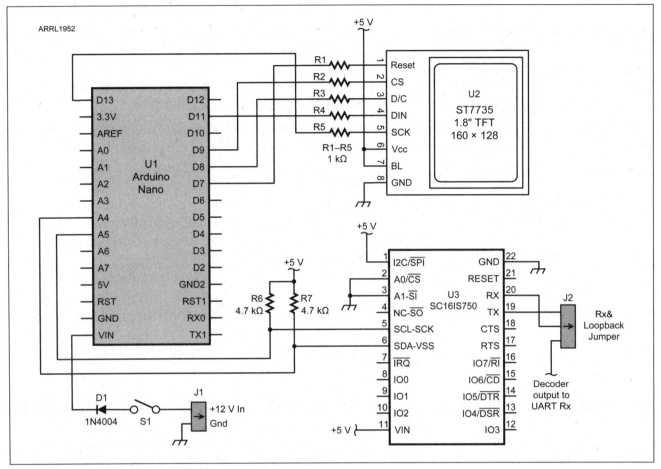

Figure 16.7 — The RTTY Reader schematic diagram.
D1 — 1N4004 diode
J1 — 2-pin DuPont male header
J2 — 3-pin DuPont male header
R1 – R5 — 1 kΩ, ⅛ W resistor
R6, R7 — 4.7 kΩ, ⅛ W resistor
S1 — SPST toggle switch
U1 — Arduino Nano
U2 — ST7735-type 1.8-inch color TFT display
U3 — SC16IS750 I²C UART Module
Enclosure — Solarbotics Mega SAFE

Next, we'll discuss the sketch used for the RTTY Reader. We'll cover testing and using the completed project once the sketch has been uploaded to the finished project.

RTTY Reader Flowchart

Figure 16.8 shows the RTTY Reader flowchart. As usual, we'll start by including the libraries needed for the ST7735 TFT display and the SC16IS750 UART. Then, we'll define the necessary constants and I/O pins needed. Next, we'll define the Baudot-to-ASCII translation array, and then declare the remaining variables needed by the sketch. Finally, we'll instantiate the TFT display and UART objects.

In the `setup()` section, we'll initialize the TFT display and display a brief startup message. Next, we'll initialize the UART and verify that it is online, then set the configuration we'll need to receive the Baudot RTTY data.

The main `loop()` is about as short as an Arduino sketch can get, con-

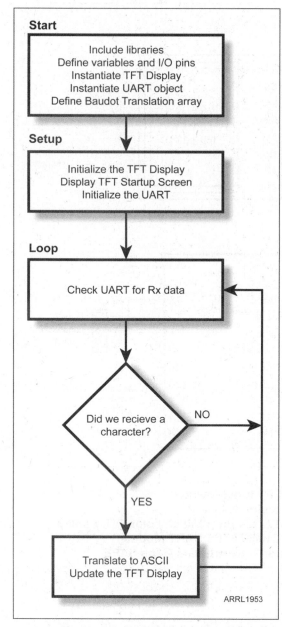

Figure 16.8 — The RTTY Reader flowchart.

sisting of a single function call to check for a received character from the UART. If a character is received, we'll translate it to its ASCII equivalent and display it on the TFT.

RTTY Reader Sketch

The RTTY Reader sketch is one that really demonstrates the power of Arduino libraries. The sketch itself is short and consists primarily of the various building blocks we've used in many of the other sketches in this book. As with all of the projects in this book, the sketch and libraries for this project can be found at **www.sunriseinnovators.com/Arduino3** and **www.kw5gp.com/Arduino3**.

The heart of this sketch is the library for the SC16IS750 UART. As of this writing in late 2020, the libraries available for the SC16IS7xx UART modules are a bit on the scarce side. The ones I found are functional and have all of the features we need to make the RTTY Reader project work. Unfortunately, the best library I was able to locate is not well documented, and many of the features needed to fully customize the UART were set up as private functions inside the library, meaning these library functions cannot be called from an external sketch. I had to modify the function declarations within the SC16IS750.h library file to change a number of the library's private internal functions to public functions, allowing them to be accessed from within the Arduino sketch. As such, until the time that a better library comes along, you'll need to install the custom SC16IS750 library from the website listed above for this project to function properly.

At this point in the sketch, we'll also define the TFT display I/O pins and the RTTY baud rate. The UART internal baud rate clock is set by a function inside the UART library. The UART's baud rate clock is set to 16 times the actual baud rate, in this case $45 \times 16 = 720$ Hz. To get this clock rate, the UART uses an internal selectable divide-by-1 or divide-by-4 prescaler, in conjunction with a 16 bit divisor register. For this project, based on the 14.7456 MHz crystal on the UART module, the library calculates a prescaler setting of 1, and a divisor setting of 20,480. Fortunately, the library handles all these calculations for us and all we have to do is give it the desired baud rate.

```
/*
  RTTY Reader by Glen Popiel - KW5GP

  Uses customized Sandbox Electronics SC16IS750 library
  Uses Adafruit ST7735 and GFX TFT display libraries
  Uses Wire library to communicate with I2C SC16IS750 UART module
  Uses SPI library to communicate with SPI TFT display

*/

//#define debug    // Enables diagnostic information

#include <Adafruit_GFX.h>     // Core graphics library
#include <Adafruit_ST7735.h>  // Hardware-specific library
#include <SPI.h>   // SPI library required for SPI devices
#include <Wire.h>  // Wire library required for I2C devices
#include <SC16IS750.h>   // SC16IS750 UART library

#define TFT_CS    10  // Assign the TFT CS to pin 10
#define TFT_RST    7  // Assign the TFT RST to pin 7
#define TFT_DC     8  // Assign the TFT DC to pin 8
#define tft_delay 10  // set the TFT command delay to 10ms

#define rtty_font_size 2   // The RTTY text display font size
#define status_font_size 1  // The font size for the status line
```

Because the RTTY Baudot character set is comprised of 5-bit data words, and the Arduino uses the 8-bit ASCII character set, we'll need to translate between the two. Because Baudot uses two sets of 32 different characters, based on whether we're in LTRS or FIGS shift, we can create two 32 character arrays, and by keeping track of the LTRS/FIGS shift, select which of the two arrays we'll use to translate the Baudot character to its ASCII equivalent. By using an array to translate between the two character sets, we can also substitute ASCII or null values for the Baudot characters we don't want to print such as the bell, LTRS and FIGS shift characters.

```
// Assign the translation values for the Baudot Letters character set
char baudot_ltrs[] = {'\0', 'E', ' ', 'A', ' ', 'S', 'I', 'U', ' ', 'D', 'R',
'J', 'N', 'F',
                     'C', 'K', 'T', 'Z', 'L', 'W', 'H', 'Y', 'P', 'Q', 'O',
'B', 'G', '\0',
                     'M', 'X', 'V', '\0'
                    };

// Assign the translation values for the Baudot Figures character set
char baudot_figs[] = {'\0', '3', '\0', '-', ' ', '\0', '8', '7', '\0', '\0', '4',
'\0', ',', '!',
                     ':', '(', '5', '+', ')', '2', '$', '6', '0', '1', '9',
'?', '&', '\0',
                     '.', '/', ';', '\0'
                    };
```

Next, we'll declare the variables we'll need in the sketch itself.

```
String status_msg;   // Variable for the current USB status message to display
String rtty_data = "";  // String variable to hold rtty text data
String rtty_char = "";  // Single character string variable to hold converted
                        // rtty char
bool system_status ;    // Flag indicating current system status - true = ok,
                        // false = bad

bool  figs = false; // The Figures/Letters toggle flag
bool unshift_on_space = true; // Enable shifting back to LTRS when a space
                              // character is received
int rx; // The received character
byte reg_data; // Variable for UART LCR Register data
```

And finally, we'll instantiate (create) the UART and TFT display objects.

```
SC16IS750 uart = SC16IS750(SC16IS750_PROTOCOL_I2C, SC16IS750_ADDRESS_BB);
 // Create the UART instance

Adafruit_ST7735 tft = Adafruit_ST7735(TFT_CS,  TFT_DC, TFT_RST);
 // Create the TFT display instance
```

In the setup() section, we'll start the Arduino IDE's Serial Monitor if debugging is enabled, and then initialize the TFT display. Next, we'll display a brief start-up message.

```
SC16IS750 uart = SC16IS750(SC16IS750_PROTOCOL_I2C, SC16IS750_ADDRESS_BB);
 // Create the UART instance

Adafruit_ST7735 tft = Adafruit_ST7735(TFT_CS,  TFT_DC, TFT_RST);
 // Create the TFT display instance
```

```
void setup()
{
#ifdef debug
  Serial.begin(9600);
#endif

  tft.initR(INITR_18BLACKTAB);    // initialize a 1.8" TFT with ST7735S chip,
                                  // black tab
  delay(tft_delay);
  tft.fillScreen(ST7735_BLACK); // Clear the display - fill with BLACK background
  delay(tft_delay);
  tft.setRotation(1); // Set the screen rotation
  delay(tft_delay);
  tft.setTextWrap(true); // Turn off Text Wrap
  delay(tft_delay);
  tft.setTextSize(3); // Set the Font Size
  delay(tft_delay);
  tft.setTextColor(ST7735_GREEN); //Set the Text Color
  delay(tft_delay);
  tft.setCursor(40, 10);   //Set the Cursor and display the startup screen
  delay(tft_delay);
  tft.print("KW5GP");
  delay(tft_delay);
  tft.setTextSize(2);
  delay(tft_delay);
  tft.setCursor(15, 65);
  delay(tft_delay);
  tft.print("RTTY Reader");

  delay(5000);   //Wait 5 seconds
  clear_display(); // Clear the startup message
```

Next, we'll start the UART at 45 baud, then verify that the UART is operational. Then, we'll update the TFT display with the UART status.

```
  uart.begin(baud_rate);   //Initialize SC16IS750 UART with defined
                           // baudrate setting
  if (uart.ping() != 1) // Check for a response from the UART
  {

#ifdef debug
    Serial.println("SC16IS750 UART not found");
#endif

    status_msg = no_uart; // set the status message to no UART found
    system_status = false;  // set the system status ok flag to false
    update_status();  // update the status line on the TFT display
    while (1);
  } else
  {
#ifdef debug
    Serial.println("SC16IS750 UART found");
#endif

    status_msg = uart_rdy; // set the status message to UART ok
    system_status = true;  // set the system status ok flag
    update_status();  // update the status line on the TFT display
  }
```

Finally, we'll set the UART's line control register (LCR) for 5-bit data words, with no parity, and one stop bit. This UART is now properly configured to receive 45 baud RTTY data.

```
#ifdef debug
  Serial.println("Start Serial Communication");
  reg_data = uart.ReadRegister(SC16IS750_REG_LCR);   // Read the UART Line
                                  // Control Register (LCR) register
  Serial.print("LCR Register = ");
  Serial.println(reg_data, BIN);
#endif

  uart.SetLine(5, 0, 0); // Set the UART Line Control Register (LCR) register
                         // fot 5 bit word, no parity, 1 stop bit

#ifdef debug
  reg_data = uart.ReadRegister(SC16IS750_REG_LCR);   // Read the UART Line
                   // Control Register (LCR) register to be sure it's set
  Serial.print("Modified LCR Register = ");
  Serial.println(reg_data, BIN);
#endif

}
```

The main `loop()` for the RTTY Reader sketch is about as simple as you can get. It comprises of a single function, the `get_rx_char()` function. This function will handle the remainder of the operations to display the received RTTY data on the TFT display.

```
void loop()
{
  get_rx_char();   // Get the received character and convert to ASCII
}
```

The `clear_display()` function is used to clear the TFT display screen by filling it with the background color, thereby erasing the display data.

```
void update_status() // Prints status messages on bottom line of TFT display
{
  if (system_status)   // True if good, false if bad
  {
    //We're all good - update status in green
    tft.setTextColor(ST7735_GREEN);  // Green on black background color required
                                     // to "clear" previous text
  } else
  {
    // There is a system error
    tft.setTextColor(ST7735_RED);  // Red on black background color required to
                                   // "clear" previous text
  }
  tft.setTextSize(status_font_size); // Set the font size to the status line
                                     // font size
  tft.setCursor(5, 118); // Move to the bottom line of the display
  tft.print(status_msg);  // Display the status

  tft.setCursor(115, 118);  // Move to the end of the status line
  tft.print("Baud:");
  tft.print(baud_rate);   // Display the baud rate

  tft.setTextSize(rtty_font_size);  // Reset the text size for the rtty
                                    // text area
  tft.setCursor(0, 5);  // Move to the top line of the display
}
```

The `update_status()` function is used to update the status information displayed on the bottom line of the TFT display. The TFT display for this project is set to display six lines of 13 characters each, with the bottom line of the TFT display reserved for status information such as the UART status and the current baud rate.

```
void update_status() // Prints status messages on bottom line of TFT display
{
  if (system_status)   // True if good, false if bad
  {
    //We're all good - update status in green
    tft.setTextColor(ST7735_GREEN);   // Green on black background color required
                                      // to "clear" previous text
  } else
  {
    // There is a system error
    tft.setTextColor(ST7735_RED);   // Red on black background color required to
                                    // "clear" previous text
  }
  tft.setTextSize(status_font_size);   // Set the font size to the status line
                                       // font size
  tft.setCursor(5, 118); // Move to the bottom line of the display
  tft.print(status_msg);   // Display the status

  tft.setCursor(115, 118);   // Move to the end of the status line
  tft.print("Baud:");
  tft.print(baud_rate);   // Display the baud rate

  tft.setTextSize(rtty_font_size);   // Reset the text size for the rtty
                                     // text area
  tft.setCursor(0, 5);   // Move to the top line of the display
}
```

The `update_text()` function will update the received RTTY text on the TFT display. The RTTY text area is six lines of 13 characters each, with the received RTTY characters scrolled within this six-line window.

```
void update_text() // Updates the RTTY text on the display
{
  tft.setTextSize(rtty_font_size);   // Set the rtty text font size
  delay(tft_delay);
  if (rtty_data.length() >= 78)   // Scroll bottom line up a line if at end
                                  // of line 6
  {
    rtty_data = rtty_data.substring(13);   // Remove the first 13 characters (the
                                           // top line)
    tft.setCursor(0, 5); // Move the cursor to the top line
    delay(tft_delay);
```

```
    tft.fillRect(0, 0, 159, 105, ST7735_BLACK); // Clear the rtty data area
    delay(tft_delay);
    tft.print(rtty_data); // Print the rtty data string variable
    tft.setCursor(0, 85);  // Set the cursor to the start of line 6
  }

  rtty_data = rtty_data + rtty_char;  // Add the character to the rtty
                                      // data string
  tft.print(rtty_char); // display the received character

#ifdef debug
  if (rtty_data.length() != 0)
  {
    Serial.print("rtty_data string length: ");
    Serial.print(rtty_data.length());
  }
#endif

}
```

The `get_rx_character()` function is called from the main `loop()` and is the only statement in the `loop()`. This function will check the UART to see if there is a received character, and if there is one, it will process and display the received character. Otherwise, the sketch will continue to loop until a character is received.

If a character is received, we'll check to see if it's a space, FIGS, or LTRS character. This sketch is designed to implement the unshift on space (USOS) RTTY feature, which will return the character set to the LTRS shift whenever a space character is received.

Next we'll check for the FIGS or LTRS character and set the figs flag accordingly.

```
void get_rx_char()   // Get a received character from the UART
{
  if (uart.available() != 0)
  {
    // Read the received character
    rx = uart.read();
    if (rx != 0)
    {
      switch (rx)
      {
        case 4: // It's a space character. Unshift if flag is enabled
          if (unshift_on_space)
          {
            figs = false; // Set FIGS/LTRS flag to LTRS
```

```
#ifdef debug
          Serial.print("   Space - unshift");
#endif


        }
      break;

    case 27:  // It's a FIGS key
      figs = true;   // Set the FIGS/LTRS flag to FIGS

#ifdef debug
          Serial.print(" FIGS");
#endif

      break;

    case 31:  // It's a LTRS key
      figs = false; // Set FIGS/LTRS flag to LTRS

#ifdef debug
          Serial.print(" LTRS");
#endif

      break;

    }

#ifdef debug
    Serial.print("    Baudot: ");
    Serial.print(rx);
#endif
```

Finally, we'll translate the received RTTY character to its matching ASCII counterpart using one of the two Baudot translation arrays selected by the figs flag, and then call the update_text() function to add the received character to the TFT display output.

Testing the RTTY Reader

The nice thing about this project is that there are several ways to tune and test the finished project. For initial testing of the decoder circuit, I used a USB sound dongle plugged into a workstation, with the headphone output of the USB dongle plugged into the audio input on the RTTY Reader. To do basic tone decoder testing, I used the online tone generator at **www.szynalski.com/tone-generator** to generate an audio signal so that I could adjust the audio input level and tune the decoder. You can measure the audio level going to the decoder circuit at the wiper of the input audio **LEVEL** potentiometer using a voltmeter on the ac setting. Remember, for the LM567 tone decoder, you'll want a maximum of 200 mV of audio. For the XR-2211 decoder, the audio input level can be any-

Figure 16.9 — The RTTY Reader display with test text generated by *MMTTY*.

where from 10 mV to 3 V. Once you have the audio level set, you can adjust the decoder **TUNING** potentiometer.

For the LM567 decoder circuit, because we're only decoding the mark tone frequency, you can set the online tone generator to 1000 Hz and adjust the **TUNING** potentiometer until the LED on the board goes out. Remember that this LED is powered by a transistor on the output of the decoder and is the inverse of the actual decoded signal. Turn the potentiometer until you have the tuning centered and you should be good to go.

For the XR-2211 version of the decoder, the tuning is slightly different. You'll actually want to tune for the center frequency, which would roughly be 1085 Hz. In either case, the tuning range is quite broad, and you have a tuning range of about 2 to 3 turns on the **TUNING** potentiometer where the tone will be properly decoded.

Once you have the tone decoder set, you can now try decoding an actual RTTY signal. I used the free *MMTTY* program available for download from **hamsoft.ca/pages/mmtty.php** for testing. After installing *MMTTY* in your PC, set the mark frequency to 1000 Hz with the **NET** button selected in the demodulator settings. With these settings, *MMTTY* will transmit the 1000 Hz mark and 1170 Hz space tones used by the RTTY Reader. Connect the computer soundcard output to the RTTY Reader input and set the *MMTTY* soundcard option to transmit to that sound card.

Now select the **RY** macro button in *MMTTY* to continuously send a series of RY characters to the RTTY Reader. (The RY character sequence was originally used to stress test mechanical RTTY machines, and it carried over to more modern RTTY decoder testing. Occasionally you still may see RYs transmitted on the air.) You can do any final tuning of the decoder while sending RYs and stop the test by selecting the *MMTTY* **TXOFF** control button. You can also press the **TX** button and type directly to the RTTY Reader. Be sure that you have the correct sound card setting for these tests. If you have your workstation connected to a transceiver, you could accidently transmit over the air instead of to the RTTY Reader. **Figure 16.9** shows the output of the RTTY Reader tested using *MMTTY*.

Once you have the RTTY Reader tuned, you're ready to connect it to the audio output of your receiver and tune in a RTTY signal. Remember that on the HF bands, RTTY is generally tuned using the LSB (lower sideband) setting on your receiver. You can use the LED on the circuit board as a tuning aid. Although RTTY software for a PC usually has a more sophisticated visual tuning indicator, with a little practice it's not hard to learn how to tune RTTY signals by ear with the help of the LED on the circuit board.

A reliable source of on-air RTTY signals you can use for testing the decoder and tuning practice is the W1AW RTTY bulletins that are transmitted daily during the week. See **www.arrl.org/w1aw-operating-schedule** for details.

Enhancement Ideas

The RTTY Reader is one of those fun projects that I could keep tinkering with and refining indefinitely. There are so many tweaks that I came up with, that it would take me forever to implement them all. But, I had to stop somewhere or else the book would never get finished, and there wouldn't be anything left for you to add yourself.

One possible enhancement would be to add the USB Keyboard project and modify the RTTY Reader to send as well as receive. A chip such as the XR-2206 monolithic function generator could be used to generate the mark and space audio tones to feed to your transmitter. This would give you a portable PC-less RTTY station you could take just about anywhere.

Another enhancement would be to display more information on the status line, such as whether you're in LTRS or FIGS shift. You could also change the font size and put more RTTY text on the screen and use more lines of the TFT display. You could even jazz up the display by using different colors and adding lines between the text and status area. You could also increase the display to a 2.2-inch TFT or larger. There's really no limit to the size of the TFT display you could use with this project.

On the XR-2211 version of the decoder, you could add a switch and an option to select either the 1000 Hz or 2125 Hz mark tone and use small reed relays to switch the two resistors and capacitor needed to make the change. You could also use this same method to change the frequency shift from 170 Hz to any other frequency shift you may choose. You could even go so far as to use the digital potentiometer chips we used in the Yaesu FH-2 keypad project to change the resistor values via software.

And finally, you can add tighter filtering on the decoder circuits. With the addition of a few components, you can tighten up the decoder filter so that it is less affected by adjacent signals. The XR-2211 and LM567 datasheets are good resources for design ideas.

I also think that adding a pushbutton switch to manually perform the LTRS shift, and another to clear the display could also be nice additions this project. Finally, you could add a piezo buzzer to one of the Arduino's I/O pins and play a tone whenever a bell character is received. (The old mechanical teleprinters had bells that rang to alert the operator of an important incoming message.)

17 In Conclusion

Once again, we come to the end of another new group of fun and useful ham radio-related Arduino projects. I hope you'll have as much fun building, using, and modifying these projects as I've had creating them. I had to stop somewhere, but I have more project ideas, concepts, and designs set aside into a folder reserved for new book ideas. So for me, the Arduino fun will continue on with those new project ideas. If you like this book and would like to see more Arduino books, be sure to write ARRL at **pubsfdbk@arrl.org** and let them know. And of course, always feel free to email me at **kw5gp@arrl.net** with your comments or suggestions.

Hopefully, this book has provided you with all of the pieces you need to create and modify your own Arduino projects. You now have all of the pieces to give your antenna rotator controller all kinds of new features, regardless of its age or manufacturer. I have several of the very old rotator controllers that don't even have position sensors. You may have seen those old units with the dial that you turned, and the controller had a solenoid that then clicked while it rotated to move the position indicator on the box. A couple of those are next up on my lab workbench, just for fun.

As you have seen, many of the projects in this book were chosen to be a proof of concept and introduce new project ideas as much as anything else. The goal was to provide you with the tools and techniques you would need to interface to the various devices in your ham shack and to enable you to create and design you own Arduino projects. Don't let this book be the end of your Arduino creation. Think up your own Arduino projects and make them happen — that's where the fun really is.

Some New Arduino Technologies

In parting, I'd like to briefly introduce you to some of the newer Arduino technologies and projects that are rapidly becoming popular in the Maker and Arduino worlds.

LIXIE Display

First up is the LIXIE. Some of you may have seen or remember the Nixie display tubes that were actually vacuum tubes that could selectively show numbers. Those original Nixie tubes are getting scarce and besides, they run on something on the order of 190 V. While we like the tubes to glow in the dark, I'm not a fan of working with high voltages. So, now we have the LIXIE, a low-voltage, LED-based alternative to the Nixie tube.

The LIXIE, as shown in **Figure 17.1**, is series of laser-cut plexiglass digits. These digits are edge-lit by an array of individually addressable RGB LEDs. They look very similar to the old-style Nixie tubes, but without the heat and high voltage. I can see the LIXIE being used to give a high-class, elegant look for Arduino and other projects.

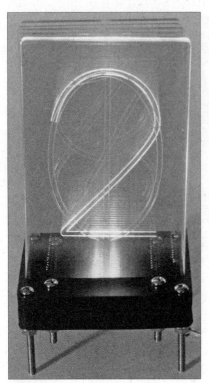

Figure 17.1 — The LIXIE display.

Speech Recognition

Speech recognition projects with the Arduino have always been at the top of my list for project ideas. The new MOVI Arduino shield by Audeme looks to be an outstanding choice for speech recognition and voice synthesis for the Arduino. It even supports Spanish and German in addition to English.

Keypads

Adafruit has really changed the game in the keypad area. The NeoTrellis M4 Express is an 8×4 keypad that's powered by the SAM D51 that uses an Arm Cortex M4 running at 120 MHz with built-in hardware digital signal processing and floating point arithmetic. This looks like it could be a very interesting and powerful platform for small portable projects.

Software Defined Transceiver

On the Arduino-powered transceiver front, HobbyPCB has created the RS-HFIQ, an 80 through 10 meter SDR transceiver. The Arduino Nano and SI5351-powered RS-HFIQ is designed to translate I/Q baseband signals to RF. It is intended to be used with a PC running SDR software and has 5 W RF output.

Arduino Pro IDE

The Arduino IDE will be getting a major upgrade with the upcoming release of the Arduino Pro IDE. As of this writing in late 2020, it's in Alpha-test, but some of the new features look outstanding, and I can't wait to try it out. In addition to having an overall upgrade to the Library Manager, Board Manager, and many other standard IDE features, it will have a dual-mode GUI, allowing you use either the classic Arduino IDE look and feel, or the new Pro Mode version. The Pro IDE will also support Arduino, Python, and JavaScript code. One of the major new features planned is an integrated debugger, something the IDE has always needed. You can find out more information about the new Pro IDE at **blog.arduino.cc** (search for Pro IDE for relevant articles).

Python Programming Language

There's also a move afoot to use the Python programming language with the Arduino. Adafruit and SparkFun, two of the major Arduino suppliers, are both developing a Python-based line of microcontrollers using CircuitPython and MicroPython. This is aimed at the Internet of Things (IoT) application side of microcontrollers.

Field Programmable Gate Arrays

The Arduino world is expanding into FPGAs (Field Programmable Gate Arrays). An FPGA is an integrated circuit that can be configured and reconfigured endlessly to implement logic functions. At first glance, this appears to be similar to the way you would upload firmware to run on the Arduino hardware. With an FPGA, however, you're programming the actual hardware itself. According to the Arduino.cc website, you can use FPGAs to implement very fast filters, audio and video processors, and other applications that microcontrollers aren't able to handle.

The Arduino MKR Vidor 4000 has a 32-bit SAMD21 Cortex-M0 CPU with 256 KB of flash memory and 32 KB of RAM onboard. In addition, it has an Intel Cyclone 10CL016 FPGA on the board. This FPGA has 15,408 logic elements. Theoretically, you can now design the hardware to do the work of what previously had to be done in software. Although FPGAs have been around for many years, Arduino is relatively new to the FPGA world and this is definitely an area to watch, especially because the plan is to have the FPGA programming

Figure 17.2—The solar-powered Skytracker, an Arduino-powered balloon payload by Bill Brown, WB8ELK. (Bill Brown, WB8ELK, photo)

functionality added directly into the Arduino Pro IDE. You can find out more about the Arduino Vidor 4000 and FPGAs at the following links:

- **www.arduino.cc/en/Tutorial/Foundations/VidorHDL**
- **www.arduino.cc/en/Tutorial/VidorGSVHDL**
- **content.arduino.cc/assets/Arduino-Vidor_c10lp-51001.pdf**

Picoballooning

And finally, the Arduino is at the heart of one of the newer ham radio-based hobbies, Picoballooning. Using something as simple as a Mylar party balloon or a larger ultra-lightweight high-altitude plastic balloon, hams are sending Arduino-powered experimental payloads with low-power VHF/UHF transmitters (**Figure 17.2**) high into the sky, sometimes as high as 45,000 feet, and tracking them using Weak Signal Propagation Reporter (WSPR) and Automatic Packet Reporting System (APRS) as they circumnavigate the world.

So, the future is looking wonderfully bright for the entire Arduino family, and I'm looking forward to playing with some of the new boards and modules. I think my biggest problem is deciding what to build next.

73,
Glen Popiel, KW5GP

Index

Note: The letters "ff" after a page number indicate coverage of the indexed topic on later pages.

µBITX V6 transceiver display: 3-33
1-Wire bus: 4-4
16×2 LCD display: 3-29

A

Accelerometer: 1-5, 2-8
   Module: 3-21
Acrylic enclosure: 3-35
AD9850 DDS module: 3-16
Adafruit: 2-1, 2-5, 3-4, 3-5, 3-7, 16-12
   12-channel LED driver module: 3-18
   128x64 OLED display: 3-31
   24-channel LED driver module: 3-19
   Bluefruit LE sniffer: 6-12
   Feather32U4: 2-10
   FeatherWing Proto: 3-14
   Grand Central M4 Express: 2-10
   Modules: 3-16ff
   Power Boost shield: 3-10
   Wave Shield: 3-8
Addressable LED: 3-28
ADS1115 16-bit A/D module: 3-26, 11-3, 11-6ff, 11-12, 12-3, 15-2ff
Altoids mint tin: 3-34
AltSoftSerial library: 14-3ff
AMS1117 voltage regulator: 11-2
Analog input: 4-3
Analog output: 4-3
Analog-to-digital converter (A/D) module: 3-26
Anderson Powerpole connector: 6-15ff
APRS: 1-5, 3-4, 3-9, 17-4
AR-40 rotator controller: 11-1ff, 12-1ff
   Modifications: 12-3ff
Arduino Create: 5-18
Arduino Every: 2-7
Arduino Integrated Development Environment (*see* IDE)
Arduino IoT Cloud: 5-18
Arduino Pro IDE: 17-2
Arduino program (sketch): 5-1ff
Arduino variant: 1-3, 1-5, 2-1

Ardweeny: 1-4, 2-3, 2-4, 4-3
Argent Data Radio Shield: 3-3
ASCII: 13-13ff, 16-2
ATmega 328 series: 2-1ff
ATmega 32U4 series: 2-5ff
ATmega processors: 1-5, 4-3
Atmel: 1-5, 2-1, 4-3
Audeme MOVI speech recognition shield: 3-15, 17-2
Audio frequency shift keying (AFSK): 16-2
Audio shield: 3-7
Austriamicrosystems: 3-25
Automatic Packet Reporting System (*see* APRS)

B

Balloon telemetry: 3-21
Banggood analog keypad: 7-10
Banzi, Massimo: 1-5
Baudot: 16-1
Block diagram: 5-2
Bluetooth: 1-3, 1-4, 2-5, 2-8, 4-8ff, 13-1
   Module: 3-24
   Sniffer: 6-12
Bluetooth Low Energy (BLE): 2-8, 2-9, 6-12
Board Manager: 5-8ff
Boarduino: 1-4
Bootloader: 5-7
Breadboard: 6-1ff
   Shield: 3-12

C

CAN Bus shield: 3-9
CDE/HyGain rotator controller: 9-1ff, 9-5ff, 9-11ff, 10-1ff, 10-6, 11-1ff, 12-1ff, 15-1ff
Cheat sheets: 6-19
Circuit simulator: 6-13
CircuitPython: 2-11, 17-3
CJMCU-116 9-axis motion/position sensor module: 3-21
Clone: 1-5
Compass dial: 10-9

Compiler: 5-17
Component tester: 6-4
Computer Aided Transceiver (CAT): 14-1ff
Construction techniques: 6-1ff, 6-14
Controller (rotator): 9-1ff, 10-1ff, 11-1ff, 12-1ff, 15-1ff
Copyleft: 1-7
Copyright: 1-3, 1-7
Cortex: 1-5
Creative Commons License: 1-8
Cuartielles, David: 1-5
Current sensor module: 3-20

D

Debugger: 6-17
Debugging methods: 5-18
Decal: 10-9
Design tools: 6-13ff
DFRobot: 3-6
Digital I/O: 4-2ff
Digital Rights Management (DRM): 1-7
Digital storage scope: 6-10
Digital-to-analog converter (D/A) module: 3-26
Direct digital synthesis (DDS)
    Module: 3-16
    Shield: 1-5
Display: 1-5, 3-29ff
    16×2 LCD: 3-29
    Adafruit 128×64 OLED: 3-31
    Addressable LED: 3-28
    E-Ink: 3-33
    ILI9341 TFT: 13-3
    LED: 3-28
        NeoPixel LED ring: 3-28
        NeoPixel LED strip: 3-28
    Nextion TFT touchscreen: 3-33
    Nokia 5110 graphic LCD: 3-29ff
    OLED display: 3-30
    RGB LED: 3-28
    SparkFun MicroView OLED: 3-31
    TFT: 3-31ff
Documentation: 5-2ff
DrDuino Explorer shield: 3-12, 6-6
DrDuino Pioneer shield: 3-13
DS3231 RTC module: 3-25
DTMF
    Generator module: 3-27
    Receiver chip: 3-28
DuPont headers and sockets: 6-14

E

E-Ink display: 3-33
Eagle: 5-2ff, 6-12
Easy VR Voice Recognition shield: 3-10
EEPROM: 1-5, 2-3, 5-17, 12-14
EMIC 2 text-to-speech module: 3-22
ENC28J60 Ethernet module: 3-21
Enclosure: 3-34, 6-19
ESP32: 2-9
ESP8266: 2-9
Espressif ESP32: 2-9
Ethernet: 2-9
    Module: 3-21
    Shield: 3-6ff
Exar XR-2211 FSK decoder: 16-5

F

Feather: 2-10
Featherwings: 2-10
Ferroelectric RAM (FRAM) module: 3-17
Field programmable gate array (FPGA): 17-3
FIGS shift: 16-2, 16-11
Flash memory: 1-3, 1-5, 2-3, 5-17
Flora: 2-5
Flowchart: 5-4
    AR-40 Rotator Controller: 11-8ff
    Keypad Entry Rotator Controller: 15-6ff
    Keypad Entry Rotator Controller
      calibration: 15-12ff
    Modified AR-40 Rotator Controller: 12-11ff
    Peltier Cooler Controller: 8-7ff
    Rotator Position Indicator: 10-4ff
    Rotator Turn Indicator: 9-10ff
    RTTY Reader: 16-9ff
    USB CW Keyboard: 13-6ff
    Yaesu CAT Display: 14-5ff
    Yaesu FH-2 Keypad: 7-5ff
FM transceiver shield: 3-2, 3-3
Franklin AS3935 lightning sensor chip: 3-25
Freeduino: 1-4
FreeRTOS: 5-19
Frequency counter: 6-7
Frequency shift keying (FSK): 16-2
Fritzing: 5-4
FTDI
    Interface: 2-4
    Module: 4-3, 4-10

G

General Public License (GPL): 1-7ff
Genuino: 1-4
Gerber file: 6-12
Global Positioning System (*see* GPS)
GNU GPL: 1-7
GPL: 1-7ff
GPS: 1-5, 2-5
    Module: 3-22
Grand Central M4 Express: 2-10
GY-906 infrared temperature sensor: 8-2ff
Gyroscope module: 3-21

H

Hall Effect current sensor: 3-20, 8-14
Ham Radio Deluxe: 10-2, 11-1ff, 11-11, 12-1ff,
                                  12-43, 15-1
HamShield: 3-3
HC-05 Bluetooth module: 3-24, 4-9
HeliOS: 5-19
Hobby Lobby enclosures: 3-35
HobbyPCB RS-HFIQ SDR: 17-2
HobbyPCB RS-UV3 Radio Shield: 3-2
HRD Rotator: 10-3, 11-13, 12-44
Human Interface Device (HID): 2-5, 3-17, 3-24,
                                    4-8ff, 13-1

I

I/O methods: 4-1ff
I/O sensor shield: 3-11
I/O shield: 6-6
I^2C: 1-4, 2-3, 3-1, 4-6
    Bus extender chip: 3-27
    Digital I/O expander chip: 3-27
ICSP: 4-5
IDE: 1-4, 2-4, 2-7, 2-9, 5-1ff, 5-6ff, 6-20
    Board manager: 5-8ff
    Libraries: 5-11ff
    Library Manager: 5-14ff
    Troubleshooting: 5-16ff
Iduino: 1-4
ILI9341 TFT display: 13-3
Instructables: 6-8
Inter-Integrated Circuit (*see* I^2C)
Interrupt: 1-5, 4-6ff, 5-19
ISD1820 voice recorder and playback module:..3-22

K

Keypad: 3-19, 7-1ff, 15-1ff
KiCad: 6-12

L

Labels: 6-19
LCD shield: 3-4
LCD5110_Basic library: 5-13
LED: 3-28
    Driver module: 3-18
    NeoPixel ring: 3-28, 9-2ff, 9-12, 10-1ff, 10-6
    NeoPixel strip: 3-28ff
    RGB: 3-28
Leonardo: 2-5, 2-6, 4-2, 4-3, 4-5ff, 5-16
Lesser GNU Public License (LGPL): 1-8
Level converter module: 3-24
LGPL: 1-8
Library: 1-4, 1-6, 5-11ff
    Installation: 5-13
Library Manager: 5-14ff
License: 1-3, 1-4, 1-6ff
Lightning sensor module: 3-25
Lilypad: 2-2, 2-4
Lilypad Arduino Simple Snap: 2-5
LinkSprite enclosure: 3-34
Linux: 1-3, 1-6
LiPo battery: 2-4, 2-5
LIXIE display: 17-2
LM567 tone decoder: 16-7
Logic analyzer: 6-10ff
LTRS shift: 16-2, 16-11
LTSpice: 6-13

M

Magnetometer: 1-5
MAX232 RS-232 driver/receiver: 14-2ff
MaxDetect: 4-4
Maxim: 4-4
MCP42010 dual digital potentiometer chip: 7-2
Mega 2560: 2-6
Mega 2560 Mini: 2-6
Mega series: 2-6
Memory issues: 5-17
Memory types: 2-3
MFJ Enterprises: 11-1, 12-1
Micro-Cap 12: 6-13
Microchip MCP23008 I^2C serial I/O expander: ..3-27
Microchip MCP23017 digital I/O expander: 3-27
Microchip MCP4725 12-bit I^2C D/A module: 3-26
Microcontroller: 1-3ff
    Types: 2-1ff
MicroPython: 17-3
Microview OLED: 2-4
Mini USB Host module: 3-17, 4-11
MLX90614 infrared temperature sensor: 8-2ff
MMTTY software: 16-19
Module: 1-4ff, 3-1ff, 3-15ff

Accelerometer: 3-21
AD9850 DDS: 3-16
Adafruit 12-channel LED driver: 3-18
Adafruit 24-channel LED driver: 3-19
ADS1115 16-bit A/D: 3-26
Analog-to-digital converter (A/D): 3-26
Bluetooth: 3-24
CJMCU-116 9-axis motion/position sensor: 3-21
Current sensor: 3-20
Digital-to-analog converter (D/A): 3-26
Direct digital synthesis (DDS): 3-16
DS3231 RTC: 3-25
DTMF generator: 3-27
DTMF receiver: 3-28
EMIC 2 text-to-speech: 3-22
ENC28J60 Ethernet: 3-21
Ethernet: 3-21
Ferroelectric RAM (FRAM): 3-17
Franklin AS3935 lightning sensor: 3-25
FTDI: 4-3
GPS: 3-22
GY-906 infrared temperature sensor: 8-2ff
Gyroscope: 3-21
HC-05 Bluetooth: 3-24
I^2C bus extender: 3-27
I^2C digital I/O expander: 3-27
ISD1820 voice recorder and playback: 3-22
LED driver: 3-18
Level converter: 3-24
Lightning sensor: 3-25
MCP42010 dual digital potentiometer: 7-2
Micro MCP23008 I^2C serial I/O expander: 3-27
Micro MCP23017 digital I/O expander: 3-27
Micro MCP4725 12-bit I^2C D/A: 3-26
Mini USB Host: 3-17, 4-11
MLX90614 infrared temperature sensor: 8-2ff
Relay: 3-18
RobotDyn nRF24L01 2-4 GHz ISM band transceiver: 3-18
RTC: 3-25
Si5351 programmable clock generator: 3-16
SparkFun I2S: 3-18
SparkFun ProDriver: 3-18
Speech: 3-23
Stepper motor driver: 3-18
TEC1-12706 Peltier: 8-2ff
Texas Instruments P82B715 I^2C bus extender: 3-27
Text-to-speech: 3-22
TinyGPS: 3-22
ublox NEO-6M GPS: 3-22
USB Host: 13-1ff
Voice recorder and playback: 3-22
WT588D speech: 3-23
Zarlink MT8870D DTMF receiver: 3-28
Motor driver shield: 1-5, 3-6
Multimeter: 6-5
Multitasking: 5-19

N

Nano: 1-1, 2-3, 4-3
Nano 33 BLE: 2-8
NeoPixel LED ring: 3-28, 9-2ff, 9-12, 10-1ff, 10-6
NeoPixel LED strip: 3-28ff
NeoTrellis keypad: 3-19, 7-10, 15-1
NeoTrellis M4 Express keypad: 17-2
Nextion TFT touchscreen: 3-33
Nixie display: 17-2
Nokia 5110 graphic LCD display: 3-29ff

O

One-Wire: 1-4, 3-1
Open Source: 1-3ff, 1-6ff
Optocoupler: 9-4
Organic LED (OLED): 2-4, 3-30
Oscilloscope: 6-8ff
    Analog: 6-9
    Digital storage: 6-10
    PC-based: 6-8
    Tektronix 475: 6-9

P

Parts suppliers: 3-36
PC-based USB storage scope: 6-8
Peltier cooler controller: 8-1ff
Perfboard: 6-14, 6-15
PhaseDock Workbench: 6-3
Picoballooning: 17-4
PJRC Teensy 4.0 and 4.1: 2-8
Power boost shield: 3-9
Power over Ethernet (PoE): 3-7
Processor: 1-1, 1-4, 1-5
Program: 1-3ff
Proprietary software: 1-7
Protoshield: 3-13ff
Prototyping: 1-5, 3-12ff, 6-1ff, 6-14
    Shield: 3-13ff, 6-15
Pulse Width Modulation (PWM): 1-4, 1-5, 2-2, 3-18, 4-2, 8-2, 8-9ff

R

Radioteletype (RTTY): 16-1ff
Random Access Memory (RAM): 1-5, 5-17
Raspberry Pi: 1-1ff
Rduino: 1-5
Real Time Clock (RTC) 1-5
    Module 3-25
Relay
    Module: 3-18
    Shield: 3-6
RGB LED: 3-28
Rights: 1-7
Rinkydinkelectronics: 5-13
RobotDyn keypad: 15-1ff
    Calibration: 15-6ff
RobotDyn nRF24L01 2-4 GHz ISM band
    transceiver module: 3-18
Rocksett high temperature adhesive: 8-6
Rotator position indicator: 10-1ff
Rotator turn indicator: 9-1ff
RS-UV3: 3-2
RTTY reader: 16-1ff

S

Satellite tracking: 3-21
SC16IS750 UART: 16-4
Schematic: 5-2ff
Scope: 6-8ff
Sensor: 1-1
Serial I/O: 4-3ff
Serial monitor: 5-8, 7-6, 13-8
Serial Peripheral Interface (*see* SPI)
Shield: 1-4, 1-5, 2-3, 3-1ff
    Adafruit FeatherWing Proto: 3-14
    Adafruit Power Boost: 3-10
    Adafruit Wave: 3-8
    Argent Data Radio Shield: 3-3
    Audeme MOVI speech recognition: 3-15
    Audio shield: 3-7
    Breadboard: 3-12
    CAN Bus: 3-9
    DrDuino Explorer: 3-12, 6-6
    DrDuino Pioneer: 3-13
    Easy VR Voice Recognition: 3-10
    Ethernet: 3-6, 3-7
    FM transceiver: 3-2, 3-3
    HamShield: 3-3
    HobbyPCB RS-UV3: 3-2
    I/O: 6-6
    I/O sensor: 3-11
    LCD: 3-4
    Mega I/O: 6-7
    Motor driver: 3-6
    Power boost: 3-9
    Protoshield: 3-13ff
    Prototyping: 3-13ff, 6-15
    Relay: 3-6
    Solarbotics XBee: 3-9
    SparkFun MP3 Trigger: 3-8
    TFT display: 3-4
    USB Host: 4-11, 13-1ff
    USB host: 3-7
    VHF/UHF transceiver: 3-3
    Voice Recognition: 3-10
    Wi-Fi: 3-7
    XBee: 3-8
Shieldlist.org: 3-2
Si5351 programmable clock generator
    module: 3-16
Simulation: 6-13
Sketch: 1-3, 1-4, 5-1ff
Sketchbook: 5-6ff
Skytracker balloon payload: 17-3
Sockets: 6-14
Software
    Arduino Create: 5-18
    Arduino IoT Cloud: 5-18
    Circuit simulators: 6-13
    Design tools: 6-13ff
    Eagle: 5-2ff, 6-12
    FreeRTOS: 5-19
    Fritzing: 5-5
    Gerber file: 6-12
    HeliOS: 5-19
    IDE: 5-1ff, 5-6ff
    KiCad: 6-12
    LTSpice: 6-13
    Micro-Cap 12: 6-13
    MMTTY: 16-19
    Simulation: 6-13
    Tinkercad: 6-13
    Virtronics Simulator for Arduino: 6-13
    Wireshark: 6-12
Solarbotics: 2-3, 2-4, 8-4
    Enclosure: 6-2
    Mega SAFE: 3-35
    XBee shield: 3-9
Soldering station: 6-5
SparkFun: 2-4, 2-5, 3-7
    ESP32: 2-9
    I2S module: 3-18

    MicroView OLED display: 3-31
    MP3 Trigger Shield: 3-8
    ProDriver module: 3-18
    Project enclosure: 3-34
    USB host shield: 4-11
Speech module: 3-23
SPI: 1-4, 2-3, 3-1, 4-4ff
ST7735 TFT display: 11-11, 15-2ff
Static RAM (SRAM): 1-5, 2-3, 5-17
Stepper motor driver module: 3-18

T

TEC1-12706 Peltier module: 8-2ff
Tektronix 475 scope: 6-9
Telemetry: 3-21
Texas Instruments P82B715 I²C bus extender chip: 3-27
Text-to-speech
    Module: 3-22
    Shield: 1-5
TFT display: 3-4, 3-31ff, 11-11, 13-3, 15-2ff
Thermal switch: 11-3
Thermoelectric: 8-1
Thermometer: 8-2
Tinkercad: 6-13
TinyGPS: 3-22
Tivoization: 1-7
Tone generator: 16-18
Tools: 6-1ff
    Bluetooth sniffer: 6-12
    Component Tester: 6-4
    Design: 6-12ff
    Frequency counter: 6-7
    Logic analyzer: 6-10ff
    Multimeter: 6-5
    Oscilloscope: 6-8ff
    Soldering station: 6-5
Torvalds, Linus: 1-6
Touchscreen: 3-5
Transceiver: 3-2
Transmitter: 1-5
Troubleshooting: 6-1ff, 6-16ff
    IDE issues: 5-16ff

U

UART: 16-3ff
ublox NEO-6M GPS module: 3-22
Universal Serial Bus (*see* USB)
Uno: 1-1, 1-4, 1-5, 2-2ff, 3-1ff, 4-2ff
USB: 13-2ff
USB communications: 4-10ff
USB controller: 5-16
USB CW keyboard: 13-1ff
USB host: 4-11, 13-2ff
    Shield: 3-7
USB serial communication: 12-14
USB-to-serial interface: 2-4

V

Variant (Arduino): 1-3, 1-5, 2-1
VHF/UHF transceiver: 3-2, 3-3
Virtronics Simulator for Arduino: 6-13
Voice recognition shield: 3-10
Voice recorder and playback module: 3-22

W

W1AW RTTY bulletin: 16-19
Weak Signal Propagation Reporter (*see* WSPR)
Wearable Arduino: 2-4, 2-5
Wi-Fi shield: 3-7
Wireshark: 6-12
WSPR: 1-5, 17-4
WT588D speech module: 3-23

X

XBee shield: 3-8

Y

Yaesu
    CAT display: 14-1ff
    FH-2 keypad: 7-1ff
    G-1000SA rotator controller: 9-8, 9-19, 10-1ff
    G-450A rotator controller: 9-7, 9-17, 10-1ff
    G-800SA rotator controller: 9-8, 9-19, 10-1ff
    Rotator controller: 9-2ff

Z

Zarlink MT8870D DTMF receiver chip: 3-28
Zigbee protocol: 3-8

Notes

Notes

Notes

Notes

Notes

Notes